Sustainable Management of Fish Genetic Resources

Uttam Kumar Sarkar • T. T. Ajith Kumar •
Neeraj Sood • Rajeev Kumar Singh •
Ravindra Kumar • Lalit Kumar Tyagi
Editors

Sustainable Management of Fish Genetic Resources

Springer

Editors
Uttam Kumar Sarkar
ICAR-National Bureau of Fish Genetic Resources
Lucknow, Uttar Pradesh, India

T. T. Ajith Kumar
ICAR-National Bureau of Fish Genetic Resources
Lucknow, Uttar Pradesh, India

Neeraj Sood
ICAR-National Bureau of Fish Genetic Resource
Lucknow, Uttar Pradesh, India

Rajeev Kumar Singh
ICAR-National Bureau of Fish Genetic Resources
Lucknow, Uttar Pradesh, India

Ravindra Kumar
ICAR-National Bureau of Fish Genetic Resource
Lucknow, Uttar Pradesh, India

Lalit Kumar Tyagi
ICAR-National Bureau of Fish Genetic Resources
Lucknow, Uttar Pradesh, India

ISBN 978-981-97-5249-2 ISBN 978-981-97-5250-8 (eBook)
https://doi.org/10.1007/978-981-97-5250-8

This Springer imprint is published by the registered company Springer Nature Singapore Pte Ltd.
The registered company address is: 152 Beach Road, #21-01/04 Gateway East, Singapore 189721, Singapore

Preface

Fisheries and aquaculture are one of the sun rise sectors supporting more than 59 million direct jobs. Global fish production reached 186.6 million metric tons (MMT) in 2023, with 87% for direct human consumption. Fish and fishery products are major trade commodities, with developing countries contributing nearly 60% of global fish exports. They are key to achieving UN Sustainable Development Goals, especially in ending hunger and promoting economic growth. Sustainable fisheries resource management is vital to address overfishing, habitat destruction, and climate change, ensuring the long-term health of fish populations and ecosystems.

India has about 10% of the world's fish diversity and is one of the 17 mega-biodiverse countries. The country is the third-largest fish and aquaculture producer and accounts for about 16% and 5% of total global inland and marine fish production, respectively. The country supports 3,240 native finfish species across various habitats, with significant marine diversity from island ecosystems such as the Andaman and Nicobar archipelago. The long coastline of 8118 km with the exclusive economic zone (EEZ) encompasses 2.02 million km^2 of coastal and offshore areas. The major inland aquatic resources include 1.2 million hectares (Mha) of brackish water area, 2.38 Mha of freshwater ponds and tanks, and 3.15 Mha of reservoirs besides about 0.19 million kilometre of rivers and canals. For millions of people, fisheries and aquaculture tend to be vital sources of food, nutrition, and income and a viable economic solution to the nation's hunger and malnutrition problems. The country's fish production has grown from 5.656 MMT in 2000–2001 to 16.248 MMT in 2021–2022, significantly contributing to the national economy and foreign exchange earnings.

Despite contributing significantly to global Fish Genetic Resources (FGR), aquatic ecosystems of the country face threats from habitat destruction, overexploitation, pollution, invasive species, and climate change. The loss of genetic diversity reduces the ability of fish populations to adapt to environmental changes, making them more vulnerable. Effective conservation measures, sustainable practices, and international cooperation are crucial for addressing these challenges and ensuring the long-term health and resilience of marine ecosystems. Sustainable management of FGR is crucial for maintaining ecological balance, supporting livelihoods, and ensuring food security. This involves utilizing genomic resources and databases to monitor genetic diversity, protect critical habitats, and develop informed aquaculture and fisheries strategies. By adopting these practices, the country can preserve its aquatic biodiversity and ensure the long-term viability of fish populations.

In this context, publishing this book, *Sustainable Management of Fish Genetic Resources*, is an appropriate measure, which aims to provide a comprehensive overview of the current state of FGR management, highlighting key challenges, innovative strategies, and practical solutions for conservation and sustainable utilization. This book contains 15 chapters and begins with exploring the specific strategies and practices adopted in the country with a vast and diverse array of fish species, to provide a contextual foundation for understanding global FGR management efforts. By examining the genetic variability in natural fish populations, we underscore the importance of genetic diversity as a cornerstone of effective conservation strategies.

Advanced methodologies and tools for assessing and managing genetic variations are discussed in detail, offering insights into how these tools can inform and enhance conservation efforts. Cutting-edge genomic techniques are presented, showcasing their role in analysing genetic variability and notifying precise conservation actions. *Ex situ* conservation focuses on maintaining genetic diversity outside natural habitats, while in situ conservation emphasizes preserving fish genetic resources within their natural environments, each approach playing a vital role in holistic conservation strategies. By implementing these conservation approaches, the stakeholders and policymakers can work together to ensure the sustainable management of fish genetic resources, contributing to biodiversity conservation and the resilience of aquatic ecosystems.

Further, the impacts of climate change on fish genetic resources are analysed, providing a forward-looking perspective on how changing environmental conditions affect fish populations and what adaptive strategies are necessary for their conservation, especially in the face of ongoing human activities and environmental changes. By implementing adaptive strategies, the investors can address the complex and dynamic challenges facing fish conservation, thereby promoting the resilience and sustainability of aquatic ecosystems for future generations. We also present a model for harmonizing conservation with livelihood opportunities, using marine ornamental aquaculture as a case study to demonstrate the sustainable practices that benefit ecosystems and local communities.

Critical issues such as habitat degradation in coral reefs and mangroves are addressed, highlighting the importance of these ecosystems for fish genetic diversity and outlining mitigation measures. Addressing habitat degradation in mangroves and coral reefs requires integrated management approaches prioritizing conservation, community engagement, and sustainable use. The use of genotoxicity biomarkers for environmental health assessment is discussed, emphasizing their role in sustainable FiGR management. Integrating the genotoxicity biomarkers into environmental health assessment allows researchers to evaluate the impact of environmental contaminants on biological systems at molecular, cellular, and organismal levels. These biomarkers provide reasonable information for risk assessment, regulatory decision-making, and development of strategies to protect human and environmental health from genotoxic hazards.

Transboundary diseases present significant challenges to wild finfish populations, and insights for managing these diseases are provided, focusing on their implications for conservation. Collaborative and multisectoral efforts are required to prevent, detect, and respond to emerging and re-emerging threats from transboundary diseases. The need for surveillance in fish disease for reporting, management, and associated technical aspects was discussed in detail to provide its emphasis on aquaculture and export promotion. The restoration of fish stocks through the conservation of genetic resources is also explored, detailing strategies to replenish depleted populations and maintain genetic diversity. Restoration of fish stocks is vital for maintaining healthy aquatic ecosystems, supporting sustainable fisheries, and ensuring food security for millions dependent on fish for nutrition and livelihoods.

Innovative applications of artificial intelligence (AI) models are discussed, showcasing how these technologies can enhance the monitoring and management of aquatic genetic resources. These advanced tools offer new possibilities for predictive modelling and adaptive management in conservation efforts. Lastly, we review the legislative framework in India, highlighting best practices and identifying challenges in implementing effective policies for sustainable management of fish genetic resources. This legislative review provides a practical perspective on how national policies can support and enhance conservation efforts.

This book is intended as a valuable resource for researchers, policymakers, and practitioners involved in the conservation and sustainable management of fish genetic resources. By offering a comprehensive view of the current status, challenges, and strategies, we hope to contribute to the ongoing efforts to preserve the genetic diversity of fish species for future generations. We extend our deepest gratitude to all the contributors and experts whose knowledge and insights have made this book possible. Their dedication and expertise are crucial in advancing the field of fish genetic resource management and ensuring the sustainability of aquatic ecosystems globally.

Lucknow, Uttar Pradesh, India — Uttam Kumar Sarkar
Lucknow, Uttar Pradesh, India — T. T. Ajith Kumar
Lucknow, Uttar Pradesh, India — Neeraj Sood
Lucknow, Uttar Pradesh, India — Rajeev Kumar Singh
Lucknow, Uttar Pradesh, India — Ravindra Kumar
Lucknow, Uttar Pradesh, India — Lalit Kumar Tyagi

Contents

1 **Sustainable Management of Fish Genetic Resources in India** 1
Uttam Kumar Sarkar, Ganesan Kantharajan, and A. Kathirvelpandian

2 **Genetic Variability in Natural Populations of Fishes** 21
Rajeev Kumar Singh, P. R. Divya, A. Kathirvelpandian, and Mahender Singh

3 **Genetic Variation and Assessment Tools for Conservation and Management of Fish Genetic Resources (FiGR)** 37
Rajeev Kumar Singh, A. Kathirvelpandian, and M. U. Rekha

4 **Structural and Functional Genomics for Assessment of Genetic Variability** 53
Vindhya Mohindra, Murali S. Kumar, Basdeo Kushwaha, P. R. Divya, Labrechai Mog Chowdhury, and Ravindra Kumar

5 **In Situ Conservation: Tools, Strategies, and Challenges** 71
Uttam Kumar Sarkar, Rejani Chandran, T. K. Teena Jayakumar, and Charan Ravi

6 **Ex Situ Conservation: Status, Strategies and Challenges** 91
Ravindra Kumar, Rajeev Kumar Singh, Murali S. Kumar, and Santosh Kumar

7 **Habitat Degradation in Coral Reef Ecosystems and Mangroves: Current Status and Management Measures** 111
T. K. Teena Jayakumar and Uttam Kumar Sarkar

8 **Climate Change and Fish Genetic Resources: Impact, Status and Perspectives** 151
Uttam Kumar Sarkar, A. Kathirvelpandian, and Ganesan Kantharajan

9 **Harmonizing Conservation with Livelihood: Aquaculture of Marine Ornamentals, A Model** 163
T. T. Ajith Kumar

10 Genotoxicity Biomarkers for Environmental Health Assessment Towards Sustainable Management of Fish Genetic Resources 173
Basdeo Kushwaha, Rashmi Srivastava, and Ravindra Kumar

11 Aquatic Animal Disease Surveillance and Important Disease Affecting Finfish and Shellfish in India 199
Neeraj Sood, P. K. Pradhan, Anutosh Paria, Chandra Bhushan Kumar, Gaurav Rathore, Uttam Kumar Sarkar, and J. K. Jena

12 Implications of Transboundary Diseases in Wild Finfish Populations: Insights for Conservation 223
Arun Sudhagar, Thangaraj Raja Swaminathan, and Neeraj Sood

13 Conservation of Genetic Resources: Restoration of Fish Stocks 237
V. S. Basheer, Charan Ravi, and Uttam Kumar Sarkar

14 Artificial Intelligence-Based Modeling for Sustainable Management of Fish Genetic Resources: Status and Opportunities 249
Achal Singh, Tanwy Dasmandal, Rejani Chandran, Ganesan Kantharajan, and Uttam Kumar Sarkar

15 National Legislations: Best Practices and Challenges in India for Sustainable Management of Fish Genetic Resources 267
Tarachand Kumawat and Lalit Kumar Tyagi

Editors and Contributors

About the Editors

Uttam Kumar Sarkar has more than 30 years of experience in fisheries science and presently serving as Director, ICAR-NBFGR. He has made significant contributions in the areas including exploring, evaluating, and characterizing the spatio-temporal pattern of fish biodiversity, biological traits, and population parameters of diversified fishes of the inland waters. He described a total of 10 new fish species across various ecosystems. He has been involved in conceptualizing and adoption of the innovative approach of state fish for the first time in India for the conservation of fish genetic resources of the country. These innovations were made to assess and understand fish assemblages, spatial diversity patterns, and habitat complexity of more than 30 Indian reservoirs and 50 wetlands, covering different climatic zones, for developing region-specific management practices. He has published 8 books, 3 policy documents, 3 fish diversity atlas, 5 databases, and DNA barcoding of 72 freshwater fishes. He has also published over 260 research papers, with above 5500 citations and supervised 13 PhDs. He is President of the Aquatic Biodiversity and Conservation Society. He has been conferred with National Academy of Agricultural Sciences (NAAS) Fellow-2024, prestigious National Award for e-Governance-2024 for the National Surveillance Programme for Aquatic Animal Diseases.

T. T. Ajith Kumar obtained his master's degree and PhD in marine biology from Annamalai University and thereafter served 27 years in teaching, research, and extension at Annamalai University and ICAR. He previously served as an Assistant Professor and, presently, is a Principal Scientist and Head of the Centre of the ICAR-National Bureau of Fish Genetic Resources, Regional Centre at Kochi, India. He has scientific interests in breeding and rearing marine ornamental organisms, contributing significantly towards integrating marine ornamental conservation and sustainable development. He has worked for the development of the community aquaculture units, marine ornamental village, and nano aquaria towards the conservation of biodiversity and livelihood upliftment. He has published more than 200 articles (citation 3,500) besides 23 books (including booklets and manuals) and 43 book chapters. Further, a team led by him discovered 17 new species of fish and shrimps and several new records to Indian waters.

Neeraj Sood is working as a Principal Scientist in the Exotics and Aquatic Animal Health Division of ICAR-National Bureau of Fish Genetic Resources, Lucknow, India. He completed his postgraduation from CCS Haryana Agricultural University, India, and doctorate from Punjab Agricultural University, Ludhiana, India, in veterinary pathology. He has over 26 years of experience working in the area of aquatic animal health and has published over 100 research papers. He has been involved in preparing the national documents on exotic aquatic species and quarantine. His areas of interest include disease diagnosis, development of fish cell lines and monoclonal antibodies, and understanding host–pathogen interaction. He is the Consortium Principal Investigator of the National Surveillance Programme for Aquatic Animal Diseases. He has also reported five new fish diseases.

Rajeev K. Singh is working as Principal Scientist and Head, Germplasm Exploration, Evaluation and Conservation Division of ICAR-National Bureau of Fish Genetic Resources, Lucknow. He earned his postgraduation and a doctorate in biotechnology. He has over 24 years of experience in the area of conservation genetics and genomics. He has characterized over 20 fish species, below the species level, which is utilized for management of genetic resources. He has sequenced and characterized the full genome of an anadromous fish and eight mitochondrial genomes. He has also been associated with developing sperm cryopreservation protocols. He has published over 100 peer-reviewed research publications; co-authored FAO's document on Genetic Management of Carps; discovered seven fish species, new to science; and supervised/co-supervised six doctorates along with over 25 postgraduate/graduate researchers.

Ravindra Kumar has 29 years of research experience in the areas of characterization and cataloguing of fish genetic resources. His area of research focuses on physical chromosome map development, whole genome/transcriptome sequencing and analyses, genomic resource databases development/maintenance, cell line development/maintenance for conservation, and sustainable use of genetic resources. He was involved in deciphering the whole genome information of *Clarias magur* and *Labeo rohita*. He has generated cytogenetic profiles of over 82 fish species from India and assessed genotoxic potential of around 15 pollutants and developed/maintaining 6 genomic resource databases/portal/webserver. He is presently working as Principal Scientist in Genomics and Computational Resources Division at ICAR-NBFGR. He has also served as Head, Molecular Biology and Biotechnology Division, ICAR-NBFGR and Director (acting). He has published 134 research papers, 02 books, and 30 popular articles with over 3216 citations.

Lalit Kumar Tyagi has over 23 years of professional experience in research, extension, and training at ICAR-National Bureau of Fish Genetic Resources, Lucknow, India. He worked on 20 research projects funded by various agencies on issues related to fisheries resources like co-management of aquatic resources, orientation of fisherfolks towards fish conservation, functioning of fishing cooperative societies, exploration and assessment of fish diversity and traditional knowledge, harmonizing biodiversity conservation and fish genetic resource, evaluation of wildlife protected areas, models for livelihood enhancement of tribal communities using fisheries enterprises and explorations, and conservation of fisheries resources in north-eastern region of India. He has 35 research publications, co-edited 2 books and a national guideline, and contributed 12 book chapters, 9 review articles, and 35 popular articles.

Contributors

V. S. Basheer ICAR-National Bureau of Fish Genetic Resources, Lucknow, Uttar Pradesh, India

Rejani Chandran ICAR-National Bureau of Fish Genetic Resources, Lucknow, Uttar Pradesh, India

Labrechai Mog Chowdhury ICAR-National Bureau of Fish Genetic Resources, Lucknow, Uttar Pradesh, India

Tanwy Dasmandal ICAR-National Bureau of Fish Genetic Resources, Lucknow, Uttar Pradesh, India

P. R. Divya ICAR-National Bureau of Fish Genetic Resources, Lucknow, Uttar Pradesh, India

J. K. Jena Indian Council of Agricultural Research, Krishi Anusandhan Bhawan-II, Pusa, New Delhi, India

Ganesan Kantharajan ICAR-National Bureau of Fish Genetic Resources, Lucknow, Uttar Pradesh, India

A. Kathirvelpandian ICAR-National Bureau of Fish Genetic Resources, Lucknow, Uttar Pradesh, India

Chandra Bhushan Kumar ICAR-National Bureau of Fish Genetic Resources, Lucknow, Uttar Pradesh, India

Murali S. Kumar ICAR-National Bureau of Fish Genetic Resources, Lucknow, Uttar Pradesh, India

Ravindra Kumar ICAR-National Bureau of Fish Genetic Resources, Lucknow, Uttar Pradesh, India

Santosh Kumar ICAR-National Bureau of Fish Genetic Resources, Lucknow, Uttar Pradesh, India

T. T. Ajith Kumar ICAR-National Bureau Fish Genetic Resources, Lucknow, Uttar Pradesh, India

Tarachand Kumawat ICAR-National Bureau of Fish Genetic Resources, Lucknow, Uttar Pradesh, India

Basdeo Kushwaha ICAR-National Bureau of Fish Genetic Resources, Lucknow, Uttar Pradesh, India

Vindhya Mohindra ICAR-National Bureau of Fish Genetic Resources, Lucknow, Uttar Pradesh, India

Anutosh Paria ICAR-National Bureau of Fish Genetic Resources, Lucknow, Uttar Pradesh, India

P. K. Pradhan ICAR-National Bureau of Fish Genetic Resources, Lucknow, Uttar Pradesh, India

Gaurav Rathore ICAR-National Bureau of Fish Genetic Resources, Lucknow, Uttar Pradesh, India

Charan Ravi ICAR-National Bureau of Fish Genetic Resources, Lucknow, Uttar Pradesh, India

M. U. Rekha ICAR-National Bureau of Fish Genetic Resources, Lucknow, Uttar Pradesh, India

Uttam Kumar Sarkar ICAR-National Bureau of Fish Genetic Resources, Lucknow, Uttar Pradesh, India

Achal Singh ICAR-National Bureau of Fish Genetic Resources, Lucknow, Uttar Pradesh, India

Mahender Singh ICAR-National Bureau of Fish Genetic Resources, Lucknow, Uttar Pradesh, India

Rajeev K. Singh ICAR-National Bureau of Fish Genetic Resources, Lucknow, Uttar Pradesh, India

Neeraj Sood ICAR-National Bureau of Fish Genetic Resources, Lucknow, Uttar Pradesh, India

Rashmi Srivastava School of Life Sciences, Department of Zoology, Babasaheb Bhimrao Ambedkar University, Lucknow, Uttar Pradesh, India

Arun Sudhagar ICAR-National Bureau of Fish Genetic Resources, Lucknow, Uttar Pradesh, India

Thangaraj Raja Swaminathan Microbiology, Fermentation & Biotechnology Division, ICAR-Central Institute of Fisheries Technology, Kochi, Kerala, India

T. K. Teena Jayakumar ICAR-National Bureau of Fish Genetic Resources, Lucknow, Uttar Pradesh, India

Lalit Kumar Tyagi ICAR-National Bureau of Fish Genetic Resources, Lucknow, Uttar Pradesh, India

1 Sustainable Management of Fish Genetic Resources in India

Uttam Kumar Sarkar, Ganesan Kantharajan, and A. Kathirvelpandian

Abstract

The conservation and management of fish genetic resources are linked with almost all Sustainable Development Goals (SDGs) and contribute to global development. India with vast aquatic resources including biodiversity hotspots in the Western Ghats, North East Region, Himalayas, and Andaman and Nicobar Islands supports rich fish diversity and contributes to the livelihood of the rural and coastal communities. The rich aquatic biodiversity is incessantly threatened by various natural and anthropogenic stressors, viz. habitat destruction, overexploitation, aquatic pollution, invasive alien species, and climate change. This chapter discusses the current status of the fish genetic resources in diverse habitats in India with the historical perspectives on new species discovery and expansion of protected areas (PAs). The role of various conservation approaches including the ex-situ and in-situ measures for the sustainable management of fish genetic resources and the significance of fishery resource management in ensuring the food and nutritional security, livelihood promotion of the native communities, economic development of the country, maintenance of ecological integrity, and health of the ecosystem are also discussed in detail in this chapter.

Keywords

Fish diversity · New species · Climate change · In-situ conservation · Ex-situ conservation

U. K. Sarkar (✉) · G. Kantharajan · A. Kathirvelpandian
ICAR-National Bureau of Fish Genetic Resources, Lucknow, Uttar Pradesh, India
e-mail: Uttam.Sarkar@icar.gov.in; kantharajan.g@icar.gov.in; A.Pandian@icar.gov.in

U. K. Sarkar et al. (eds.), *Sustainable Management of Fish Genetic Resources*,
https://doi.org/10.1007/978-981-97-5250-8_1

1.1 Introduction

India is one among the 17 mega-biodiverse countries consisting of four biodiversity hotspots including the Western Ghats, North East Region, Himalayas, and Andaman and Nicobar Islands, which supports 7–8% of the world's biodiversity (Venkataraman and Sivaperuman 2018). The vast aquatic resources include a coastal length of 8118 km, a continental shelf area of 0.5 mil km^2, an exclusive economic zone (EEZ) of 2.02 mil km^2, rivers and canals (2,52,431 km), reservoirs (4.03 Mha), tanks and ponds (9.2 Mha), beels/oxbow lakes (0.48 Mha), and brackish water (10.65 Mha) and support rich biodiversity as well as livelihood (DoF 2020). A rich diversity of finfishes (3218), molluscs (3271), crustaceans (2934), echinoderms (765), and coral species (>200) are also reported from India. The conservation and sustainable management of aquatic genetic resources are vital as it contributes to the livelihood and provision of various ecosystem services, including fish production, food web dynamics, nutrient cycling, ecosystem resilience, diversity, and recreational services.

The diverse aquatic germplasm offers a lot of scope for aquaculture species diversification in India. However, the fish genetic resources of the country are being incessantly threatened by various natural and anthropogenic stressors including habitat destruction, overexploitation, aquatic pollution, invasive alien species, and climate change. The conservation and management approaches adopted in the country are broadly classified into ex-situ (outside the natural habitats) and in-situ (within the natural habitats) conservation including spatial protection measures, temporal fishing regulation, species-specific management measures, declaration of state fish, fish ranching for natural stock enhancement, live gene bank, tissue and DNA bank, genetic characterisation, conservation aquaculture, and co-management. The concept of 'state fish' is one such approach which involves the stakeholders and policymakers to boost the conservation and sustainable utilisation of the fish species and create awareness of fish genetic resource conservation among the general public (Sarkar et al. 2024). This chapter discusses the current status of fish diversity, the expansion of the protected area (PA) network in India *vis-à-vis* the world along with threats and significance of fish genetic resources and recommendations for their sustainable management and utilisation.

1.2 Status of Fish Genetic Resources in India

Globally, the number of valid species is 36,779, of which 18,675 are found in freshwater. The top five families include Gobiidae (2000 species), Cyprinidae (1785 species), Cichlidae (1747 species), Characidae (1260 species), and Loricariidae (1060 species) (as of 02 April 2024; Fricke et al. 2024). The IUCN Red List status of class Actinopterygii (25,675 species) indicates globally 82 species are extinct; 11 species extinct in the wild; 739 species critically endangered; 1210 species endangered; 1410 species vulnerable; 765 species near threatened; 16,556 species least concern; and 4902 species are data deficient (IUCN 2024).

The Aquatic Genetic Resource Information System of India (AqGRISI) lists a total of 3218 native finfish species belonging to 1038 genera, 256 families, and 57 orders from India. This included about 1500 species exclusively marine and about 950 species exclusive to freshwater ecosystems besides brackish water habitats (https://aqgrisi.nbfgr.res.in/#/). In India, there are 2757 species assessed under the class Actinopterygii by the IUCN Red List which includes 18 critically endangered species, 96 endangered species, 116 vulnerable species, 44 near-threatened species, 2150 least concern species, and 333 data-deficient species. The critically endangered species distributed in India are *Glyptothorax kudremukhensis* (Kudremukh Glyptothorax), *Puntius deccanensis* (Deccan barb), *Schistura papulifera*, *Neolissochilus bovanicus* (Bovany barb), *Pethia pookodensis* (Pookode Lake barb), *Glyptothorax kashmirensis*, *Puntius madhusoodani*, *Tor remadevii* (Hump-backed mahseer), *Triplophysa kashmirensis*, *Hypselobarbus pulchellus* (Haragi), *Hypselobarbus thomassi* (Red Canarese barb), and *Garra arunachalami* (IUCN 2024).

1.2.1 Freshwater Fish Diversity

The vast spread and diverse freshwater ecosystems in India support over 950 fish species. Rivers are the major freshwater ecosystem with dynamic habitat characteristics that support numerous fish species. The fish species reported in the major river systems of India in recent years are given in Table 1.1. Apart from major rivers, tributaries of the major rivers also support sizeable fish diversity, viz. Gomti River in Ganga basin (56 species), Ib River (55 species) and Tel River (71 species) in Mahanadi basin (55 species), Pranhita River in Godavari basin (53 species), and Kabini River in Cauvery basin (136 species) (Sarkar et al. 2010; Chandran et al. 2019, 2022; Thampy et al. 2021; Kantharajan et al. 2022). Further, small

Table 1.1 Fish diversity reported in major rivers in India based on recent literature

Sl. No	River system	Fish diversity			Source
		Species	Genera	Family	
1	Ganga (Gangotri–Hooghly)	143	72	32	Sarkar et al. (2012)
	Ganga (Harsil—Frazerganj)	190	97	65	ICAR-CIFRI (2020)
2	Brahmaputra (North East India)	229	–	–	Vishwanath (2017)
	Brahmaputra (Barak and Assam)	217	–	36	Bhattacharjya et al. (2004)
3	Mahanadi and tributaries	121	85	43	Tyagi et al. (2021)
4	Godavari	152	84	37	Jadhav et al. (2024)
5	Krishna	127	–	–	Das et al. (2017)
6	Cauvery	146	–	52	ICAR-CIFRI (2019)
7	Tapti (Multai to Kamrej)	80	53	25	ICAR-CIFRI (2020)
8	Narmada and tributaries	196	126	51	Bhakta et al. (2020)
9	Yamuna (Dakpathar to Allahabad)	143	–	–	Sharma et al. (2017)

independent rivers such as the Thamirabarani River (125 species) and Bharathappuzha River (117 species) originating from the Western Ghats region support a rich fish diversity (Bijukumar et al. 2013; Mogalekar 2019).

The rich diversity also reported from various wetlands in India is associated with the diversity reported from the rivers it is connected (Sarkar et al. 2020a, b). 'Saraiyaman' is a protected natural oxbow lake in the Ganga basin that supports 58 fish species belonging to 7 orders, 20 families, and 40 genera (Joshi et al. 2022a, b). In another study by Pathak et al. (2020) in Surha Lake, 66 fish species belonging to 23 families have been reported. A total of 185 species were reported from the 3 selected Ramsar wetlands in Kerala, including Vembanad Lake, Ashtamudi Lake, and Sasthamkotta Lake (Divya et al. 2020). Nandur Madhmeshwar wetland in the Godavari basin in Maharashtra supports 39 fish species, while the Kolleru Lake in Andhra Pradesh, one of the largest freshwater lakes in India, supports 51 species (Barman 2004; Wagh et al. 2018).

The reservoirs are the man-made habitats that support a rich diversity of fish species (Khedkar et al. 2014). Sarkar et al. (2015) compiled a list of 117 fish species from Indian reservoirs exhibiting a rich diversity of food fishes (96.5%). The Chandil reservoir in Subarnarekha (42 species) (Lianthuamluaia et al. 2019) supports more than 60% of the total fish species reported in the respective river system indicating its potential habitat for fish genetic resources, though negative impacts on fish diversity are also reported in India as well as the world.

1.2.2 Marine Fish Diversity

The marine fish diversity of India comprises 2492 species belonging to 941 orders 240 families. The most species-rich families are Gobiidae (190), Pomacentridae (92), and Labridae and Serranidae (each 85 species) (Joshi et al. 2017). The rich diversity of marine fish in India is reported from the coastal waters of the Andaman and Nicobar archipelago (1431 species), followed by the east coast (1121 species) and west coast of India (1071 species). About 91 marine fish species are endemic to India's coastal waters (Joshi et al. 2017).

1.2.3 Deep-Sea Fish Diversity

The deep sea is the largest and unique biome on the Earth and supports rich biodiversity. The deep-sea ecosystem (>200 m depth) covers >95% of the global ocean volume, and its biodiversity is poorly explored (Danovaro et al. 2020). The abundance and distribution of the deep-sea ecosystem depend on various factors including habitat heterogeneity and nutrient flux; hence the heterogeneity occurs at various spatial scales including macro- (e.g. continental margins), meso- (e.g. cold-water reefs and canyons), and micro-scale. The distribution pattern of demersal fishes in the deep sea of south-west region of Indian EEZ (200–1100 m depth; 7–15^0 N Lat) based on exploratory trawl surveys conducted during 1998–2002 and 2005–2007 by

FORV Sagar Sampada of the Ministry of Earth Sciences, Government of India, revealed the occurrence and distribution of 149 deep-sea fish species belonging to 123 genera, 70 families, and 24 orders at 164 stations (Venu 2013).

1.2.4 Island Fish Diversity

The island fishery resources in India support rich fish diversity and offer scope for the sustainable development. The coastal Andaman and Nicobar Islands cover a length of 1962 km, a continental shelf area of 35,000 km^2, and an exclusive economic zone (EEZ) of about 6,00,000 km^2. A total of 1439 fish species belonging to 577 genera, 165 families, and 33 orders were recorded in the Andaman and Nicobar archipelago including 400 food fishes. The fragile coral reef ecosystem of Andaman archipelago supports a whopping diversity of >1000 species (Rajan and Sreeraj 2013). The Lakshadweep archipelago in the Arabian Sea consists of 12 atolls, 5 submerged banks, 3 reefs, and diverse coastal habitats. The recent checklist reports the distribution of 856 species of 432 genera, 144 families, and 43 orders from the Lakshadweep Islands which also includes 154 commercial fish species that contribute to the fisheries of these islands (Rajan et al. 2021).

1.2.5 Brackish Water Fish Diversity

The estuaries formed by the major river on the east coast, viz. Hooghly-Matlah, Mahanadi, Godavari, Krishna, and Cauvery, and on the west coast, viz. Narmada and Tapti, are important brackish water aquatic resources in India. Apart from that, coastal lakes and lagoons like Chilika, Pulicat, and Vembanad support a majority of the brackish water fish species reported in India. Among the 317 species reported from the Chilika Lake, 43.85% of species are of brackish water fish species (139) (Mohanty et al. 2015). The updated checklist of fishes from the estuaries of Tamil Nadu reports the occurrence of 330 species (Mogalekar et al. 2017). Bharadhirajan et al. (2015) report the distribution of 59 fish species from the Coleroon estuary in the Cauvery basin. Lal et al. (2020) reported 105 finfish and shellfish species from the Ulhas River estuary along the central west coast of India. Some of the common brackish water finfish species include *Etroplus suratensis*, *Mugil cephalus*, *Chanos chanos*, *Tenualosa ilisha*, *Lates calcarifer*, *Elops machnata*, *Liza macrolepis*, *L. parsia*, *Epinephelus tauvina*, *Escualosa thoracata*, *Siganus javus*, and *Eleutheronema tetradactylum*. The commercial important shellfish resources include *Metapenaeus dobsoni*, *Penaeus indicus*, *Penaeus monodon*, *Penaeus semisulcatus*, *Scylla serrata*, *S. tranquebarica*, *Villorita cyprinoides*, *Perna viridis*, *P. indica*, and *Crassostrea madrasensis*.

1.2.6 Cold-Water Fish Diversity

India has vast cold-water fishery resources comprising upland streams, rivers, high and low altitudinal freshwater and brackish water lakes and reservoirs in the Indian Himalayan region comprising an area of 594,400 km^2. These waterbodies are mostly located above 900 m above mean sea level, and the temperature range is limited to 10–20 °C. The cold-water fish diversity of India comprises 258 species belonging to 21 families and 76 genera, of which the maximum number of species reported from the Nort Eheast Himalaya region (255 species) (Singh and Akhtar 2015). Some of the commercially important cold-water fishery resources include mahseer (*Tor*, *Naziritor*, *Neolissochilus* species), snowtrout (*Schizothoraichthys*, *Schizothorax* species), Indian hill trout (*Barilius bendelisis*), minor carps, *Garra* species, and Pengba (*Osteobrama belangeri*).

1.3 Discovery of New Species: Global *Vis-à-Vis* India

The discovery of a new species and understanding the biology of an already described species are essential for the conservation and sustainable utilisation of resources through aquaculture. The new species discovery made in the world during 2005–2023 revealed that a total of 7702 fish species have been newly discovered which also includes 4778 freshwater fishes contributing to about 62% of the total fish discoveries made (Fricke et al. 2024) (Fig. 1.1). The increase in the new species discovery in recent years is attributed to the increased exploration in many countries, expansion of areas for exploration, more funding, and applications of advanced molecular tools through an integrative approach in ichthyotaxonomy.

During the last decade (2013–2022), a total of 268 new finfish discoveries and 70 new records have been made in India, according to the Annual Animal Discoveries

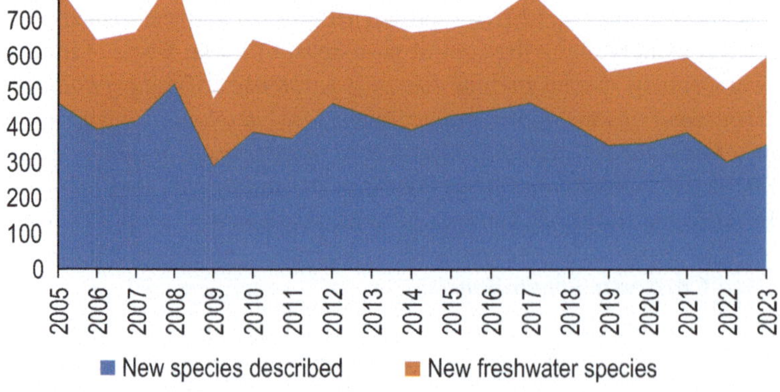

Fig. 1.1 Trend of global new fish species discovery (data source: Fricke et al. 2024)

published by the Zoological Survey of India. The highest of 38 fish species was discovered in 2019 followed by 36 species in 2013. The spatial distribution map of new species discovery indicates that more fish discoveries have been made from the Western Ghats and North East Region in India (Fig. 1.2). The targeted exploratory surveys may be carried out in Central India and the Northern Himalayan region to document fish species.

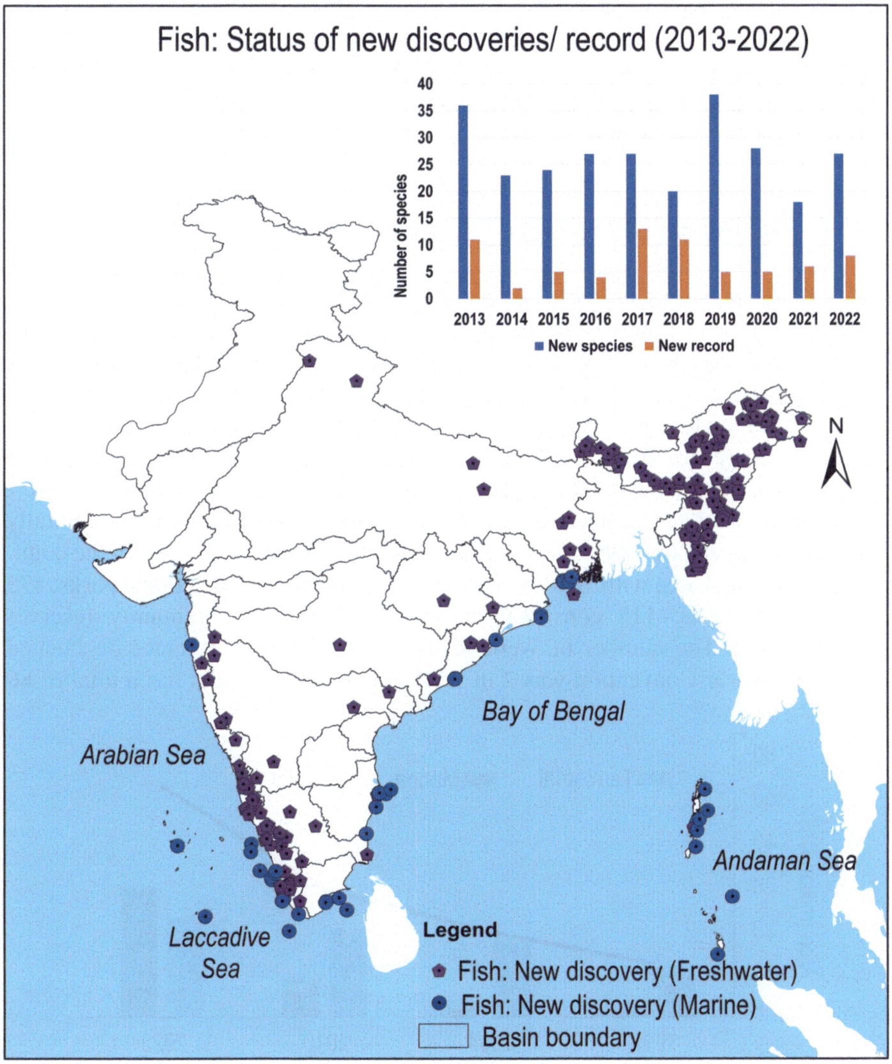

Fig. 1.2 Trend and spatial distribution of new finfish species discoveries (2013–2022) made in India

1.4 Exploration of Fish Genetic Resources

The advent of advanced molecular tools based on an integrated taxonomy approach and the expansion of fish explorations to unexplored areas enables the research communities to understand the existence of undescribed biota and facilitate the discovery of new species. The ICAR-National Bureau of Fish Genetic Resources initiated exploration in various unique freshwater and marine ecosystems to unravel the ichthyofauna diversity. Such programmes led to the discovery of more than 60 new fish species by ICAR-NBFGR and catalogued the diversity of various river basins including Ganga, Mahanadi, Godavari, Cauvery, and Luni River basins and numerous streams in Western Ghats and North East India, coastal habitats in Sundarbans, Gulf of Mannar, and island ecosystem in Lakshadweep and Andaman and Nicobar. Further explorations are underway to catalogue the unique fish diversity in deep-sea ecosystems and caves in North East India and Lakshadweep Islands.

1.5 Expansion of Protected Area Network

The positive impact of protected areas (PAs) on the conservation of fish genetic resources in terms of increasing the diversity, richness, abundance, and support of the threatened fish species is reported from diverse ecoregions in India (Sarkar et al. 2013; Kantharajan et al. 2022). Globally the area under the protected area network has seen a tremendous increase (Fig. 1.3). The area under the marine protected area (MPA) has increased from 0.5% of the global marine area in 2004 to 7.74% of coastal and marine areas in 2021 (CBD 2022).

The area under the protected area (PA) network of India increased gradually from 3.34% in 1988 to 4.89% in July 2026 and to approximately 5.32% of the country's geographical area with 1014 designated PAs, including 106 national parks, 573 wildlife sanctuaries, 115 conservation reserves, and 220 community reserves (NWDC 2023). The number of Wetlands of International Importance designated under the Ramsar Convention was 2 in 1981 and at present India has a total of 80

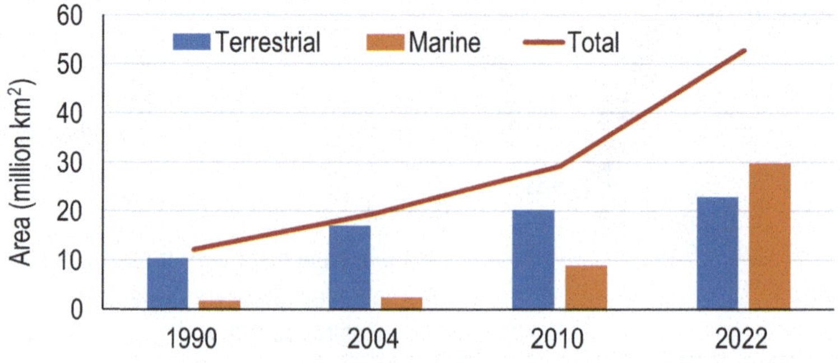

Fig. 1.3 Expansion of the area under the global protected area network (data source: https://www.protectedplanet.net/en Protectedplanet 2024)

spans to a total area of 1.33 Mha. During the year 2021–2022 itself, 33 wetlands have been added to the Ramsar network in India for biodiversity conservation. It includes some of the prominent habitats including Chilika Lake, Pulicat Lake, Pichavaram mangrove wetlands, and Sundarbans along the east coast of India and Vembanad Lake and Ashtamudi Lake and Thane Creek along the west coast of India.

Further, increased conservation and restoration measures have also led to an overall increase in the mangrove cover. For instance, the mangrove cover in the country in 2021 is 4992 km^2, which is about 23.4% higher than the area reported during the year 1987 (4046 km^2), according to the State of Forest Report published by the Ministry of Environment, Forest and Climate Change, Government of India (ISFR 2021).

1.6 Threats to Fish Genetic Resources

The fish genetic resources of the country face a lot of anthropogenic and man-made threats which hamper their survivability and biological processes. The intensity of the threat and the impact varies between the ecosystem and species.

Habitat destruction and fragmentation. The destruction and degradation of aquatic habitats like rivers, wetlands, mangrove areas, and lakes are mainly attributed to the change in the land use and land cover due to unplanned construction. The construction of the dam and change in the rainfall pattern result in the fragmentation of the river flood-plain, decrease or increase in the water spread, flow conditions, and reductions in the deep pools which support the fish diversity during the non-monsoon seasons. The dams constructed against the Bhagirathi River in the Garhwal region resulted in the fragmentation of riverine habitat and affected the distribution and assemblage structure of rheophilic and indigenous fish species including snow-trout, mahseer, *Glyptothorax*, *Garra*, *Barilius*, and *Psilorhynchus* (Agarwal et al. 2018). Further, construction of the Farakka Barrage in the Ganga River has significantly disturbed the migration of anadromous fish, hilsa (Mukherjee and Suresh 2007), while the Thanneermukkom Bund in Vembanad Lake impacted the fishery of freshwater prawn, *Macrobrachium rosenbergii* (Suseelan 1987).

Overexploitation. The unsustainable harvesting of fishery resources without considering the future, disturbs the aquatic ecosystem balance and leads to the collapse of the fishery. Illegal, unreported, and unregulated (IUU) fishing and juvenile fishing are a major global problem threatening sustainable fisheries. Usage of low mesh-sized fishing gear in rivers and wetlands impacts the growth and other reproductive traits which affects the natural recruitment of fish species. This is a major cause of concern for fish with slow growth rates, low reproductivity potential (e.g. sharks and rays), and higher economic value (e.g. hilsa). A study on the commercial trawlers operating along central Kerala revealed that the by-catch contributed 28% to the total catch with a rich diversity of 67 fish species (Madhu et al. 2017).

Aquatic pollution. The inflow of source and non-point source pollutants degrades the quality of the water bodies and affects the ichthyofaunal resources in numerous ways. The enrichment of organic pollutants and nutrients causes eutrophication which leads to the formation of algal blooms, dead zones, and subsequent fish mortality. Further, the fish species living in polluted habitats are expected to be more immunologically suppressed than those living in good habitat conditions. Pollutants like pesticides, heavy metals, industrial effluents, hydrocarbons, and sewage alter the optimum habitat conditions for sensitive fish species. This further leads to the elimination of such species from the ecosystem and paves the way for the establishment of resistant exotic species. The localised impact of the inflow of sewage and other pollutants on fish assemblage patterns and catch is reported in the Ganga River system (Sarkar et al. 2012). Further, the negative impact of contaminants of emerging concern (CEC) including microplastics, pharmaceuticals, nanoparticles, antibiotics, and pesticides on the survival, growth, and reproductive biology of fish poses a major threat to aquatic biodiversity. The occurrence of plastics in the mangrove habitats threatens biodiversity by altering the habitat structure and food web interactions (Kantharajan et al. 2018). A study on beach litter, man-made objects which are discarded directly or indirectly on the coastal beaches in 254 selected beaches in India, reported the highest quantity along the Goa coast (205.75 g/m^2), and the values reported in island ecosystems of Andaman and Lakshadweep were higher than the value reported in many coastal states in the mainland India (Kaladharan et al. 2017). The emergence of antimicrobial resistance (AMR) negatively impacts the aquatic ecosystem and human health. The transfer of AMR genes from the aquaculture system to natural waterbodies disturbs the wild fish genetic resources and potentially links with human health (Watts et al. 2017).

Invasive species. The dominance of exotic fishes in natural waterbodies causes the extirpation of indigenous fish diversity due to their aggressive nature and simple biological traits (Singh et al. 2013). The non-native fish species contend with the indigenous species for food and habitat and may facilitate the transmission of new parasites or diseases in the aquatic ecosystem. This affects the native aquatic biodiversity and livelihood of dependent local communities (Joshi et al. 2021). The higher abundance and range extension of *Cyprinus carpio* in the Ganga River is attributed to the change in the riverine habitat characteristics from the lotic to the lentic system which has threatened the native fish species (Sarkar et al. 2012). A recent study in Yamuna River noted the abundant occurrence and gradual expansion of African magur, *Clarias gariepinus*, and the occurrence of a wide range of sized individuals indicating the establishment risk of this fish which is highly dangerous to native fish fauna (Joshi et al. 2022a, b).

Climate change. Climate change is a significant driver of predominantly negative change in farmed aquatic genetic resources and their wild relatives, particularly in tropical regions. It affects fish production by altering the biological process at individual, community, and ecosystem levels (FAO 2019). The variations in climatic parameters can extremely impact the fish genetic resources by altering their geographic distribution, community structure, abundance, loss of species, and change in population biology, breeding, migration, and physiological behaviour of fish species (Sarkar and Das 2021). A threat assessment of future flow and water temperature extremes (3.2 °C warmer) in ~11,500 riverine fish species indicates that 36% of them will have over half of their present-day geographic range exposed to the climatic extreme with the highest expected threat impact in tropical and subarid regions (Barbarossa et al. 2021).

Climate has a strong controlling influence on physical, chemical, and biological processes in freshwater ecosystems. The change in the rainfall affects the riverine flow and thus influences the life cycle of the dependent fish communities. The correlation between the decreasing trend in the rainfall and the increasing trend in the temperature correlated with the decreased fish production in the flood-plain wetlands of Lower Ganga basin indicates the enviro-climatological impact on fish diversity factors (Sarkar et al. 2020a, b). An apparent shift occurred in the distribution of some fishes from the Ganges (Sarkar et al. 2012), with several fish species previously only reported from the warmer middle and lower sections of the Ganges in the 1950s, now being recorded from upper sections of the river. Alterations in temperature and rainfall led to the breeding failure of the Indian major carps and a consequent decline in fish spawn availability in the Ganga River (Vass et al. 2009). The alteration in the feeding intensity and food composition due to heatwave phenomena was noted in some fish species as revealed by its low gastro-somatic index (GaSI) in India (Sarkar and Das 2021). The moderate to high reproductive vulnerability observed in *Nandus nandus*, Gangetic leaf fish, in flood-plain wetlands of the major river basins in India reveals the impacts of regional climate change on the population attributes and breeding phenology (Lianthuamluaia et al. 2023).

The impact of climate change on marine fisheries is also prominent in India as evidenced by the expansion of boundaries of the small pelagics, i.e. *Sardinella longiceps* and *Rastrelliger kanagurta*, towards northern latitudes due to the warming of the sea surface. Further, the rise in the sea surface temperature (SST) has also led to the vertical expansion of the Indian mackerel towards downward. The apparent shift in the spawning season of threadfin breams *Nemipterus japonicus* and *N. mesoprion* from warmer to relatively cooler months was reported along the Indian coast. The rise in the SST above the optimal level causes bleaching in the coral reef habitats in India. This affects the entire associated flora and fauna of the species-rich ecosystem and subsequently disturbs the livelihood of the dependent communities. The sea-level rise is another major issue which affects the intertidal habitats including the mangrove ecosystems in India (Rao and Vivekanandan 2008).

1.7 ICAR-NBFGR Efforts for Sustainable Fish Genetic Resource Management

Exploration, cataloguing, and documentation of fish diversity from diverse ecoregions and habitats are paramount for the conservation and management of fish genetic resources in the country. Developing open-access biodiversity portals/databases is essential for data sharing and information dissemination to support decision-making and policy implementation (Krishnan et al. 2022). The Aquatic Genetic Resource Information System of India (AqGRISI) portal (AqGRISI 2024; https://aqgrisi.nbfgr.res.in/) of the National Bureau of Fish Genetic Resources (NBFGR), Lucknow, of the Indian Council of Agricultural Research (ICAR) provides information on systematics, biology, type specimen, disease, habitat, distribution, patent, fish nutrition, and bibliography on fish resources reported in India.

The protected areas (PAs) are considered an important step towards conserving and managing threatened fish species in their natural habitat. However, fish genetic resources are poorly represented in the existing PA network which is mainly devoted to conserving the terrestrial flora and fauna, and to some extent, it offers protection to charismatic aquatic fauna like Gangetic dolphin and gharials (Gupta et al. 2014). However, the positive impact of terrestrial PAs on fish genetic resource conservation has been reported in India in various ecoregions. This emphasis is on the prioritisation of selected terrestrial PAs from the existing network based on their habitat potential to threatened fish fauna for extending their scope for conservation.

The conservation and protection of the threatened aquatic genetic resources in their natural habitat are also ensured through various acts and legal measures available at the national level including the Wild Life Protection Act, 1972; Indian Fisheries Act, 1897; Biological Diversity Act, 2002; etc. which are amended from time to time. Apart from that, the international-level organisations/instruments including the Convention on Biological Diversity (CBD), Ramsar Convention, Convention on Migratory Species (CMS), and Convention on International Trade in Endangered Species of Wild Fauna and Flora (CITES) are also involved in the conservation of fish genetic resources.

The novel concept of the ‘state fish’ proposed in the year 2006 involves the native community and other stakeholders in the conservation and management of fish genetic resources towards sustainable development. At present 22 Indian states declared their state fish considering their preference and popularity among local communities; ecological, commercial, cultural, and historical significance; culinary values; endemism; aquaculture; and conservation potential (Sarkar et al. 2024).

Captive breeding programmes are a principal tool to recompense declining wild fish populations and enhance yield. The ICAR-NBFGR has developed breeding and seed production technology for more than 62 indigenous food and ornamental fish and shrimp species including *Chitala chitala*, *Ompok bimaculatus*, *Hemibagrus punctatus*, *Horabagrus nigricollaris*, *Stenopus hispidus*, *Thor hainanensis*, and *Ancylocaris brevicarpalis*.

Ranching is another important in-situ measure which enhances the natural populations of threatened and endangered species achieved by releasing captive-bred fish seeds that are reared under artificially controlled conditions (Jena and Gopalakrishnan 2012). The ICAR-NBFGR produces quality fish seeds from the wild brood stocks collected from the native environment to support the ‘river ranching programme’ initiated by the Government of India under Pradhan Mantri Matsya Sampada Yojana (PMMSY) scheme to rebuild the indigenous fish stocks. Further, the ranching of endemic and threatened fish species including the ‘critically endangered’ *Hemibagrus punctatus*, ‘endangered’ *Horabagrus nigricollaris*, and ‘vulnerable’ *Horabagrus brachysoma*, in the pristine habitats of the Western Ghats region is regularly undertaken to safeguard India’s aquatic biodiversity. Understanding the distribution, abundance, and biological life history traits of the fish species is essential for developing the captive propagation techniques to support the ranching programmes.

The involvement of the community in the conservation process is regarded as an important step towards the management of aquatic genetic resources and the livelihood generation of native populations. The documentation of indigenous knowledge and applications of the native people's expertise improve the outcomes of the conservation process with reduced time and effort.

Cryopreservation of fish gametes is a vital ex-situ conservation measure which aids in the long-term preservation of fish sperm for conservation and aquaculture. ICAR-NBFGR developed species-specific cryopreservation protocols for about 39 fish species including commercially important endemic fishes in the country. Further, it effectively addresses the issues of insufficient milt production or variation in the maturity of two sexes for captive breeding of cultivable fish species.

Tissue banking enables the long-term storage of material of fish species. The DNA and tissue repository of the National Fish Museum and Repository maintains the >19,000 tissue accessions collected from commercially important and endemic fish resources of biodiversity hotspot areas such as the Western Ghats, North Eastern states, Andaman and Nicobar Islands, and other parts of the country.

The establishment of live gene banks for the endemic, threatened, and important fish species is considered an important approach to the conservation of fish species in its native ecoregion. This also facilitates to study of the life history traits and reproductive biology of the fish species. ICAR-NBFGR has established live fish germplasm resource centres in Agatti, Lakshadweep (island ecosystem); Airoli, Maharashtra (coastal biodiversity of central west coast of India); Guwahati, Assam (North East Region); Kochi, Kerala (Western Ghats); Lamphelpat, Manipur (North East Region); Lucknow, Uttar Pradesh (Ganga basin); Nagarjuna Sagar, Telangana (Deccan Plateau); and Pichavaram, Tamil Nadu (coastal biodiversity of south-east coast of India).

DNA barcoding is considered an important tool which supports the conservation of aquatic biodiversity through taxonomic characterisation and discovery of new species, identification of cryptic and polymorphic species, curbing the illegal trade, and forensic identification of *Clarias magur*, *Pampus chinensis*, the endangered and wildlife-protected whale shark (*Rhincodon typus*), and sea cow (*Dugong dugon*). Further, species-specific DNA barcodes were developed for over 600 Indian fish species for accurate identification and fishery resource management purposes.

The genomic resources of fish species are useful for developing effective strategies for aquaculture and fisheries management. Population genetic structure for 30 finfish and shellfish species was ascertained using molecular markers. Microsatellite markers were developed for over 50 important Indian fish species for genetic variation and differentiation studies. Deciphered the complete mitochondrial genomes of 28 fish species. Decoded the genomes of various fish species like magur, rohu, and hilsa.

Fish cell lines are considered an important resource for ex-situ conservation and are widely applied for virus isolation, disease diagnosis, ecotoxicology, cytology, developmental biology, and gene expression studies (Murali et al. 2020).

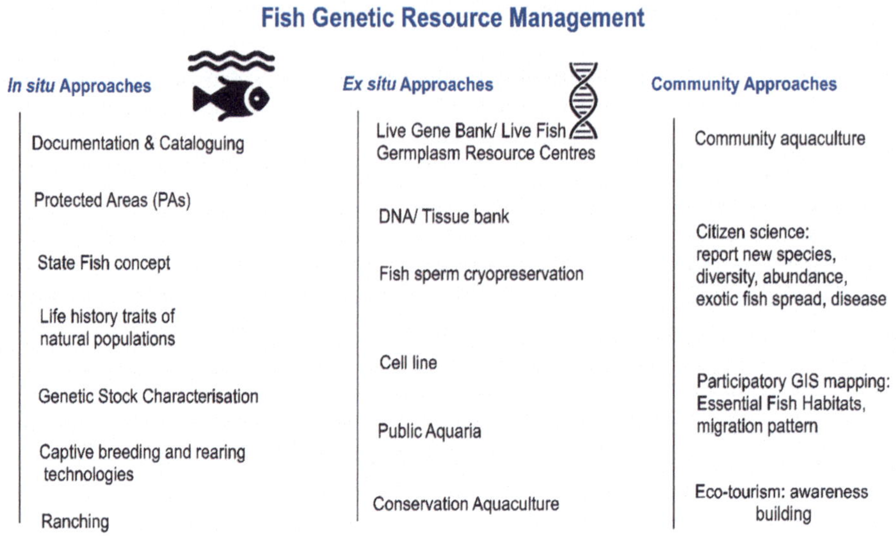

Fig. 1.4 Approaches for the sustainable management of fish genetic resources

ICAR-NBFGR established the National Repository of Fish Cell Lines (NRFC), Lucknow, in 2010, a designated repository under the Biological Diversity Act, 2002 (Subsection (1) of Sec. 39), for the fish genetic resources category. It is one of the largest collections of authenticated fish cell lines in the world and presently holds 81 cell line accessions developed from 16 tissues of 39 fish species belonging to 32 genera (29 freshwater, 3 brackish water, and 7 marine species). The most common approaches used for the sustainable management of fish genetic resources in India are depicted in Fig. 1.4.

1.8 Significance of Sustainable Fish Genetic Resource Management

Fish genetic resources contribute directly or indirectly to the United Nations Sustainable Development Goals (SDGs). Fish resources offer livelihood to the dependent community (SDG 1) and rich sources of various macro- and micronutrients which play a vital role in food security (SDG 2) and ensure well-being (SDG 3). Fisheries have a vast scope for advancing the status of gender equality (SDG 5), and the fish business accelerates the growth of the country's economy (SDG 8). The 'life below water' (SDG 14) is one of the directly linked goals which aim for the conservation and sustainable utilisation of the ocean and marine resources for development. The concept of 'blue economy' involves the sustainable use of ocean resources for the livelihood promotion and economic development of the country, while preserving the health of the marine ecosystems.

1.8.1 Food and Nutritional Security

Fish is a prime source of protein; essential nutrients including omega-3 fatty acids (PUFA), calcium, phosphorus, and vitamins A, D, B, and C; and minerals, viz. iron, zinc, iodine, magnesium, and potassium which are essential for the development of the infants and support the health of adult populations by reducing the risks of malnutrition and noncommunicable diseases (NCDs) including heart disease. India is the third largest fish-producing country, accounting for about 8% of global fish production with 17.54 mil t in the year 2022–2023, which encompasses the inland fish production of 13.1 mil t (PIB 2023a). The marine and inland capture fisheries support the food security of the quantifiable population due to the low domestication level in aquaculture. Hence aquatic biodiversity management and sustainable utilisation are relevant in the present context for achieving food and nutritional security. The small indigenous fishes (SIFs) encompass fin- and shellfish species attaining 25–30 cm maximum length in their adult stage, regarded as the rich source of various nutrients to the rural mass. The major SIFs reported in the wetlands of the Ganga basin include minnows (*Salmostoma*, *Rasbora*, *Amblypharyngodon*, and *Chela*), barbs (*Puntius* sp.), schilibids (*Ailia coila* and *Eutropiichthys*), clupeids (*Gonialosa manmina* and *Gudusia chapra*), and some bagrid catfishes (Sarkar et al. 2010).

1.8.2 Livelihood Promotion

About 59 million people are employed globally in fisheries and aquaculture production systems (FAO 2022). In India, fisheries offer livelihood, employment, and entrepreneurship support to >28 million fishers and fish farmers at the primary level (PIB 2023b). Inland fisheries are significant to poverty alleviation through nutritional and livelihood security for rural populations by supporting the small-scale and subsistence fisheries in rivers, wetlands, and lakes. The involvement of women in fisheries is prominent as they constitute about half of the employed workforce globally. In India, about 12.4 million women are engaged in various activities, especially in marketing, fish processing, seaweed cultivation, and ornamental fish culture. ICAR-NBFGR has established the model of 'conservation aquaculture with community participation' in Lakshadweep Island in which the high-valued marine ornamental fish species are bred in captivity and reared by the marginalised island community which ensures the long-term sustainability of the marine resources and livelihood promotion of the native community (Kumar et al. 2020; Rekha et al. 2024).

1.8.3 Economic Development

The fisheries sector contributes to ~1.09% of India's gross value added (GVA) and > 6.724% to agriculture GVA in 2022–2023 (PIB 2023b). The seafood export of India reached an all-time high value of 17,35,286 MT by quantity, worth USD 8.09

billion in 2022–23, in which the capture fisheries alone contributed to about 37.13% of the export value, while the remaining share was contributed by the aquaculture sector. The ornamental fish was exported to the tune of 305 tonnes with a value of USD 4.76 million in the year 2022–2023 (MPEDA 2023). India with rich indigenous and endemic fish diversity has an important role to play in the global aquarium trade. The development of aquaculture technologies for the fish species being exported and potential ornamentals will lead to the avoidance of wild collection, thus ensuring the ecological balance in the native ecosystem. ICAR-NBFGR has explored the hidden biodiversity of the Lakshadweep Islands and standardised the breeding technology for ornamental shrimp, viz. *Thor hainanensis* and *Ancylocaris brevicarpalis*, and upscaled seed production of indigenous clown fishes, *Amphiprion nigripes* and *A. clarkii*, which are reared by the beneficiaries of the native community for marketing (Rekha et al. 2024). The exploration of the hidden biota in deep-sea and other extreme aquatic habitats will pave the way for the discovery of potential bioactive compounds and the development of value-added fishery products which will uplift the economic development of the country. The ecotourism and sport fishery activities such as camping, angling, coastal walking, etc. are based on the existence of flagship species like mahseer, trout, trevally, and marlins and rich diversity in the natural environment. Hence the conservation and management of these fishes are mandatory to sustain the livelihood of the native community tagged with this sector and support the economic development at the regional level.

1.8.4 Ecological Integrity and Health

Fish is an integral component of the ecosystem and is seen as a part of the food web, nutrient cycling, ecosystem resilience, and engineering process. The rich fish diversity of an ecosystem reveals the ecological integrity. The distribution and relative abundance of fish from diverse ecological niches indicate the healthy state of the ecosystem. Fish is a part of the energy transfer system (food chain, food web, and food pyramid) and nutrient cycling process, thus facilitating the movement of energy in the system. The diversity of fish species at multiple trophic levels maintains the predator-prey relationships and ensures the homeostasis of the ecosystem; the change in the relative abundance of a particular species may lead to the dominating ecosystem which may hamper the ecosystem structure and processes and ecosystem service potential.

1.9 Conclusion and the Way Forward

The sustainable management of fish genetic resources lies in the collective effort of individual stakeholders, policymakers, and resource managers. The applications of advanced tools, viz. integrative taxonomy, open-access databases, eDNA approach, spatial analytics (geographic information system and remote sensing including drones), artificial intelligence, and machine learning are expected to upsurge the role of traditional in-situ and ex-situ approaches. Further, the mapping and quantification of ecosystem services associated with fish genetic resources are mandatory to claim the benefits of conservation approaches being implemented in the country and to reinforce the legal measures for sustainable management. Considering the looming climate change impacts on fish genetic resources, identifying climate resilience stocks is warranted for sustainable fishery development.

References

Agarwal NK, Singh G, Singh H et al (2018) Ecological impacts of dams on the fish diversity of Bhagirathi River in Central Himalaya (India). J Coldwater Fish 1(1):74–84

AqGRISI (2024) Aquatic Genetic Resource Information System of India. https://aqgrisi.nbfgr.res.in/#/. Accessed 6 Apr 2024

Barman RP (2004) The fishes of the Kolleru Lake, Andhra Pradesh, India with comments on their conservation. Rec Zool Surv India 103(1–2):83–89

Barbarossa V, Bosmans J, Wanders N et al (2021) Threats of global warming to the world's freshwater fishes. Nat commun 12(1):1701. https://doi.org/10.1038/s41467-021-21655-w

Bhakta D, Solanki S, Vadhel N et al (2020) Finfish diversity of river Narmada and its tributaries. Proc Zool Soc 73:325–337

Bharadhirajan P, Murugan S, Gopalakrishnan A et al (2015) Finfish diversity in Coleroon Estuary, Southeast coast of India. Indian J Geo-Mar Sci 46(10):1968–1985

Bhattacharjya BK, Choudhury M, Sugunan VV (2004) Icthyofaunistic resources of Assam with a note on their sustainable utilization. In: Mahanta PC, Tyagi LK (eds) Proceedings of the participatory approach for fish biodiversity conservation in north East India. ICAR-NBFGR, Lucknow

Bijukumar A, Philip S, Ali A et al (2013) Fishes of river Bharathapuzha, Kerala, India: diversity, distribution, threats and conservation. J Threat Taxa 5(15):4979–4993

CBD (2022) Global status, gaps and opportunities of protected areas and other area based conservation measures. Published by Convention on Biological Diversity, Montreal, p 6

Chandran R, Tyagi LK, Jaiswar AK et al (2019) Diversity and distribution of fish fauna in the Ib River, a tributary of Mahanadi, India. Indian J Fish 66(1):92–98

Chandran R, Mandal S, Trivesh SM et al (2022) Fish diversity and habitat ecology of Tel River, a tributary of Mahanadi River. J Inland Fish Soc India 54(1):03–15

Danovaro R, Fanelli E, Aguzzi J et al (2020) Ecological variables for developing a global deep-ocean monitoring and conservation strategy. Nat Ecol Evol 4(2):181–192

Das AK, Manna RK, Rao DSK et al (2017) Status of the river Krishna: water quality and riverine environment in relation to fisheries. Aquat Ecosyst Health Manag 20(1–2):160–174

Divya PR, Syamkrishnan MU, Kumar RG et al (2020) Handbook on fishes & shellfishes of Ramsar sites in Kerala. ICAR-NBFGR, Lucknow, p 220

DoF (2020) Handbook on Fisheries Statistics—2020. Published by Department of Fisheries, Ministry of Fisheries Animal Husbandry & Dairying, Government of India, New Delhi

FAO (2019) The state of the World's aquatic genetic resources for food and agriculture. FAO Commission on Genetic Resources for Food and Agriculture assessments, Rome

FAO (2022) The state of world fisheries and aquaculture (2022) Towards blue transformation. Rome, FAO. https://doi.org/10.4060/cc0461en

Fricke R, Eschmeyer WN, Fong JD (2024) Species by family/subfamily. http://researcharchive.calacademy.org/research/ichthyology/catalog/SpeciesByFamily.asp. Accessed 06 Apr 2024

Gupta N, Raghavan R, Sivakumar K et al (2014) Freshwater fish safe zones: a prospective conservation strategy for river ecosystems in India. Curr Sci 107(6):949–950

ICAR-CIFRI (2019) Annual report, April–December 2019. ICAR-Central Inland Fisheries Research Institute, Kolkata

ICAR-CIFRI (2020) Annual report, January–December 2020. ICAR-Central Inland Fisheries Research Institute, Kolkata

ISFR (2021) Mangrove cover in India state of Forest Report-2021. Forest survey of India (FSI). Ministry of Environment Forest & Climate Change, Government of India, New Delhi

IUCN (2024) The IUCN red list of threatened species. Version 2023-1. https://www.iucnredlist.org. Accessed 6 Apr 2024

Jadhav S, Karuthapandi M, Jaiswal D et al (2024) Freshwater fishes of river Godavari. Zoological Survey of India, Kolkata

Jena JK, Gopalakrishnan A (2012) Aquatic biodiversity management in India. Proc Natl Acad Sci India Sect B Biol Sci 82:363–379

Joshi KK, Thobias PA, Varsha MS (2017) Present status of ichthyofaunal diversity of Indian seas. Manual on summer school on advanced methods for fish stock assessment and fisheries management. ICAR-Central Marine Fisheries Research Institute, Cochin

Joshi KD, Basheer VS, Kumar A et al (2021) Alien fish species in open waters of India: appearance, establishment and impacts. Indian J Anim Sci 91(3):167–173

Joshi KD, Kumar A, Srivastava SM et al (2022a) Mounting threat of invasive alien fish species in ganga river system and open water resources of the basin. Indian J Anim Sci 92(2):268–271

Joshi KD, Pathak AK, Kumar S et al (2022b) Fish community diversity assessment of protected Saraiyaman wetland in the Ganga River basin, India. Commun Ecol 23(1):63–73

Kaladharan P, Vijayakumaran K, Singh VV et al (2017) Prevalence of marine litter along the Indian beaches: a preliminary account on its status and composition. J Mar Biol Assoc India 59(1):19–24

Kantharajan G, Pandey PK, Krishnan P et al (2018) Plastics: a menace to the mangrove ecosystem of megacity Mumbai, India. ISME/GLOMIS Electronic J 16(1):1–5

Kantharajan G, Yadav AK, Chandran R et al (2022) Impact of terrestrial protected areas on the fish diversity and habitat quality: Evidence from tropical river Pranhita, India. J Nat Conserv 68:126187. https://doi.org/10.1016/j.jnc.2022.126187

Khedkar GD, Lutzky S, Rathod S et al (2014) A dual role of dams in fragmentation and support of fish diversity across the Godavari River basin in India. Ecohydrology 7(6):1560–1573

Krishnan P, Mukherjee R, Hari MS et al (2022) Streamlining science-policy linkage for fisheries management in South Asia: lessons from regional initiatives. In: George G, Acharya GD, Hossain MB (eds) Cross learning for addressing emergent challenges of aquaculture and fisheries in South Asia. SAARC Agriculture Centre, Dhaka

Kumar T, Charan R, Jayakumar TT et al (2020) Framework for participatory linkage of marine ornamentals germplasm conservation to livelihoods: is community aquaculture an inclusive option? Aquacult Mag 24(4):3–9

Lal DM, Sreekanth GB, Soman C et al (2020) Fish community structure as an indicator of the ecological significance: a study from Ulhas River estuary, Western Coast of India. J Environ Biol 41(4):745–754

Lianthuamluaia L, Mishal P, Panda D et al (2019) Understanding spatial and temporal patterns of fish diversity and assemblage structure *Vis-a-Vis* environmental parameters in a tropical Indian reservoir. Environ Sci Pollut Res 26:9089–9098

Lianthuamluaia L, Kumari S, Sarkar UK et al (2023) Improving approaches and modeling framework for assessing vulnerability of Asian leaf fish in the major river basin floodplains of India in changing climate. Ecol Inform 73:101926

Madhu VR, Raphel L, Jeevan J et al (2017) Status of bycatch from commercial trawlers operated off Central Kerala. Fish Tech 54:162–169

Mogalekar HS (2019) Fishes from Tamiraparani river system, Tamil Nadu. Indian J Anim Sci 89(3):340–346

Mogalekar HS, Canciyal J, Jawahar P et al (2017) Estuarine fish diversity of Tamil Nadu, India. Indian J Geo-Mar Sci 46(10):1968–1985

Mohanty R, Mohapatra A, Mohanty S et al (2015) Ichthyofaunal diversity of Chilika Lake, Odisha, India: an inventory, assessment of biodiversity status and comprehensive systematic checklist (1916-2014). Check List 11(6):1

MPEDA (2023) Marine products export development authority-annual report 2022–23. Published by MPEDA, Ministry of Commerce and Industry, Government of India, Kochi
Mukherjee A, Suresh V (2007) Impact of Farakka barrage on Hilsa fisheries: an overview. In: Goswami UC (ed) Natural and anthropogenic hazards on fish and fisheries. Narendra Publishing House, Delhi, pp 153–160
Murali SK, Kushwaha B, Mishra AK et al (2020) National repository of fish cell lines: accelerating fish cell line research in fisheries and aquaculture. Curr Sci 119:589–590
NWDC (2023) National Wildlife Database Cell. https://wii.gov.in/nwdc_aboutus. Accessed 07 Apr 2024. (Updated 27 Nov 2023)
Pathak AK, Sarkar UK, Abidi R (2020) Status of fish diversity in Surha Taal, a natural lake in the floodplains of river Ganga, in Ballia District, Uttar Pradesh, India. Indian J Fish 67(3). https://doi.org/10.21077/ijf.2020.67.3.70744-18
PIB (2023a) Press releease—Ministry of Fisheries, Animal Husbandry & Dairying, Government of India. https://pibgovin/PressReleaseIframePageaspx?PRID=1936017. Posted 28 June 2023
PIB (2023b) Press release—Ministry of Fisheries, Animal Husbandry & Dairying, Government of India. https://pibgovin/PressReleaseIframePageaspx?PRID=1986155. Posted 14 Dec 2023
Protectedplanet (2024). https://www.protectedplanet.net/en. Accessed 30 May 2024
Rajan PT, Sreeraj CR (2013) Fish fauna of Andaman and Nicobar Islands: a review. In: Venkataraman K, Sivaperuman C, Raghunathan C (eds) Ecology and conservation of tropical marine faunal communities. Springer, Heidelberg
Rajan R, Rajan PT, Mishra SS et al (2021) Fishes of Lakshadweep archipelago: new records, review and a revised checklist. Mar Biodivers Rec 14(1):14
Rao P, Vivekanandan E (2008) Impact of climate change on Indian marine fisheries. CMFRI, Cochin, pp 32–37
Rekha MU, Ajith Kumar TT, Sarkar UK (2024) Potentials and prospects of marine aquaculture a window view from Lakshadweep Island. Indian Farming 74(1):31–34
Sarkar UK, Das BK (2021) Research advances in climate and environmental change impacts on inland fisheries of India: status, vulnerability and mitigation strategies. Aquat Ecosyst Health Manag 24(3):7–17
Sarkar UK, Gupta BK, Lakra WS (2010) Biodiversity, ecohydrology, threat status and conservation priority of the freshwater fishes of river Gomti, a tributary of river Ganga (India). Environmentalist 30:3–17. https://doi.org/10.1007/s10669-009-9237-1
Sarkar UK, Pathak AK, Sinha RK et al (2012) Freshwater fish biodiversity in the river ganga (India): changing pattern, threats and conservation perspectives. Rev Fish Biol Fish 22(1):251–272
Sarkar UK, Pathak AK, Tyagi LK et al (2013) Biodiversity of freshwater fish of a protected river in India: comparison with unprotected habitat. Rev Biol Trop 61(1):161–172
Sarkar UK, Sharma J, Mahapatra BK (2015) A Review on the fish communities in the Indian reservoirs and enhancement of fisheries and aquatic environment. J Aquac Res Development 6:297. https://doi.org/10.4172/2155-9546.100029
Sarkar UK, Bakshi S, Lianthuamluaia L et al (2020a) Understanding enviro-climatological impact on fish biodiversity of the tropical floodplain wetlands for their sustainable management. Sustain Water Resour Manag 6:1–12
Sarkar UK, Mishal P, Borah S et al (2020b) Status, potential, prospects, and issues of floodplain wetland fisheries in India: synthesis and review for sustainable management. Rev Fish Sci Aquac 29(1):1–32
Sarkar UK, Kathirvelpandian A, Kantharajan G et al (2024) The concept of "state fishes of India": current status, knowledge gaps, and strategic plans for conservation and sustainable utilization. J Fish Biol 104:1675. https://doi.org/10.1111/jfb.15729
Sharma AP, Das MK, Vass KK et al (2017) Patterns of fish diversity, community structure and ecological integrity of river Yamuna, India. Aquat Ecosyst Health Manag 20(1–2):30–42
Singh AK, Akhtar MS (2015) Coldwater fish diversity of India and its sustainable development. In: Singh P (ed) Biodiversity for sustainable development. UP Biodiversity Board, Lucknow

Singh AK, Srivastava SC, Kumar D et al (2013) Exotic fish diversity, invasion and its impacts on aquatic biodiversity and ecosystems in Uttar Pradesh. Water and Biodiversity-Uttar Pradesh State Biodiversity Board, Lucknow, pp 129–139

Suseelan C (1987) Impact of environmental changes and human interference on the prawn fishery resources. Mar Fish Infor Serv Tech Exten Ser 73:1–4

Thampy DR, Sethu MR, Paul MB et al (2021) Ichthyofaunal diversity in the upper-catchment of Kabini River in Wayanad part of Western Ghats, India. J Threat Taxa 13(2):17651–17669. https://doi.org/10.11609/jott.6159.13.2.17651-17669

Tyagi LK, Chandran R, Mandal S (2021) Checklist of fishes of Mahanadi River, India. J Indian Fish Assoc 48(1):1–11

Vass KK, Das MK, Srivastava PK et al (2009) Assessing the impact of climate change on inland fisheries in river ganga and its plains in India. Aquat Ecosyst Health Manag 12(2):138–151

Venkataraman K, Sivaperuman C (2018) Biodiversity hotspots in India. In: Sivaperuman C, Venkataraman K (eds) Indian hotspots: vertebrate faunal diversity, conservation and management, vol 1. Springer, Cham, pp 1–27. https://doi.org/10.1007/978-981-10-6605-4_1

Venu S (2013) Deep-Sea fish distribution along the south-west region of Indian EEZ. In: Venkataraman K, Sivaperuman C, Raghunathan C (eds) Ecology and conservation of tropical marine faunal communities. Springer, Berlin, Heidelberg. https://doi.org/10.1007/978-3-642-38200-0_17

Vishwanath W (2017) Diversity and conservation status of freshwater fishes of the major rivers of Northeast India. Aquat Ecosyst Health Manag 20(1–2):86–101

Wagh P, Kurhade S, Jadhav S et al (2018) Fish fauna of Nandur-Madhmeshwar wetland, Maharashtra, India. J Threat Taxa 10(7):11973–11979

Watts JE, Schreier HJ, Lanska L et al (2017) The rising tide of antimicrobial resistance in aquaculture: sources, sinks and solutions. Mar Drugs 15(6):158

Genetic Variability in Natural Populations of Fishes

2

Rajeev Kumar Singh, P. R. Divya, A. Kathirvelpandian, and Mahender Singh

Abstract

Intraspecific variation encompasses variability at the genetic as well as phenotypic level and is found within and among populations. The documentation of below-species-level (intraspecific) variations is important for conservation and management. The FAO's Commission on Genetic Resources for Food and Agriculture (CGRFA) has underscored the significance of biodiversity conservation for posterity. Advancements in molecular markers have enabled the delineation of genetic diversity (intra-and interspecific), which has applications in stock characterization, parentage assignment, and genetic rescue/stock enhancement programs. This chapter highlights the importance and need for revealing intraspecific genetic variations, their quantification, and neutral as well as adaptive divergence. Evolutionary forces that shape the patterns of genetic diversity are also described. Threats emanating from various anthropogenic causes, especially loss of natural genetic diversity due to the unintended mixing of farm-bred stock are worth mentioning in the current scenario. The genetic information, below-species-level plays a significant role in resource management, registration of genetic stocks and the elite germplasm.

Keywords

Intraspecific variation · Stock characterization · Divergence · Evolutionary forces · Management

R. K. Singh (✉) · P. R. Divya · A. Kathirvelpandian · M. Singh
ICAR-National Bureau of Fish Genetic Resources, Lucknow, Uttar Pradesh, India
e-mail: Rajeev.Singh1@icar.gov.in; Divya.R@icar.gov.in; A.Pandian@icar.gov.in; Mahender.singh2@icar.gov.in

U. K. Sarkar et al. (eds.), *Sustainable Management of Fish Genetic Resources*, https://doi.org/10.1007/978-981-97-5250-8_2

2.1 Introduction

Populations, groups of interbreeding individuals, that exchange members with other populations very rarely, are considered as management units. Population genetics is the discipline related to genetic differences within/across populations and their dynamics. These differences arise due to mutations in germline cells of the individuals. For fisheries management, the principles of population genetics play a vital role. Genetic variability manifests in two distinct forms: intraspecific and interspecific. Intraspecific variability denotes the variations in genetic characteristics among individuals of the same species. Intraspecific variability is of paramount importance as it ensures a rich gene pool within a species, reinforcing its capacity to navigate evolving environments and challenges. However, this diversity is often undervalued due to limited sample sizes or restricted geographic sampling, leading to a misinterpretation of the true magnitude of genetic diversity within a species. Conversely, interspecific genetic variability encompasses the differences between different fish species, which is crucial for comprehending the evolutionary relationships, and can provide insights into their shared genetic history.

2.2 Genetic Variation in Natural Populations

Marine species. The natural populations of marine species are typically panmictic or have random mating, with an occasional subdivided or structured populations, known as demes. Genetic variation in these populations results from the interplay between local populations or genetic diversity between populations. The study conducted by Gandra et al. (2021) provided insights into the genetic differentiation of more than 170 fish species across the globe, revealing that marine species had lower genetic differentiation than others. This discovery has significant implications for our understanding of marine ecosystems. However, other schools of thought emphasize that marine species with high dispersal have more subdivisions than previously believed. The application of molecular techniques and multi-locus approaches have provided a new understanding of population divergence (Thomas 1996). According to another study conducted by Martinez et al. (2018), marine fishes had significantly higher genetic diversity than freshwater fishes. It is worth noting that there is a positive relationship between genetic diversity and sea surface temperature for marine species (Manel et al. 2020). The age at maturity and fecundity were negatively correlated with genetic variation in marine and freshwater fishes.

Freshwater species. The genetic diversity of freshwater species varies primarily across the regional basins and has been negatively correlated with the average river slope. However, it is interesting to observe that the genetic diversity was not lower in threatened fishes compared to non-threatened ones, contrary to the researchers' expectations (Martinez et al. 2018). The study suggests that life history characteristics and habitats are important factors influencing the patterns of genetic diversity in fishes and should be considered when prioritizing species for conservation. In the current scenario of climate change, it is noted that tropical species have less variation in their tolerance to heat than the temperate species. However, among tropical species, both freshwater and island ecosystems exhibit less tolerance variation, which indicates that they are more vulnerable to the harmful effects of thermal stress.

2.2.1 Genetic Variation Within Populations

Genetic variation is fundamental for the short as well as long-term survival of any species as it enables the populations to adapt to changing environmental conditions. Genetic variability is assessed as two major approaches: (1) heterozygosity and (2) allele/haplotype diversity, of which the heterozygosity is mostly used for comparing genetic variation. The parameters used to assess genetic variation within natural populations are described below:

H-statistics. There are two different measures of heterozygosity, observed (Hobs) and expected (Hexp). Hobs and Hexp represent the observed and expected proportion of heterozygotes, given that the population is random mating. While comparing the two values, Hexp can better estimate the genetic status of the population (Nei 1973). For a given locus, a proportion of heterozygous individuals is known as heterozygosity, which ranges from 0 to 1. Lower values of H (=0) demonstrate that all individuals possess the identical allele, while higher H (=1) represents all individuals with entirely different alleles. For mitochondrial genes, the corresponding parameter is 'h' that shows that two individuals (randomly drawn) possess different haplotypes. As per an estimate, lower heterozygosity (0.05) was determined for allozymes, whereas microsatellites exhibited much higher (0.60) values (DeWoody and Avise 2000).

Allele/haplotype diversity. Allelic or haplotype diversity is another important estimate of genetic variability. Depending on effective population size and evolutionary history, the allele number, and their frequency values are also related to the sample size. Allelic richness measures the long-term evolutionary potential of populations, and therefore, it is also an important estimate. While quantifying, frequency of the most common microsatellite or allozyme allele is usually set at 0.95 or 0.99. Genetic variation can also be measured by the proportion of loci that possess alternative alleles. Compared to natural populations, small isolated populations are severely affected owing to genetic drift as well as inbreeding. This results into the reduction in allelic diversity and heterozygosity, which may further affect the fitness. Out of several markers, microsatellites are considered the most effective for estimating allelic diversity in a population. These markers can assess the allele diversity loss in a stock/population.

Nucleotide diversity. Nucleotide diversity estimates the average divergence in DNA sequence between individuals. Lower values (=0) demonstrate that all members are genetically identical; however, this exceed 0.05 very rarely within fish populations. A low value of genetic diversity indicates that colonization might have been initiated by a small number of individuals, while higher levels are indicative of older populations.

Hardy-Weinberg Equilibrium. The Hardy-Weinberg principle states that the allele and genotype frequencies remain in equilibrium, given that the population is large and randomly mating. This principle is based on certain assumptions such as large population, random mating, the absence of natural selection, no gene flow or migration, no mutation, and autosomal locus. In case of violation of these assumptions, deviation from Hardy-Weinberg proportions is experienced.

Effective population size (Ne). This parameter displays the number of individuals that pass their genes to the next generation. Ne is determined by the extent of nucleotide diversity. Higher levels are exhibited by a stable population.

2.2.2 Genetic Variation Among Populations

The components of genetic variation, within a single population and among different populations, can be assessed by the hierarchical analysis of partitioning. This statistical approach partitions the total genetic variation of a species into different components. Another important approach is the pairwise coefficient of genetic differentiation. The details are given below:

F-statistics. The coefficient of genetic differentiation (F_{ST}) represents the proportion of intraspecific genetic variability, which ranges from zero to one. Zero represents the absence of genetic structure, a high extent of gene flow, and that the populations are frequently interbreeding. In such a situation, the population is considered as a single panmictic without any substructure. The higher F_{ST} indicates low or absence of gene flow and substructuring in the population, while the values close to zero show the absence of genetic differentiation. A modified form known as GST is used for haploid data such as mtDNA. Rho statistics (RST) represents the expectations about how microsatellites mutate (Slatkin 1995). Smaller values of RST are indicative of a significant restriction on gene flow between populations.

Partitioning of genetic variance. Analysis of molecular variance (AMOVA) is most popularly used to detect population differentiation. This is employed for haplotypes, as well as allelic datasets. In panmictic populations, most of the variance is expected to arise from within samples. However, if the analysis shows that most of the variation is among samples (within populations) or among populations, it becomes obvious that the population is substructured.

Principal Component Analysis. Principal component analysis (PCA) is based on multiple variables. This preserves the covariance while reducing the data's dimensionality. PCA is performed to identify the main components of variation in a dataset. Major variation is explained by principal component 1 (PC1), which is followed by PC2. This process continues (cumulatively) till all the available variation is explained. In the marker dataset, PCA identifies major axes of variation in allele frequencies and subsequently produces the coordinates.

Discriminant Analysis of Principal Components. Discriminant analysis of principal components (DAPC) (Jombart et al. 2010) is also a multivariate statistical tool, which clusters the genetically related individuals. In this approach, log transformed data is used for identifying clusters in a gene pool. The populations are structured on the basis of principal component and discriminant analyses. It is a powerful tool compared to other Bayesian clustering methods.

Population structure. Population genetic structure represents the interplay between ecological factors and genetics. Understanding the genetic characteristics of a population is important for conservation. Groups of individuals, showing significantly different haplotype/allele frequencies represent shallow population structure, while deep population structure refers to accumulated differences in sequences over time.

2.3 Tools to Assess Genetic Variability

Various marker systems have been developed ranging from allozymes to single nucleotide polymorphisms. Every marker has its own specialty; however, the usage depends on the questions that are being addressed. Generally, the molecular markers are classified as type I or type II markers: type I is associated with genes of known function, while type II is associated with anonymous genomic segments. As such, allozyme markers were considered important type I markers, in the past. Expressed sequence tag (EST) markers are also type I markers because they represent transcripts of genes. Random amplified polymorphic DNA (RAPD), amplified fragment length polymorphism (AFLP), and microsatellite markers are type II markers. Single nucleotide polymorphism (SNP) markers are mostly type II unless developed from expressed sequences. Substantial genetic variability exists even between different populations within the same species. From conservation point of view, such intraspecific differences are important to document, and genetic markers can successfully decipher the extent of genetic variability. Moreover, it becomes more informative when these markers are used in combination. For example, the combined use of mitochondrial gene and nuclear markers can potentially assess the female-versus-male-mediated gene flow. Different markers and their applications have been discussed in another chapter.

2.4 Genetic Characterization

The FAO's Commission on Genetic Resources for Food and Agriculture (CGRFA) highlights the significance of conserving biodiversity for nutritional security. International obligations such as SDG 14 mentions about sustainable utilization of the genetic resources. For conservation, we need to identify causes of stock depletion. These causes include destructive, unregulated fishing practices and illegal overfishing. The FAO also emphasizes documenting genetic information on stocks and strains. All of this necessitates documenting the native fish diversity below the species level (Singh et al. 2022).

Since its inception, ICAR-NBFGR has focused on deciphering intraspecific genetic variation of fish species, which are important for food and trade. Currently, about 41 prioritized finfish/shellfish species from Indian waters have been genetically characterized by NBFGR (alone or in partnership). The majority of the species were characterized in a flagship program, "Outreach Activity on Fish Genetic Stocks", which was implemented in two phases (Phase I, 2008–2014; Phase II, 2015–2020). These two plan-fund programs were awarded by the Indian Council of Agricultural Research (ICAR) to NBFGR as the need was felt to register our native fish germplasm to comply with international obligations. The species' that were characterized in these multi-institutional programs belonged to diverse ecological regimes, including fresh, brackish, and marine waters. Subsequently, other funding agencies such as the Department of Biotechnology and the Department of Science and Technology had also funded for the work on similar lines.

Genetic stock identification of prioritized species, markers used and significant inferences are summarized in Table 2.1.

2.5 Adaptive Genetic Diversity

Genetic studies using neutral or type II molecular markers have been commonly used for making informed decisions and formulating conservation strategies; however, this lacks the information about the adaptive evolution. To preserve adaptive genetic variation, geneticists and population biologists advocated an alternate approach that could identify the gene pools, experiencing selection pressure. With high-throughput sequencing, identification of large numbers of outlier loci (in genes) has become possible. Adaptive or outlier loci (SSR and SNP) fall beyond the range of expected variation compared to the neutral loci. Analysis of outlier loci provides estimates of adaptive divergence. These loci, associated with production traits are a promising tool for addressing emerging issues such as climate change and overexploitation.

In order to determine adaptive divergence, adaptive genes are identified after determining adaptive variation in large populations. The markers that are associated with genes of known functions are generally used. Population variability parameters are determined, and the candidate outlier loci are correlated with prevailing environmental variables.

2.6 Role of Evolutionary Forces

Mutation. Mutations are unexpected alterations in the DNA sequence of an organism. They play a crucial role in shaping the genetic variability of populations. Mutations can occur due to various factors, such as errors during DNA replication, exposure to mutagens, or environmental stressors. These genetic changes can lead to emergence of new traits and genetic diversity, ultimately driving the evolution of fish species. Point mutations, insertions and deletions (indels), repeat expansions, and chromosomal rearrangements, all leading to genetic variability in an organism/population.

Table 2.1 Genetic stock identification of important species from Indian waters

Sl. no.	Species	Habitat	Markers	Inference	References
1	*Priacanthus hamrur*	Marine water	RAPD; mitochondrial cyt b	Lack of genetic subdivision	Vidya et al. (2017)
2	*Pampus candidus*	Marine water	Microsatellite, ATPase 6/8	Three distinct stocks in Indian waters	Divya et al. (2015), Mohitha et al. (2021)
3	*Rachycentron canadum*	Marine/brackish water	Microsatellite, ATPase 6/8	Three distinct stocks in Indian waters	Joy et al. (2016) Divya et al. (2019)
4	*Coilia dussumieri*	Marine/ brackish/ freshwater	Microsatellite, ATPase 6/8	Two distinct stocks, east and west population	Kathirvelpandian et al. (2014a, b)
5	*Rastrelliger kanagurta*	Marine water	Microsatellite, cyt b	Lack of genetic subdivision in the Indian mainland and distinct from the Andamans population	ICAR NBFGR Annual Report (2015)
6	*Scomberomorus commerson*	Marine water	Microsatellite, D-loop	Lack of genetic subdivision	Radhakrishnan et al. (2018), Vineesh et al. (2016, 2017), Joy et al. (2016)
7	*Chanos chanos*	Marine/ brackish/ freshwater	ATPase 6/8 Microsatellite, cyt b	Low but significant structuring	Sri Hari et al. (2019), Jose et al. (2023)
8	*Sillago malabarica*	Marine/brackish water	Microsatellite, cyt b	Low but significant structuring	ICAR Annual Report (2020)
9	*Etroplus suratensis*	Brackish water/ freshwater	Microsatellite, cyt b	Four distinct stocks in Indian waters	Chandrasekar et al. (2019), Balasubramaniam et al. (2022)
10	*Lutjanus argentimaculatus*	Marine/ brackish/ freshwater	Microsatellite, ATPase 6/8, cyt b	Three distinct stocks in Indian waters	Vineesh et al. (2023)
11	*Eleutheronema tetradactylum*	Marine/ brackish/ freshwater	Microsatellite, ATPase 6/8, cyt b, COI	Distinct genetic structures (FAO fishing areas 51, 57, and 61)	Ismail et al. (2019)

(continued)

Table 2.1 (continued)

Sl. no.	Species	Habitat	Markers	Inference	References
12	*Sardinella longiceps*	Marine water	Microsatellite, SNP, D-loop, COI	Major subdivisions between Mumbai and Mangalore, Oman Sea and Indian waters	Sebastian et al. (2017)
13	*Trichiurus lepturus*	Marine/brackish water	COI	Major subdivisions between east and west coast of India	Lakshmi (2023)
14	*Sphyrna lewini*	Marine/brackish water	COI, D-loop	Lack of genetic subdivision in Indian waters	Sukumaran et al. (2020)
15	*Scoliodon laticaudus*	Marine/brackish water	COI, D-loop, ND2	Two distinct genetic subdivisions in Indian waters	Sukumaran et al. (2023)
16	*Carcharhinus longimanus*	Marine water	D-loop	Lack of genetic subdivision in Indian waters	Sreelekshmi et al. (2020)
17	*Chitala chitala*	Freshwater	Cyt b, ATPase6/8, microsatellite	High gene flow in Gangetic rivers, four natural genetic stocks	Dutta et al. (2020)
18	*Tor tor*	Freshwater	Cyt b, ATPase6/8, microsatellite	Four genetic stocks	Sah et al. (2021)
19	*Silonia silondia*	Freshwater	Cyt b, ATPase6/8, microsatellite	Three management units	Mandal et al. (2020)
20	*Labeo rohita*	Freshwater	Microsatellite, cyt b	10–11 genetic stocks	Luhariya et al. (2012)
21	*Labeo catla*	Freshwater	Microsatellite, cyt b	9–11 genetic stocks	Das et al. (2012)
22	*Cirrhinus mrigala*	Freshwater	Microsatellite, cyt b	8–10 genetic stocks	Chauhan et al. (2007)
23	*Labeo calbasu*	Freshwater	Microsatellite, ATPase6/8	Low to moderate level of differentiation	Singh et al. (2012, 2016)
24	*Tenualosa ilisha*	Marine/brackish/freshwater	Cyt b, ATPase6/8, 12SrRNA, 16S rRNA	Two distinct genetic stocks	Mohindra et al. (2019)
25	*Notopterus notopterus*	Freshwater	ATPase6/8	Common ancestry of Mahanadi population	Gupta et al. (2013)

26	*Lates calcarifer*	Marine/ brackish/ freshwater	Microsatellite, cyt b ATPase6/8	Moderate to high level of differentiation	Singh et al. (2023)
27	*Channa marulius*	Freshwater	Cyt b, ATPase6/8, microsatellite	4–5 genetic stock	Pathak et al. (2018)
28	*Ompok bimaculatus*	Freshwater	Microsatellite	Substructured population	Chowdhury et al. (2023)
29	*Osteobrama belangeri*	Freshwater	ATPase6/8	Substructured population	Maisnam et al. (2017)
30	*Labeo dero*	Freshwater	Allozymes, microsatellite	Five genetic stocks	Mohindra et al. (2005), Chaturvedi et al. (2011)
31	*Labeo gonius*	Freshwater	ATPase6/8, D-loop	Greater divergence in east coast (Himalayan rivers)	Roy et al. (2024)
32	*Tor khudree*	Freshwater	ATPase6/8, cyt b	High genetic diversity, recent demographic expansion	Das et al. (2024)
33	*Sperata seenghala*	Freshwater	Microsatellite	Five distinct genetic stocks	Acharya et al. (2019)
34	*Neolissochilus hexagonolepis*	Freshwater	Microsatellite	Two populations with high genetic diversity	Sharma et al. (2019)
35	*Labeo fimbriatus*	Freshwater	ATPase 6, D-loop, cyt b, microsatellite	Two genetic stocks	Swain et al. (2022)
Shellfishes					
36	*Thenus orientalis*	Marine	RAPD; cyt b, COI	Lack of genetic subdivision	Jeena et al. (2016)
37	*Parapenaeopsis stylifera*	Marine	Microsatellite	Two distinct genetic subdivisions in Indian waters	Chowdhury et al. (2021)
38	*Perna viridis*	Marine/brackish	RAPD, microsatellite; cyt b	Genetic structuring	Divya et al. (2012, 2020, 2022)

(continued)

Table 2.1 (continued)

Sl. no.	Species	Habitat	Markers	Inference	References
39	*Panulirus polyphagus*	Marine	COI, D-loop	Lack of genetic subdivision	CMFRI (2018)
40	*Macrobrachium rosenbergii*	Brackish/ freshwater	ATPase 6/8 genes, microsatellite	Substructured population	Kumar et al. (2017)
41	*Metapenaeus dobsoni*	Marine/brackish	RAPD	Lack of genetic subdivision in Indian waters	Mishra et al. (2009)

Genetic recombination. Genetic recombination is a natural process that occurs during cell division when genetic material from two individuals is combined to create offspring with a unique combination of genes. During this process, genetic material is exchanged between multiple chromosomes or different regions of the same chromosome. Recombination is an essential mechanism in DNA repair and replication for both prokaryotic and eukaryotic organisms. It is the primary means by which diversity is introduced into populations.

Genetic drift is likely to occur in all-natural populations. Even the larger populations experience the effects of genetic drift. Hence, it is important to understand this event. Genetic drift is defined as a random change in allele frequencies from generation to generation. This effect was first demonstrated by R. A. Fisher and Sewall Wright in small populations. In general, the effects of genetic drift are larger on small populations. The changes in allele frequencies can affect the genetic traits over several generations. Additionally, genetic drift can cause a new population to have genetic characteristics different from those of its original population, which suggests that it may contribute to the evolution of new species.

Gene flow. Gene flow is the movement of genes between populations, resulting in the transfer of genetic variation. Gene flow maintains consistency within a species. When gene flow is restricted, the species can diverge, leading to new species. This process, called speciation, contributes to biodiversity.

2.7 Founder and Wahlund Effects

Founder effect. In a new population, founded by a small number of individuals, the allele frequencies are changed. This change is due to the fact that a small number of individuals represent only a small proportion of the original genetic diversity. This also results in the loss of genetic variation. Founder events and population bottlenecks cause a greater effect on the number of alleles in a population, compared to heterozygosity.

Wahlund effect. Under the Wahlund effect, the expected heterozygosity (Hexp) of a locus gets reduced compared to Hardy-Weinberg expectations. This deficit in Hexp occurs when the population is substructured.

2.8 Threats to Natural Genetic Diversity

There are many reasons for the loss of genetic diversity in natural fish populations. Overfishing is one of the main causes. Many studies have shown that when fish populations decline due to overfishing, their genetic diversity also decreases (Sadler et al. 2023), but some studies have yielded different inferences. The differences in the habitats and life history traits of the fish are some of the reasons for this contradiction. The methodologies used to estimate the effect of fishing on genetic diversity also differ among studies. Anthropogenic disturbances, such as climate change, also exacerbate the loss of genetic diversity. Gene flow through immigration, availability of refugia from fisheries, or mutation, can help mitigate the loss of genetic diversity. Therefore, fisheries management should consider the loss of genetic diversity when managing recovery of populations. Farm escapees can also be a source of a significant threat to the genetic health of natural populations. These escapees interbreed with wild conspecifics and put them at risk. Several researchers have mentioned the genetic risks emanating from intentional/accidental release of hatchery-bred individuals in the wild (Waples et al. 2016). This results in the loss of genetic diversity, within and among populations. Theoretically, the genetic diversity loss has a direct effect on the adaptive potential of the species. This issue aggravates more when farmed individuals are raised from a small number of parents. In this context, Ryman-Laikre effect assesses the reduction in effective population size wherein the captive and wild populations thrive together. This results in an increase in inbreeding and occurrence of genetic drift, ultimately leading to genetic homogenization among the native populations (Waples et al. 2016).

2.9 Significance of Stock Characterization

Resource management. Conserving biodiversity is important for the sustainable utilization of natural resources. It is essential to delineate population substructure and identify the genetic stocks. Based on the genetic patterns, the unique stock may be promoted for scientific management.

Domestication. Characterization can determine distinct natural stock that can be used to derive domesticated strains. Domestication is driven by several genetic processes. These include inbreeding, genetic drift (leading to changes in the gene pool), natural selection (certain traits favoring a captive environment), and the effects of selection pressure.

Conservation and biodiversity. Deciphering genetic parameters and patterns of genetic diversity enables policymakers to make informed decisions for conservation of unique genetic variants (distinct stocks), and document the genetic health of a population. The characterization can be of particular importance for endangered species, that can prevent their extinction by adopting suitable measures. Genetically heterogeneous stocks can be gene banked, and used for restocking natural bodies. In marine populations, genetic stock identification is the most effective when the species are philopatric, and dispersal is restricted among populations with significant levels of differentiation.

Selective breeding and aquaculture. Genetically diverse populations are important for creating base populations, which is crucial for selective breeding programs. It is widely accepted to assemble a base population drawn from diverse stocks so as to capture comprehensive genetic variation. Such information can help in enhancing fish stocks by identifying superior traits (growth, resistance, etc.).

Adaptation and evolution. Fishes inhabit diverse environments (salinity, temperature, etc.). Migratory fishes have the potential to adapt to the environment. Using genome sequences, markers with adaptive potential may be identified and related to environmental variables. Evolutionary relationships among fish species can be delineated using comparative genomics.

References

Acharya AP, Pavan-Kumar A, Babu PG et al (2019) Population genetics of Indian giant river-catfish, *Sperata seenghala* (Sykes, 1839) using microsatellite markers. Aquat Living Resour 32:4

Balasubramaniam S, Soman M, Katneni VK et al (2022) Mitochondrial DNA based diversity studies reveal distinct and sub structured populations of pearlspot, *Etroplus suratensis* (Bloch, 1790) in Indian waters. J Genet 101(3):PMID35129136

Chandrasekar S, Sivakumar R, Mathialagan R et al (2019) Population genetic structure of *Etroplus suratensis* Bloch, 1790 in South India: preliminary evidence of founder haplotypes shared among populations. J Asia-Pac Biodivers 12(3):376–381

Chaturvedi A, Mohindra V, Singh RK et al (2011) Population genetic structure and phylogeography of cyprinid fish, *Labeo dero* (Hamilton) inferred from allozyme and microsatellite DNA marker analysis. Mol Biol Rep 38(5):3513–3529

Chauhan T, Lal KK, Mohindra V et al (2007) Genetic differentiation in the population of Indian major carp, *Cirrhinus mrigala* (Hamilton-Buchanan, 1882): evidence from allozyme and microsatellite markers. Aquaculture 269:135–149

Chowdhury LM, Kathirvelpandian A, Divya PR et al (2021) Genetic characterization of Kiddi shrimp, *Parapenaeopsis stylifera* (H. Milne Edwards, 1837) Along the Indian coast using microsatellite markers. Fish Res 244:106128

Chowdhury LM, Chaturvedi S, Mandal S et al (2023) Development of novel microsatellite markers for population differentiation and detection of natural selection in wild populations of butter catfish, *Ompok bimaculatus* (Bloch, 1794). Mol Biol Rep 50:2435–2444

Das R, Mohindra V, Singh RK et al (2012) Intraspecific genetic diversity in wild *Catla catla* (Hamilton, 1822) population assessed through mtDNA cytochrome b sequences. J Appl Ichthyol 28:280–283

Das G, Das SP, Sahoo L et al (2024) High genetic differentiation and genetic diversity in endangered Mahseer *tor khudree* (Sykes, 1839) as revealed from concatenated *ATPase 6/8 and Cyt b* mitochondrial genes. Biochem Genet. https://doi.org/10.1007/s10528-023-10623-2

DeWoody J, Avise J (2000) Microsatellite variation in marine, freshwater and anadromous fishes compared with other animals. J Fish Biol 56:461–473

Divya PR, Thomas PC, Gopalakrishnan A et al (2012) Stock structuring in Asian green mussel *Perna viridis* population along the Indian coast based on shell morphometrics and RAPD markers. Indian J Anim Sci 82(7):775–778

Divya PR, Gopalakrishnan A, Basheer VS et al (2015) Mitochondrial ATPase 6/8 genes to infer the population genetic structure of silver pomfret fish *Pampus argenteus* along the Indian waters. Mitochondrial DNA B Resour 26(2):189–194

Divya PR, Linu J, Mohitha C et al (2019) Deciphering demographic history and fine-scale population structure of cobia, *Rachycentron canadum* (Pisces: Rachycentridae) using microsatellite and mitochondrial markers. Mar Biodivers 49:381–393

Divya PR, Jency PME, Joy L et al (2020) Population connectivity and genetic structure of Asian green mussel, *Perna viridis* along Indian waters assessed using mitochondrial markers. Mol Biol Rep 47(7):5061–5072

Divya PR, Joy L, Poulose S et al (2022) Genetic stock identification in *Perna viridis* (Linnaeus1758) from the Indian peninsula by using microsatellite markers. Mol Biol Rep 49(4):3357–3363

Dutta N, Singh RK, Pathak A et al (2020) Mitochondrial DNA markers reveal genetic connectivity among populations of Osteoglossiform fish *Chitala chitala*. Mol Biol Rep 47:8579–8592

Gandra M, Assis J, Martins MR, Abecasis D (2021) Reduced global genetic differentiation of exploited marine fish species. Mol Biol Evol 38(4):1402–1412
Gupta A, Lal KK, Mohindra V et al (2013) Genetic divergence in natural populations of bronze featherback, *Notopterus notopterus* (Osteoglossiformes: Notopteridae) from five Indian rivers, analyzed through mtDNA ATPase6/8 regions. Meta Gene 12(1):50–57
ICAR-CMFRI (2018) Annual Report 2017–18. Central Marine Fisheries Research Institute, Kochi
ICAR-NBFGR (2015) Annual report 2015. ICAR-National Bureau of Fish Genetic Resources, Lucknow
ICAR-NBFGR (2020) Annual Report 2020. ICAR-National Bureau of Fish Genetic Resources, Lucknow
Ismail S, Vineesh N, Peter R et al (2019) Identification of microsatellite loci, gene ontology and functional gene annotations in Indian salmon (*Eleutheronema tetradactylum*) through next-generation sequencing technology using illumina platform. Ecolog Genet Genom 11:100038
Jeena NS, Gopalakrishnan A, Kizhakudan JK et al (2016) Population genetic structure of the shovel-nosed lobster *Thenus unimaculatus* (Decapoda, Scyllaridae) in Indian waters based on RAPD and mitochondrial gene sequences. Hydrobiologia 766:225–236
Jombart T, Devillard S, Balloux F (2010) Discriminant analysis of principal components: a new method for the analysis of genetically structured populations. BMC Genet 11:94
Jose DM, Divya PR, Lal KK (2023) Panmictic stock structure of milkfish (*Chanos chanos*, Forssål 1775) from Indian waters determined using mtDNA marker. J Genet 102:3
Joy L, Mohitha C, Divya PR, Gopalakrishnan A et al (2016) Weak genetic differentiation in cobia, *Rachycentron canadum* from Indian waters as inferred from mitochondrial DNA ATPase 6 and 8 genes. Mitochondrial DNA A DNA Mapp Seq Anal 27(4):2819–2821
Kathirvelpandian A, Gopalakrishnan A, Lakra WS et al (2014a) Microsatellite markers to determine population genetic structure in the Golden anchovy, *Coilia dussumieri*. Biochem Genet 52(5–6):296–309
Kathirvelpandian A, Gopalakrishnan A, Lakra WS et al (2014b) Mitochondrial ATPase 6/8 genes reveal genetic divergence in the *Coilia dussumieri* (Valenciennes, 1848) populations of north east and northwest coasts of India. Mol Biol Rep 41(6):3723–3731
Kumar R, Gopalakrishnan A, Divya PR et al (2017) Population genetic structure of *Macrobrachium rosenbergii* (Palaemonidae) from Indian waters using mitochondrial ATPase 6/8 gene. Mitochondrial DNA A DNA Mapp Seq Anal 28(4):602–605
Lakshmi M (2023) Population structure and genetic variability of the largehead hairtail *Trichiurus lepturus* linnaeus 1758 from Indian waters. PhD Thesis, Mangalore University, Mangaluru, Karnataka
Luhariya RK, Lal KK, Singh RK et al (2012) Genetic divergence in wild population of *Labeo rohita* (Hamilton, 1822) from nine Indian rivers, analyzed through MtDNA cytochrome b region. Mol Biol Rep 39(4):3659–3665
Maisnam M, Gupta A, Lal KK et al (2017) Genetic divergence in natural and farm populations of Pengba fish, *Osteobrama belangeri* (Valenciennes, 1844), an endemic fish of north-East India derived from mtDNA *ATPase 6/8* gene. Mitochondrial DNA B Resour 2(2):658–661
Mandal S, Singh A, Priyanka S (2020) Genetic and morphological assessment of a vulnerable large catfish, *Silonia silondia* (Hamilton, 1822), in natural populations from India. J Fish Biol 98:430–444
Manel S, Guerin PE, Mouillot D et al (2020) Global determinants of freshwater and marine fish genetic diversity. Nat Commun 11:692
Martinez AS, Willoughby JR, Christie MR (2018) Genetic diversity in fishes is influenced by habitat type and life-history variation. Ecol Evol 8(23):12022–12031
Mishra PS, Chaudhari A, Krishna G et al (2009) Genetic diversity in *Metapenaeus dobsoni* using RAPD analysis. Biochem Genet 47(5–6):421–426
Mohindra V, Anshumala SRK et al (2005) Allozyme markers for population structure analysis in *Labeo dero* (Hamilton-Buchanan, 1822). Asian Fish Sci 18(1& 2):71–76

Mohindra V, Divya BK, Kumar R et al (2019) Genetic population structure of a highly migratory Hilsa Shad, *Tenualosa ilisha*, in three river systems, inferred from four mitochondrial genes analysis. Environ Biol Fish 102:939–954

Mohitha C, Divya PR, Joy L et al (2021) Spatial structuring of silver pomfret *Pampus candidus* in northern Indian Ocean using microsatellite DNA. Animal Gene 19:200110

Nei M (1973) Analysis of gene diversity in subdivided populations. Proc Natl Acad Sci USA 70(12):3321–3323

Pathak A, Singh RK, Mohindra V et al (2018) Development and characterization of novel microsatellite markers in great snakehead, *Channa marulia* (Hamilton, 1822). Meta Gene 18:143–148

Radhakrishnan DP, Nedumpally V, Kathirvelpandian A et al (2018) Population structure of Spanish mackerel *Scomberomorus commerson* (Lacepede 1800) in the northern Indian Ocean determined using microatellite markers. Aquat Living Resour 31:22

Roy S, Behera BK, Rout VLR et al (2024) Genetic characterization of minor carp (*Labeo gonius*) from Indian rivers revealed through mitochondrial ATPase 6/8 and Dloop region analysis: implications for conservation and management. Front Mar Sci 11:1345649

Sadler DE, Watts PC, Uusi-Heikkilä S (2023) The riddle of how fisheries influences genetic diversity. Aust Fish 8(10):510

Sah P, Mandal S, Singh RK et al (2021) Development of novel microsatellite marker panel in threatened tetraploid mahseer, *tor tor* (Hamilton 1822) for insights into its genetic diversity and population structure. Meta Gene 28:100880

Sebastian W, Sukumaran S, Zacharia PU et al (2017) Genetic population structure of Indian oil sardine, *Sardinella longiceps* assessed using microsatellite markers. Conserv Genet 18:951–964

Sharma L, Ali S, Siva C et al (2019) Genetic diversity and population structure of the threatened chocolate mahseer (*Neolissochilus hexagonolepis* McClelland 1839) based on SSR markers: implications for conservation management in Northeast India. Mol Biol Rep 46:5237–5249

Singh RK, Lal KK, Mohindra V et al (2012) Genetic diversity of Indian major carp, *Labeo calbasu* (Hamilton, 1822) populations inferred from microsatellite loci. Biochem Syst Ecol 44:307–316

Singh RK, Lal KK, Mohindra V et al (2016) Characterization of mitochondrial ATPase 6/8 genes in wild *Labeo calbasu* (Hamilton, 1822) and mapping of natural genetic diversity. Mitochondrial DNA A DNA Mapp Seq Anal 27(5):3078–3078

Singh RK, Divya PR, Mohindra V et al (2022) Diversity of fish genetic resources below the species level, characterization and applications in resource management. Indian J Plant Gen Res 35:301–304

Singh RK, Kumar R, Bhordia A et al (2023) Development of genome-wide novel microsatellite markers and genetic variability analysis of Asian seabass, *Lates calcarifer* (Bloch, 1790). Aquacult Inter 31:3055–3070

Slatkin M (1995) A measure of population subdivision based on microsatellite allele frequencies. Genetics 139:457–462

Sreelekshmi S, Sukumaran S, Kishor TG et al (2020) Population genetic structure of the oceanic whitetip shark, *Carcharhinus longimanus*, along the Indian coast. Mar Biodivers 50:78

Sri Hari M, Abidi ZJ, Kathirvelpandian A (2019) Lack of genetic differentiation in milkfish, *Chanos chanos* (Forsskål, 1775), revealed by mitochondrial ATPase 6/8 genes. Mitochondrial DNA A DNA Mapp Seq Anal 30(3):511–516

Sukumaran S, Wilson S, Lakshmi M et al (2020) Molecular analyses reveal a lack of genetic structuring in the scalloped hammerhead shark, *Sphyrna lewini* (Griffith & Smith, 1834) along the Indian coast. Mar Biodivers 50(18):1–6

Sukumaran S, Sebastian W, Zacharia PU et al (2023) Population genetic research on the Spadenose shark, *Scoliodon laticaudus* (Chondrichthyes, Carcharhinidae), reveals the presence of two significantly differentiated clades along the Indian coast. Mar Biodivers 53(16):1–9

Swain SK, Sahu BP, Das SP et al (2022) Population genetic structure of fringe-lipped carp, *Labeo fimbriatus* from the peninsular rivers of India. 3 Biotech 12:300

Thomas JH (1996) Population genetics of marine species: the interaction of natural selection and historically differentiated populations. J Exp Mar Biol Ecol 200(1–2):67–83

Vidya R, Gopalakrishnan A, Biradar RS et al (2017) Mitochondrial DNA marker reveals shallow genetic structuring in *Priacanthus hamrur* (Forsskål, 1775) along the Indian coast. J Mar Biol Assoc India 59(2):29–35

Vineesh N, Kathirvelpandian A, Divya PR et al (2016) Hints for panmixia in *Scomberomorus commerson* in Indian waters revealed by mitochondrial ATPase 6 and 8 genes. Mitochondrial DNA A DNA Mapp Seq Anal 27:2822–2824

Vineesh N, Divya PR, Kathirvelpandian A et al (2017) Four evolutionarily significant units among narrow-barred Spanish mackerel (*Scomberomorus commerson*) in the Indo-West Pacific region. Mar Biodivers 48:2025–2032

Vineesh N, Shihab I, Akhilesh KV et al (2023) Is *Lutjanus argentimaculatus* genetically connected along the Arabian Sea and bay of Bengal? A study using microsatellite markers. Aquaculture 562:738863

Waples RS, Hindar K, Karlsson S et al (2016) Evaluating the Ryman-Laikre effect for marine stock enhancement and aquaculture. Curr Zool 62(6):617–627

Genetic Variation and Assessment Tools for Conservation and Management of Fish Genetic Resources (FiGR)

3

Rajeev Kumar Singh, A. Kathirvelpandian, and M. U. Rekha

Abstract

Genetic variation is fundamental to biodiversity, and crucial for long-term survival and ecological resilience of any living species or population. Understanding genetic diversity at molecular levels aids in ecological comprehension and practical applications, particularly in fisheries management, aquaculture, and conservation. This chapter describes genetic variation, mechanisms (mutations, recombination, and genetic drift), and their impact on fish populations' adaptability to environmental changes. It delves into genetic variation factors, such as natural selection, gene flow, and population demographics. Furthermore, the implications of low genetic variability, susceptibility to environmental stressors and reduced fitness, are discussed. Various tools to assess genetic variation, ranging from morphological assessments to molecular markers (protein, nuclear, and mitochondrial) are detailed. Various prioritized fish species, across the globe, have been characterized for genetic divergence using these marker systems. Such genetic data has applications in multiple ways. Future perspectives emphasize the potential of genomics in revolutionizing genetic improvement and sustainable practices. This comprehensive understanding of genetic variation and its applications underscores its significance in maintaining healthy, resilient fish populations and overall aquaculture development.

Keywords

Genetic variation · Biodiversity · Adaptation · Fisheries management · Aquaculture · Conservation

R. K. Singh (✉) · A. Kathirvelpandian · M. U. Rekha
ICAR-National Bureau of Fish Genetic Resources, Lucknow, Uttar Pradesh, India
e-mail: Rajeev.Singh1@icar.gov.in; A.Pandian@icar.gov.in; Rekha.MU@icar.gov.in

U. K. Sarkar et al. (eds.), *Sustainable Management of Fish Genetic Resources*,
https://doi.org/10.1007/978-981-97-5250-8_3

3.1 Introduction

Evolution is a dynamic and continuous process, through which the characters of a species change at different rates and patterns. This also gives rise to new species as well. Life on earth began approximately 3.5–4.0 billion years ago, and since then, all living taxa have been evolving and contributing to biodiversity. Genetic variation is the cornerstone of biodiversity, which enables populations to adapt to changing environments. Such genetic variation is essential for the long-term survival of any species. It refers to the diversity between/among individuals, populations, and groups. Knowledge of the genetic makeup of fish populations enhances our ecological understanding for practical applications. From an ecological standpoint, genetic diversity is crucial for comprehending their evolutionary history, adaptation potential, and resilience to environmental changes. Evaluating genetic variation in fish populations is indispensable for devising suitable measures for conservation and management. The extent of genetic variation has direct implications for fisheries management, aquaculture, and conservation efforts. Genetic data can be meaningful in sustainable harvesting practices and breeding programs to enhance desirable traits in aquaculture species (Ruzzante et al. 2019).

3.2 Genetic Variation

Genetic variation, the foundation of evolutionary processes, refers to the diversity in DNA sequences within populations, among groups that result in the change of allele, gene, and haplotype frequencies (Kimura 1983; Nei 1987). It serves as the raw material for natural selection, which is useful in adapting to changing environments, thereby ensuring the persistence of species over time (Lewontin 1974). Genetic variation arises from various mechanisms, such as mutations, recombination, and genetic drift (Hartl and Clark 2007). These processes individually or in combination contribute to the genetic diversity.

3.3 Origin of Genetic Variation

Genetic variation introduces diversity into populations' gene pools and, therefore, is central to numerous biological phenomena, including disease susceptibility, physiological traits, and ecological interactions (Hedrick 2001). The variations at the genetic level can arise in multiple ways; however, the primary mechanism is mutation, which is considered as the ultimate source (Hartl and Clark 2007).

Mutation is defined as the heritable change(s) in the genetic material of an individual. These changes are random, such as nucleotide substitutions, insertions, deletions, or duplications, occurring during DNA replication or repair processes (Lewontin 1974). While most mutations are neutral or deleterious, some may confer adaptive advantages, leading to their persistence in populations through natural selection (Kimura 1983). The basic types of mutation are as follows:

(a) Based on the size of the mutation, there are point mutations and frameshift mutations. In point mutation, only one nucleotide pair is replaced and the total number of nucleotide pairs remains unchanged in a gene. These substitutions can be synonymous, i.e., they do not alter the encoded amino acid, or non-synonymous, where they result in altered amino acids. Such alteration at the protein level can potentially affect its function (Nachman and Crowell 2000). The addition (insertion) or deletion of one or few nucleotide pairs may lead to a frameshift mutation. These changes can produce nonfunctional (or with changed properties) peptides/proteins due to changes in reading frames.

(b) Another type of mutation is based on chromosome structure, which includes inversion, duplication, and translocation. These are considered as the large-scale alterations in DNA sequence. With the availability of high-throughput sequencing, genome-wide assessment has opened a new dimension leading to the identification of copy number variations (CNVs). CNVs, which refer to the change the number of copies of specific DNA segments (Conrad et al. 2010). These variations can influence gene dosage, gene expression levels, and phenotypic diversity within populations.

Recombination is a crucial process that occurs during meiosis when homologous chromosomes exchange their genetic material (Hartl and Clark 2007). This process shuffles existing genetic variation and generates new combinations of alleles. Recombination increases genetic diversity within populations and can facilitate the spread of advantageous alleles, which may contribute to adaptation and evolutionary change.

Another type of genetic variation is epigenetic modifications. These changes do not alter the genotype, and the underlying DNA sequence remains the same. These modifications, including DNA methylation and histone modifications can influence gene activity, chromatin structure, and cellular identity. Gene expression is affected due to the presence of RNAs, which can alter the access of the transcriptional system to the helix. Epigenetic variation can arise in response to environmental cues, developmental processes, and stochastic events, ultimately contributing to phenotypic plasticity and adaptation.

3.4 Factors Affecting Genetic Variation

Several factors affect genetic variation and play an essential role in shaping the evolutionary trajectories and ecological dynamics of the species. One of the primary factors is natural selection, whereby advantageous traits confer higher fitness and are thus more likely to become prevalent over time. Genetic drift, a stochastic process, is important in shaping genetic variation within populations (Hartl and Clark 2007). This phenomenon refers to random changes in allele frequencies over generations due to sampling error in finite populations (Kimura 1983). Particularly pronounced in small populations, genetic drift can lead to the fixation or loss of alleles, reducing genetic variation over time (Nei 1987).

Migration causes mutations to spread to the subpopulations and creates limits for genetic divergence. It also facilitates the adaptation of a stock/unit to local environments, as some alleles/haplotypes may experience a positive correlation with certain environmental variables. Gene flow, the movement of individuals and their genetic material between populations, can either homogenize genetic variation or maintain genetic diversity within populations (Slatkin 1987). Restricted gene flow due to geographic barriers can lead to population differentiation and increased genetic variabilities among populations.

Population size and demographic history also significantly influence genetic variation (Frankham et al. 2017). Demographic events such as bottlenecks and founder effects can drastically reduce genetic variation within populations by selectively removing alleles, leading to genetic homogenization (Allendorf et al. 2010).

3.5 Assessment of Variability

Various assessment tools have been developed to quantify the genetic variability and genetic structure within the populations. In morphological assessments, external features and internal structures are examined to document phenotypic variabilities within/between populations. Morphometric analyses, such as TRUSS (transformation of uniformly scaled shape) morphometry, quantify shape and size variation to evaluate divergence at phenotypic level.

3.5.1 Phenotypic Variability

Morphological assessments (morphometric characters and otolith analysis) are important in deciphering variabilities within/between fish populations. Morphometric characteristics, such as body shape, fin morphology, and meristic counts, provide insights into phenotypic diversity. Otolith analysis focuses on examining the ear stones to provide information on fish age, growth rates, and ecological conditions for reconstructing population dynamics and connectivity (Campana et al. 2000).

TRUSS morphometry is a powerful tool for quantifying phenotypic diversity. Developed as a geometric morphometric technique, this tool offers a robust approach for comparative analysis of shape differences. Unlike traditional morphometric methods that rely on linear measurements, TRUSS morphometry uses landmark coordinates to capture shape variabilities. This framework is particularly useful in studies, where shape variations hold ecological significance or reflect adaptive responses to environmental conditions.

Integrating the TRUSS morphometry with other analytical techniques enhances its utility in understanding the factors that lead to shape variation through evolutionary processes within fish populations. New tools, such as thin-plate splines and Procrustes analysis, have been incorporated for body-shape visualization, with more precision (Chandran et al. 2022). Recently, the morphometric analyses and molecular markers have been analyzed together, which allowed the identification of genotype-phenotype associations and the exploration of adaptive divergence among populations (Pujolar et al. 2011).

3.5.2 Genetic Variability

Molecular markers are the proven tools for assessing genetic variation. These are protein-based, cytogenetic, cytoplasmic, and nuclear DNA markers (Avise 2004). Among the recently used molecular markers are microsatellites, single nucleotide polymorphisms (SNPs), and mitochondrial DNA sequences, which provide information on allele/haplotype frequencies, population structure, historical demographic events, and genetic constraints (Allendorf et al. 2010). Molecular markers can be classified into type I and type II markers. Type I markers are associated with genes of known function, while type II markers are linked with unidentified genomic regions (Liu and Cordes 2004). The application of the marker system needs to be done on a case-to-case basis, depending on the issue to be addressed. Table 3.1 includes the most prevalent genetic markers and their usage in delineating genetic variation in economically significant fish species.

3.5.2.1 Allozymes

Isozymes are structurally different molecular forms of an enzyme encoded by one or more loci (Markert and Moller 1959). Isozymes exhibiting different alleles at the same locus are designated as “allozymes” or “alloenzymes” (Starck 1998). Amino acid differences in polypeptide chains (different allelic forms) of an enzyme represent changes in the underlying DNA sequences. Depending on the nature of the amino acid changes, the resulting protein products may migrate at different rates (due to charge and size differences) upon electrophoresis. Differences in the allele frequencies are used to quantify genetic variation and population structure (Singh et al. 2010). Allozymes have been used as species-specific loci (Mohindra et al. 2007) and in detecting hybrids (Lal et al. 2007). Other significant applications of allozymes include the tracking of inbreeding and parentage analysis. Some studies have correlated allozyme markers with performance traits (McGoldrick and Hedgecock 1997). Their usage in linkage mapping has been demonstrated in studies of salmonids (May and Johnson 1990) and poeciliids (Morizot et al. 1991).

3.5.2.2 Nuclear DNA Markers

Major nuclear DNA markers are random amplified polymorphic DNA (RAPD), amplified fragment length polymorphisms (AFLPs), microsatellites, and single nucleotide polymorphisms (SNPs). These marker systems have been widely used for analyzing genetic diversity and population structure in fish species.

3.5.2.3 Random Amplified Polymorphic DNA (RAPD)

RAPD markers have been considered the popular markers across the commodities. In this technique, the genomic DNA is amplified using short nucleotide sequence primers, which can detect high levels of polymorphisms. These loci represent

Table 3.1 Applications of popular genetic markers in important fishes belonging to different eco-regions

Sl. no.	Species	Common name	Marker	Applications	Reference
	Marine				
1	*Thunnus albacares*	Yellowfin tuna	SNP	Population genomics	Pecoraro et al. (2018)
2	*Scomberomorus commerson*	Spanish mackerel	Microsatellite DNA	Validated SSR panel for genetic variability	Joy et al. (2020)
3	*Katsuwonus pelamis*	Skipjack tuna	mtDNA D-loop	Population structure	Menezes et al. (2012)
4	*Euthynnus affinis*	Little tuna or kawakawa	mtDNA D-loop	Population structure	Kumar et al. (2012)
5	*Clupea harengus*	Atlantic herring	Microsatellites, allozymes, and mt DNA	Neutral and adaptive divergence	André et al. (2011)
6	*Scomber scombrus*, *Scomber japonicus*	Atlantic mackerel, chub mackerel	Mitochondrial D-loop	Population genetic structure	Zardoya et al. (2004)
7	*Salmo salar*	Atlantic salmon	Microsatellites	Landscape genetics	Dionne et al. (2008)
8	*Carcharhinus limbatus*, *Carcharhinus sorrah*, *Sphyrna lewini*	Blacktip shark, spot-tail shark, scalloped hammerhead	mtDNA D-loop and microsatellites	Genetic parameters of exploited shark species	Spaet et al. (2015)
9	*Engraulis encrasicolus*	European anchovy	Microsatellites	Genetic structure in seas of Turkey	Tuncay and Bardakci (2014)
	Brackish water				
10	*Mugil curema*	White mullet	Microsatellites, mtDNA	Genetic patterns	Cossu et al. (2021)
11	*Chanos chanos*	Milkfish	mtDNA D-loop, and cytochrome b	Genetic variability and demographic expansion	Santos et al. (2019)
12	*Lates calcarifer*	Asian sea bass	Microsatellite	Genome-wide SSR and genetic structure	Yue et al. (2009), Singh et al. (2023)

(continued)

Table 3.1 (continued)

Sl. no.	Species	Common name	Marker	Applications	Reference
13	*Lateolabrax maculatus*	Chinese sea bass	SNPs	Genetic and adaptive divergence	Zhao et al. (2018)
14	*Dicentrarchus labrax*	European sea bass	SNPs and microsatellites	Range-wide genetic structure	Souche et al. (2015)
	Freshwater				
15	*Labeo rohita*	Rohu	mtDNA	Natural genetic divergence	Luhariya et al. (2012)
16	*Cirrhinus mrigala*	Mrigal	Microsatellite	Population genetics	Chauhan et al. (2007)
17	*Labeo catla*	Catla	Cytochrome b	Population genetics	Das et al. (2012)
18	*Labeo calbasu*	Kalbasu	Microsatellite, mtDNA	Natural genetic divergence	Singh et al. (2012)
19	*Channa marulius*	Great snakehead	Microsatellite	Validated SSR panel for genetic variability	Pathak et al. (2018)
20	*Hypophthalmichthys molitrix*, *Hypophthalmichthys nobilis*	Silver carp, bighead carp	Microsatellite	Phylogeography and population genetics	Farrington et al. (2017)
21	*Cyprinus carpio*	Common carp	Microsatellite	Genetic variability in Carpathian Basin	Tóth et al. (2020)

numerous sites across the genome. These markers have been used mainly as species-specific diagnostic tools (Mohindra et al. 2007a) and also for deciphering population genetic structure (Mamuris et al. 2002). RAPD loci are dominant markers and scored as present/absent. The limitations of RAPD include the inability to distinguish between homozygotes and heterozygotes. Low reproducibility due to the low annealing temperature used in the PCR amplification further limits the application of this marker.

3.5.2.4 Amplified Fragment Length Polymorphism (AFLP)

AFLP is a PCR-based, multi-locus fingerprinting technique. It is the selective PCR amplification of restriction fragments from digested DNA. It includes base substitutions at PCR primer binding sites, similar to RAPD. In AFLP, the adaptors are ligated to digested fragments of genomic DNA. The amplified fragments are resolved through gel electrophoresis. The high polymorphism of AFLP is useful in generating species-specific profiles or detecting variations between individuals. For instance, Young et al. (2001) generated 133 polymorphic AFLP loci, of which 23 were diagnostic to distinguish rainbow trout, coastal cutthroat trout, and their hybrids. These markers have successfully been used in fishes like *Scomberomorus niphonius* (Shui et al. 2008), among many others.

3.5.2.5 Microsatellite DNA

Microsatellites are tandemly repeated sequences, abundantly distributed across genomes and known to demonstrate high levels of polymorphism. They are codominant markers of relatively small size, which are easily amplified. Microsatellites, known as simple sequence repeats (SSRs), range from 1 to 6 base pairs, e.g., ACA or GATA (Litt and Luty 1989). The microsatellite repeats are generated during crossing over (meiosis) or replication slippage.

Microsatellite markers have been used extensively in fisheries research, such as linkage mapping, parentage, kinships, and stock structure analysis. These markers can detect demographic bottlenecks, where populations might have experienced genetic constraints influencing the distribution of genetic variation. Loss of genetic diversity may reduce the potential of small populations to respond to selective pressures and increased inbreeding and consequently may reduce population viability. Microsatellite markers also have applications in aquaculture to assess loss of genetic variation in hatcheries.

3.5.2.6 Single Nucleotide Polymorphism

Single nucleotide polymorphisms (SNPs) are single base changes in a DNA sequence/gene. Most SNPs are biallelic and are less variable than the microsatellite markers. However, the lack of allele number is compensated by their high abundance in the genome. SNPs are codominant markers that evolve through the simple mutation and infinite allele models. SNP markers are less prone to homoplasy, unlike microsatellites. This makes them more reliable for various applications. They have been utilized across various disciplines, such as population ecology, evolutionary studies, and conservation genetics.

A significant advantage of SNP markers is the high reproducibility among different laboratories, enabling newly collected data to be easily compared with the reference data. Moreover, SNP markers can be used to genotype poor-quality samples, including historical or degraded samples (Morin and McCarthy 2007). Recent advancements in next-generation sequencing technologies, along with enhanced computational capabilities, have significantly increased the feasibility of high-throughput SNP detection, particularly in non-model organisms. These technological advancements facilitate cost-effective genotyping of large panels of SNPs, further enhancing the accessibility and utility of data.

3.5.2.7 Mitochondrial DNA Markers

The mitochondrial genome is maternally inherited and exhibits unique evolutionary dynamics. The mitogenome is circular, encoding 13 proteins, two ribosomal RNAs, and 22 transfer RNAs. Typically, the teleost mitochondrial genome contains 16,000–17,000 base pairs, a highly variable control region D-loop, which is crucial for replication initiation (Harrison 1989). The mitochondrial genome evolves rapidly due to a lack of recombination and a high mutation rate, making it highly informative for population genetics studies. Several mitochondrial genes have been found useful in population-level investigations, such as population structure, historical demography, and phylogeography in various fish species. Currently, concatenated genes of the mitochondrial genome (Dutta et al. 2020) are being used for determining neutral and adaptive divergence.

3.5.2.8 Cytogenetic Markers

Cytogenetic markers, such as chromosomal morphology, banding/staining patterns, provide insights into genome organization and chromosomal evolution in fish species. Karyotype analysis enables researchers to identify species-specific chromosomal characteristics, detect chromosomal abnormalities, and infer phylogenetic relationships among the taxa. These markers have been applied in aquaculture and fisheries management, where they can be used to assess genetic integrity, identify hybridization events and monitor the impact of environmental stressors on the fish populations (Bardakci and Skibinski 2001).

Techniques such as chromosome banding, fluorescence in situ hybridization (FISH) and comparative genomic hybridization (CGH) allow researchers to visualize chromosomal abnormalities, gene mapping and interspecific comparisons. For instance, FISH has been crucial in identifying specific chromosomal regions associated with traits of interest, aiding in selective breeding programs and conservation (Bertollo et al. 2017). Additionally, cytogenetic markers have been used to assess the impact of environmental stressors on fish populations, revealing patterns of genomic instability and adaptation (Molina et al. 2014).

3.5.2.9 Combined Use of Markers

Currently, the trend is to use multiple markers to address and resolve a biological issue. Integrating multiple genetic tools provides a comprehensive understanding of genetic variation within and among the populations. Some examples of combined markers' usage include *Chitala chitala* (Mandal et al. 2009) and *Gama orientalis* (Xuan et al. 2021), where mitochondrial and microsatellite markers were employed.

Combining morphological variabilities with molecular markers can elucidate the relationship between phenotype and genotype facilitating the identification of adaptive traits and population-specific adaptations (Pritchard et al. 2000; Mandal et al. 2020). Furthermore, integrating genetic data with ecological and demographic information facilitates the policymakers to develop effective conservation and management strategies for target species (Chandran et al. 2022). There are other examples where different tools have been used in combination to reach a conclusion. For instance, Biswal et al. (2018) employed integrated morpho-molecular tools to resolve taxonomic ambiguities between *Systomus sarana sarana* (Hamilton, 1822) and *S. sarana subnasutus* (Valenciennes). Currently, the trend of integrative taxonomy is in place for delimitating a new species. The literature is growing, where morphological and molecular tools are used as complementary to each other.

In summary, molecular markers have revolutionized our understanding of genetic variation and evolutionary processes. These markers offer diverse applications in fisheries management, aquaculture and conservation biology, enabling researchers to address fundamental questions regarding population dynamics, adaptive evolution, and biodiversity conservation.

3.6 Implications of Low Genetic Variability

Low genetic variability in fish populations can greatly impact their long-term survival and ecological functioning. Reduced genetic diversity can compromise the adaptive capacity of the fish populations, making them more susceptible to environmental stressors, such as pollution, habitat degradation, and climate change. This vulnerability to changing conditions can increase the risk of population decline or extinction (Reed and Frankham 2003). Inbreeding depression is a phenomenon where mating occurs between the individuals with similar genetic backgrounds. This reduces within family/population variations and increases among family/population variations. This can result in reduced fitness and viability among the offsprings, and can lead to potential population decline (Frankham et al. 2017).

Furthermore, reduced genetic diversity can weaken the resilience of fish populations to novel pathogens and parasites. Studies have shown that populations with low genetic variability may lack the necessary immune responses to combat emerging diseases, leading to widespread mortality and decline in populations (Laurentino et al. 2020). Reduction in genetic variability can have cascading effects on ecosystem dynamics and functioning. Fish play crucial roles in maintaining ecosystem balance through predation, herbivory, and nutrient cycling. Reduced genetic diversity within fish populations may disrupt these ecological processes, leading to altered community structures and diminished ecosystem services (Hughes et al. 2008).

Large natural populations typically harbor substantial genetic variability, while threatened species often exhibit reduced levels. The Japanese eel, *Anguilla japonica*, exhibits extremely low genetic variability, as evidenced by studies (Yu et al. 2020), and is regarded as belonging to a single gene pool. The Indian scad, *Decapterus russelli*, a widely distributed species, was found to have a single panmictic population (Jose et al. 2023). Other examples include the swordfish *Xiphias gladius* in the Indian Ocean, where spatiotemporal analysis of multi-marker systems resulted in a single stock (Muths et al. 2013). These events are likely due to small founder populations or bottleneck events in their history. This phenomenon isn't exclusive to fishes; large animals are also susceptible to reduced genetic diversity. For instance, the indiscriminate hunting of northern elephant seals nearly drove them to extinction. As these seals recovered from severely reduced populations, they exhibited minimal variability in allozyme loci (Bonnel and Selander 1974).

3.7 Application in Fisheries Management and Aquaculture

3.7.1 Conservation and Management

Genetic variation is a fundamental component of maintaining healthy fish populations, as it influences the adaptability, resilience, and overall fitness of a species in response to environmental changes and stressors. Understanding the genetic variation is essential for effectively managing and conserving the fish populations.

Genetic improvement programs are dependent on the genetic variation. This variability is essential for creating base populations in selective breeding programs. Selective breeding can improve traits such as disease resistance, growth rates, and reproductive success. Such programs have been successful in enhancing the disease resistance of farmed fish species such as Atlantic salmon (*Salmo salar*) and rainbow trout (*Oncorhynchus mykiss*) (Gjedrem et al. 2012).

Endangered fish populations often suffer from reduced genetic diversity due to habitat loss, overexploitation, and pollution. Low genetic diversity can increase the risk of inbreeding depression, reduced reproductive success, and decreased adaptability to changing environmental conditions. The genetic tools can assess the genetic diversity of endangered fish populations and develop strategies to maintain genetic variation. Captive breeding programs, genetic rescue, and translocation of individuals between populations can prevent the loss of genetic diversity and improve the long-term survival prospects of endangered fish species (Frankham et al. 2017).

3.7.2 Adaptation to Changing Environments

Fish species inhabit diverse aquatic ecosystems with unique environmental conditions such as temperature, salinity, etc. As these ecological variables undergo natural or anthropogenic changes, the fish populations must adapt to survive and propagate. Genetic variation provides the raw material for evolutionary processes to act upon, allowing fish populations to evolve and adapt to new environmental challenges. For example, in response to increasing water temperatures due to climate change, fish populations with genetic variants that confer heat tolerance may have a higher likelihood of survival and reproduction. Over time, these advantageous genetic variants can become more prevalent in the population, leading to adaptation to environmental conditions. Genetic variation in heat shock proteins can influence the thermal tolerance of fish species (Feder and Hofmann 1999).

3.7.3 Resistance to Biotic and Abiotic Factors

Genetic variation in fish populations is critical for resistance toward biotic (living organisms) and abiotic (nonliving environmental factors) parameters. The intraspecific genetic diversity has a positive correlation with the ability to resist diseases and harmonize with stressors such as temperature fluctuations, pollution, and habitat degradation.

3.8 Future Perspectives

Integrating genomics in fisheries management has the potential to revolutionize the sector. Despite the efforts, unlike other commodities, these are at the initial levels in fisheries. Large-scale discovery of genome-wide makers has improved the quality of population-level studies. The use of high computational power has further elevated this area of research. Along with the neutral markers, the outlier loci are also being identified to determine the adaptive divergence among the fish populations, and identify associated selection pressure. Furthermore, genomic data can strengthen the sectorial growth. The genes associated with traits such as disease resistance, growth rates and reproduction may be identified and used in breeding programs. The researchers can selectively breed individuals with these advantageous genetic traits. This can help to improve the overall health and productivity of fish populations, ultimately benefiting both commercial fisheries and ecosystem health. Using such informative data, policymakers can formulate strategies and make more informed decisions to promote sustainable fishing practices and conservation efforts.

References

Allendorf FW, Hohenlohe PA, Luikart G (2010) Genomics and the future of conservation genetics. Nat Rev Genet 11(10):697–709

André C, Larsson LC, Laikre L et al (2011) Detecting population structure in a high gene-flow species, Atlantic herring (*Clupea harengus*): direct, simultaneous evaluation of neutral vs putatively selected loci. Heredity 106:270–280

Avise JC (2004) Molecular markers, natural history, and evolution, 2nd edn. Sinauer Associates, Sunderland

Bardakci F, Skibinski DO (2001) Assessment of genetic diversity in the Mediterranean mussel (*Mytilus galloprovincialis*) populations from the Adriatic and Aegean seas using PCR-based DNA markers. Mar Biol 139(2):241–251

Bertollo LAC, Cioffi M, Galetti PM Jr et al (2017) The role of chromosomal rearrangements in the speciation of neotropical fishes. In: Genome dynamics. Karger Publishers, Basel, pp 279–309

Biswal JR, Singh RK, Lal KK et al (2018) Molecular and morphological evidences resolve taxonomic ambiguity between *Systomus sarana sarana* (Hamilton, 1822) and *S. sarana subnasutus* (Valenciennes) and suggest elevating them into distinct species. Mitochondrial DNA Part B 3(2):838–844

Bonnel ML, Selander RK (1974) Elephant seals: genetic variation and near extinction. Science 184(4139):908–909

Campana SE, Thorrold SR, Jones CM et al (2000) Otoliths, increments, and elements: keys to a comprehensive understanding of fish populations? Can J Fish Aquat Sci 57(1):30–38

Chandran R, Singh A, Singh RK et al (2022) Phenotypic variation of *Chitala chitala* (Hamilton, 1822) from Indian rivers using truss network and geometric morphometrics. Peer J 10:e13290

Chauhan T, Lal KK, Mohindra V et al (2007) Evaluating genetic differentiation in wild populations of the Indian major carp, *Cirrhinus mrigala* (Hamilton–Buchanan, 1882): evidence from allozyme and microsatellite markers. Aquac 269(1–4):135–149

Conrad DF, Pinto D, Redon R et al (2010) Origins and functional impact of copy number variation in the human genome. Nature 464(7289):704–712

Cossu P, Mura L, Scarpa F et al (2021) Genetic patterns in *Mugil cephalus* and implications for fisheries and aquaculture management. Sci Rep 11(1):2887

Das R, Mohindra V, Singh RK et al (2012) Intraspecific genetic diversity in wild *Catla catla* (Hamilton, 1822) population assessed through mtDNA cytochrome b sequences. J Appl Ichthyol 28:280–283

Dionne M, Caron F, Dodson JJ et al (2008) Landscape genetics and hierarchical genetic structure in Atlantic salmon: the interaction of gene flow and local adaptation. Mol Ecol 17(10):2382–2396

Dutta N, Singh RK, Pathak A et al (2020) Mitochondrial DNA markers reveal genetic connectivity among populations of Osteoglossiform fish *Chitala chitala*. Mol Biol Rep 47:8579–8592

Farrington HL, Edwards CE, Bartron M et al (2017) Phylogeography and population genetics of introduced silver carp (*Hypophthalmichthys molitrix*) and bighead carp (*H. Nobilis*) in North America. Biol Invasions 19:2789–2811

Feder ME, Hofmann GE (1999) Heat-shock proteins, molecular chaperones, and the stress response: evolutionary and ecological physiology. Annu Rev Physiol 61(1):243–282

Frankham R, Ballou JD, Briscoe DA (2017) Introduction to conservation genetics. Cambridge University Press, Cambridge

Gjedrem T, Robinson N, Rye M (2012) The importance of selective breeding in aquaculture to meet future demands for animal protein: a review. Aquac 350:117–129

Harrison RG (1989) Animal mitochondrial DNA as a genetic marker in population and evolutionary biology. Trends Ecol Evol 4(1):6–11

Hartl DL, Clark AG (2007) Principles of population genetics. 4th edition Sinauer associates. Ecoscience 14(4):544–545

Hedrick PW (2001) Conservation genetics: where are we now? Trends Ecol Evol 16(11):629–636

Hughes AR, Inouye BD, Johnson MT et al (2008) Ecological consequences of genetic diversity. Ecol Lett 11(6):609–623

Jose A, Sukumaran S, Roul SK et al (2023) Genetic analyses reveal panmixia in Indian waters and population subdivision across Indian Ocean and indo-Malay archipelago for *Decapterus russelli*. Sci Rep 13:22860

Joy L, Paulose S, Divya P et al (2020) Microsatellite marker development in Spanish mackerel Scomberomorus commerson using third generation sequencing technology. Mol Biol Rep 47:10005–10014

Kimura M (1983) The neutral theory of molecular evolution. Cambridge University Press, Cambridge, p 367

Kumar G, Kunal SP, Menezes et al (2012) Single genetic stock structure of kawakawa (*Euthynnus affinis*) (cantor, 1849) along the Indian coast inferred from sequence analyses of mitochondrial DNA D-loop region. Conserv Genet 13:1119–1131

Lal KK, Singh RK, Mohindra V et al (2007) Molecular detection of a *Labeo calbasu* x *Labeo rohita* hybrid, from the river Ganges. Acta Zool Sin 53(4):773–776

Laurentino TG, Sanches A, Lima SMQ (2020) Genetic diversity and resistance to natural diseases in Brazilian farmed tilapia. Aquacul 524:735262

Lewontin RC (1974) The genetic basis of evolutionary change. Columbia University Press, pp 319–336

Litt M, Luty JA (1989) A hypervariable microsatellite revealed by in vitro amplification of a dinucleotide repeat within the cardiac muscle actin gene. Am J Hum Genet 44(3):397–401

Liu ZJ, Cordes JF (2004) DNA marker technologies and their applications in aquaculture genetics. Aquac 238(1–4):1–37

Luhariya RK, Lal KK, Singh RK et al (2012) Genetic divergence in wild population of *Labeo rohita* (Hamilton, 1822) from nine Indian rivers, analyzed through MtDNA cytochrome b region. Mol Biol Rep 39(4):3659–3665

Mamuris Z, Stamatis C, Triantaphyllidis C (2002) Genetic differentiation among Greek populations of the European sea bass (*Dicentrarchus labrax L.*) revealed by random amplified polymorphic DNA (RAPD) markers. Mar Biol 141(2):181–189

Mandal A, Kuldeep KL, Vindhya M (2009) Population structure of Indian feather-back fish, *Chitala chitala* using allozyme, random amplified polymorphic DNA (RAPD) and microsatellites. Biochem Genet 47(3–4):216–234

Mandal S, Singh A, Sah P et al (2020) Genetic and morphological assessment of a vulnerable large catfish, *Silonia silondia* (Hamilton, 1822), in natural populations from India. J Fish Biol 98:430–444

Markert CL, Moller F (1959) Multiple forms of enzymes: tissue, ontogenetic, and species-specific patterns. Proc Natl Acad Sci 45(5):753–763

May B, Johnson K (1990) Composite linkage map of salmonid fishes. In: O'Brien SJ (ed) Genetics maps, 5th edn. Cold, Spring Harbor, Cold Spring Harbor, pp 4151–4159

McGoldrick DJ, Hedgecock D (1997) Genetic markers in aquaculture. Aquac 111(1–4):43–63

Menezes MR, Kumar G, Kunal SP (2012) Population genetic structure of skipjack tuna (*Katsuwonus pelamis*) from the Indian coast using sequence analysis of mitochondrial DNA D-loop region. J Fish Biol 80(8):2198–2212

Mohindra V, Singh RK, Palanichamy M et al (2007) Genetic identification of three species of genus *Clarias* using allozyme and mitochondrial DNA markers. J Appl Ichthyol 1:104–109

Mohindra V, Khare P, Lal K et al (2007a) Molecular discrimination of five *Mahseer* species from Indian peninsula, by RAPD analysis. Acta Zool Sin 53(4):725–732

Molina WF, Neto JG, Martinez PA et al (2014) Evaluation of environmental quality: use of cytogenetic biomarkers in fish. Braz J Biol 74(2):475–482

Morin PA, McCarthy M (2007) Highly accurate SNP genotyping from historical and degraded DNA. Genome Res 17(11):1588–1595

Morizot DC, Duteau NM, Baril CT (1991) Genetic maps and their implications for management of poeciliid fishes. J Fish Biol 39(A):1–21

Muths D, Le Couls S, Evano H et al (2013) Multi-genetic marker approach and Spatio-temporal analysis suggest there is a single Panmictic population of swordfish *Xiphias gladius* in the Indian Ocean. PLoS One 8(5):e63558

Nachman MW, Crowell SL (2000) Estimate of the mutation rate per nucleotide in humans. Genetics 156(1):297–304

Nei M (1987) Molecular evolutionary genetics. Columbia University Press, New York, p 512

Pathak A, Singh RK, Mohindra V et al (2018) Development and characterization of novel microsatellite markers in great snakehead, *Channa marulia* (Hamilton, 1822). Meta Gene 18:143–148

Pecoraro C, Babbucci M, Franch R et al (2018) The population genomics of yellowfin tuna (*Thunnus albacares*) at global geographic scale challenges current stock delineation. Sci Rep 8(1):13890

Pritchard JK, Stephens M, Donnelly P (2000) Inference of population structure using multilocus genotype data. Genetics 155(2):945–959

Pujolar JM, Jacobsen MW, Als TD et al (2011) Assessing patterns of hybridization between North Atlantic eels using diagnostic single-nucleotide polymorphisms. Heredity 107(2):147–156

Reed DH, Frankham R (2003) Correlation between fitness and genetic diversity. Conserv Biol 17(1):230–237

Ruzzante DE, Mariani S, Bekkevold D et al (2019) Biocomplexity in a highly migratory pelagic marine fish, Atlantic herring. Proc Natl Acad Sci 116(27):13536–13541

Santos BS, Basiao ZU, Quilang JP (2019) Genetic diversity and patterns of demographic expansion in natural populations of milkfish, *Chanos chanos* (Forsskål, 1775), in The Philippines. Mitochondrial DNA Part A 30(2):312–324

Shui BN, Xu H, Zhao WD (2008) Genetic diversity analysis of chub mackerel, *Scomber japonicus* using amplified fragment length polymorphism markers. J Fish China 32(4):458–463

Singh RK, Lal KK, Mohindra V et al (2010) Assessing genetic differentiation in geographic populations of *Labeo calbasu* (Hamilton, 1822) using allozyme markers. Biochem Genet 48(9):760–778

Singh RK, Lal KK, Mohindra V et al (2012) Genetic diversity of Indian major carp, *Labeo calbasu* (Hamilton, 1822) populations inferred from microsatellite loci. Biochem Syst Ecol 44:307–316
Singh RK, Kumar R, Bhordia A et al (2023) Development of genome-wide novel microsatellite markers and genetic variability analysis of Asian seabass, Lates calcarifer (Bloch, 1790). Aquacult Int 31(5): 3055–3070
Slatkin M (1987) Gene flow and the geographic structure of natural populations. Science 236(4803):787–792
Souche EL, Hellemans B, Babbucci M et al (2015) Range-wide population structure of European sea bass (*Dicentrarchus labrax*). Biol J Linn Soc 116(1):86–105
Spaet JL, Jabado RW, Henderson AC et al (2015) Population genetics of four heavily exploited shark species around the Arabian peninsula. Ecol Evol 5(12):2317–2332
Starck M (1998) Allozymes: a handbook of genetic markers for environmental studies. Oxford University Press, Oxford
Tóth B, Khosravi R, Ashrafzadeh MR et al (2020) Genetic diversity and structure of common carp (*Cyprinus carpio L.*) in the Centre of Carpathian Basin: implications for conservation. Genes 11(11):1268
Tuncay SS, Bardakci F (2014) Population genetics of European anchovy (*Engraulis encrasicolus* L.) in the seas of Turkey based on microsatellite DNA. Front Mar Sci
Xuan L, Wu L, Jiang J et al (2021) Genetic differentiation and demographic history of *Coilia mystus* and *Coilia nasus* from Chinese waters based on mitochondrial and microsatellite markers. BMC Evol Biol 21(1):1–15
Young WP, Wheeler PA, Coryell VH et al (2001) A detailed linkage map of rainbow trout produced using doubled haploids. Genetics 158(1):289–300
Yu L, Liu Y, Liu J (2020) Gene-associated microsatellite markers confirm panmixia and indicate a different pattern of spatially varying selection in the endangered Japanese eel *Anguilla japonica*. J Ocean Limnol 38:1572–1583
Yue GH, Zhu ZY, Lo LC et al (2009) Genetic variation and population structure of Asian seabass (*Lates calcarifer*) in the Asia-Pacific region. Aquac 293(1–2):22–28
Zardoya R, Castilho R, Grande C et al (2004) Differential population structuring of two closely related fish species, the mackerel (*Scomber scombrus*) and the chub mackerel (*Scomber japonicus*), in the Mediterranean Sea. Mol Ecol 13(7):1785–1798
Zhao Y, Peng W, Guo H et al (2018) Population genomics reveals genetic divergence and adaptive differentiation of Chinese sea bass (*Lateolabrax maculatus*). Mar Biotechnol 20:45–59

Structural and Functional Genomics for Assessment of Genetic Variability

4

Vindhya Mohindra, Murali S. Kumar, Basdeo Kushwaha, P. R. Divya, Labrechai Mog Chowdhury, and Ravindra Kumar

Abstract

Implementation of the Convention on Biodiversity (CBD) at the global level and enactment of the Biological Diversity Act 2002 by the Parliament of India have accelerated the efforts for mainstreaming of our natural genetic resources. The biodiversity conservation essentially requires information on genetic diversity at inter-and intraspecies level and for planning conservation and management strategies and decisions, based on scientific data. Recent years have witnessed the emergence of a variety of genomics tools, which have facilitated characterization and assessment of genetic diversity at the genomics level. In addition, these have led to better understanding of genome structure and functions, including underlying mechanisms for biological traits. In this chapter, the advancements in the structural and functional genomics studies in fishes have been reviewed and their potential applications for generation and assessment of our endemic fish diversity, including improved varieties developed through selection. This would enable us to encounter the changing global scenario, following conservation of biodiversity and their sustainable use.

Keywords

Genomics · Transcriptomics · Mitogenome · Chromosome mapping · Conservation

V. Mohindra (✉) · M. S. Kumar · B. Kushwaha · P. R. Divya · L. M. Chowdhury · R. Kumar
ICAR-National Bureau of Fish Genetic Resources, Lucknow, Uttar Pradesh, India
e-mail: Mohindra.Vindhya@icar.gov.in; S.Murali@icar.gov.in; basdeo.kushwaha@icar.gov.in; Divya.R@icar.gov.in; Labrechai.Chowdhury@icar.gov.in; Ravindra.Kumar1@icar.gov.in

U. K. Sarkar et al. (eds.), *Sustainable Management of Fish Genetic Resources*,
https://doi.org/10.1007/978-981-97-5250-8_4

4.1 Introduction

India is bestowed with vast fish genetic resources (FGR), spread over water resources covering rivers and canals, minor and major reservoirs, ponds and lakes, and derelict water bodies, in addition to brackish water and saline/alkaline-affected areas, etc. These open water resources offer great opportunities for enhanced production, along with livelihood development. The inland open water fishery sector consists mainly of capture fisheries, culture-based fisheries, natural stocking and harvesting, and enclosure culture (cage and pen culture). The country's vast coastline covers 2.02 million km^2 of exclusive economic zone (EEZ) and continental shelf area of 0.53 million km^2 (https://nfdb.gov.in), and from these marine resources, India has an estimated fishery potential of 4.41 million tonnes (https://dof.gov.in/marine-fisheries). Thus, to attain the potential of production, mainstreaming of fish genetic resources (FAO 2022) is essential, for conservation and sustainable use of genetic resources, for food and nutrition security and to respond to future challenges.

With the new technological advancements, assembling whole genomes and transcriptomes, mining for trait-specific genes and associated molecular markers have become important research areas in agriculture science, including fisheries. The critical considerations for the development of genomic technologies require information of the high-quality whole genome assemblies, which can serve as reference genomes and are the backbone for annotations for variations between and below the species (Kumar et al. 2022a, b). Through genome re-sequencing, multiple reference genomes or pangenomes (Mwaura et al. 2023) provide insights into structure and organization of genes, identification of core gene sets, as well as discovery of the genomic variations, which in turn would facilitate the development of genome-wide markers, to estimate admixture, genomic variability, and for their associations with a particular trait. Such genomic data aid to define species, strains and genomic variations, divergence, and differentiation.

4.2 Structural Genomics

4.2.1 Genomes: Draft and Chromosome Level

A genome constitutes a complete set of genetic information, which an organism needs to function; and genomics is the study to describe the complete set of genes, including other types of DNA of an organism, along with their functions and their interactions, along with the environment. Total genomic information in animals, including fish, is contained in an enormous nuclear genome, along with a small mitochondrial genome, which is several thousands of million base pairs.

The last two decades have witnessed tremendous advances in genome sequencing technologies and genomic studies in vertebrate and non-vertebrate species. Fishes represent half of the vertebrates on the globe, and some fish species are among those whose genomes were initiated, viz. zebrafish in 2001 and *Takifugu* in 2002 (Aparicio et al. 2002), in the beginning attempts of genome sequencing programmes. Genome research accelerated tremendously due to advancements in massively parallel sequencing in 2005 (Heather and Chain 2016), which resulted in a many-fold reduction in sequencing cost followed by the development of computational and assembly tools. Later in recent years, the applications of Hi-C technology (Belton et al. 2012); long reads, PacBio systems (Chen-Shan et al. 2013); and ultra-long-read sequences, Nanopore GridION systems (Tyler et al. 2018), made revolutionary advancements in genome assemblies, reaching up to the chromosome and gapless telomere-to-telomere level (Sun et al. 2024).

For genomics studies, bioinformatic skills are the basic requirement, as it also requires prior information for analysis and interpretation of output in a piece of conclusive information. For a core genome research of a species/group of organisms, a beginner should have an idea regarding the genome size or nuclear DNA content, chromosome numbers, prevalent technologies of sequencing, bioinformatic and computational tools, and most importantly, the aim of conducting the genome research on a specific species or a group of animal or plant species. Some of the important attributes of genomic information in fish are highlighted.

4.2.2 Chromosome Numbers and Genome Size

Genome size and chromosome number information are necessary to generate complete and chromosome-level assembly. Genome size is an important feature of genome research as it represents the amount of DNA present in the complete haploid set of chromosomes in the nucleus of a cell. It is also recognized as a 'C value'. There is a huge variation in genome size or nuclear DNA content and chromosome numbers not only in fishes but also in plants, animals, and prokaryotes. The evolution of genome size and chromosome numbers in organisms is complex to understand as there is no natural rule pattern for fixed or controlled amount of nuclear DNA at taxon/genus/species and closely related species level in particular (Eddy 2012).

The genome size in fish or a group of fish species is regularly estimated by researchers for various study purposes including to understand genome and evolutionary aspects. The Animal Genome Size Database (https://www.genomesize.com/search.php) represents the genome size records of animals, including fish species. Access to this data revealed that genome size information is available for 2186 (26 jawless, 199 cartilaginous, 1942 ray-finned, and 19 lobe-finned) fishes. The genome size of fishes varies enormously as low as 0.35 pg in *Tetraodon nigroviridis* to as high as 132.83 pg in lungfish, *Protopterus aethiopicus*.

4.2.3 Status of Genome Sequencing and Assemblies in Fishes

Genome sequencing initiated with microbial genome sequencing and the human genome's completion in 1999 revolutionized genome research that impacted every sphere of biological sciences including fisheries and aquaculture. In 1997, the first workshop on aquaculture genomics in the United States (FAO 2021) targeted genome research on six species. The first fish genome sequenced and published was that of ray-finned fish, *Fugu rubripes* (Aparicio et al. 2002), and since then, genome research has been initiated for several fish species. In the beginning, the genome assembly releases were limited to a few species probably due to high sequencing cost and sequencing time. A boom in the fish genome happened after the introduction of low-cost, less time-consuming high-throughput sequencing platforms. Fish genomics is transforming very rapidly as evidenced by the increasing availability of numbers of high-quality genome assemblies at the chromosome level and the availability of the large number of sequencing datasets, and the utilization of these resulted in landmark discoveries (Ahmed et al. 2022). A total of 2519 teleost (fish) genomes have been released, as per summary of the NCBI genome datasets (https://www.ncbi.nlm.nih.gov/datasets/genome/?taxon = 32,443) accessed as of date 04.04.2024 (Fig. 4.1). It is evidenced that in the first quarter of 2024, 172 genome assemblies were released, and the last 4 years (2020–2023) alone witnessed a release of 1905 genomes, while in the last 20 years (2000–2019) only 442 genomes were released. The development of technology of sequencing of long-read sequences and assembly and computational bioinformatic tools including Hi-C data correlations for assembly at chromosome level has made remarkable progress in genome research. This led to the release of 680 fish genome assemblies at chromosome level, whereas a total of 1649 genomes are at scaffold level and only some of them, 182, are at contig level (Fig. 4.2), and almost a similar trend could be seen in the case of animal and plant genome assembly releases.

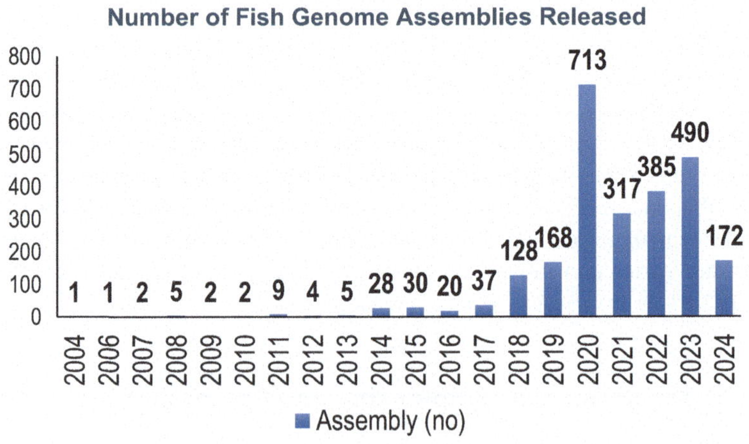

Fig. 4.1 Bar chart depicting the growth of fish genome assemblies published

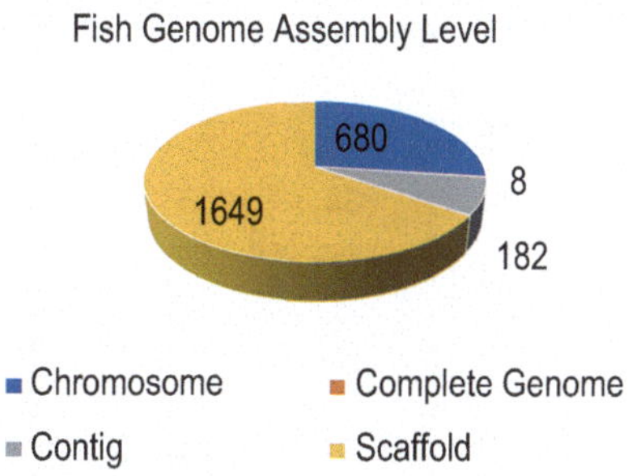

Fig. 4.2 Genome assembly completeness level among fish species

4.2.4 Mitochondrial Genome: Mitogenome

Mitochondria, the powerhouse of the eukaryotic cells, are energy-transducing organelles, in which energy molecule ATP is produced through a process of oxidative phosphorylation. Its genome is extranuclear and compact and consists of a closed circular double-stranded DNA molecule, approximately 15–18 kilobase pairs in size (Anderson et al. 1981). Fish mitogenome, like other vertebrates, has been reported to encode 37 genes (13 protein-coding genes, 22 transfer RNAs, and 2 ribosomal RNAs) (Satoh et al. 2016). It has been reported that these genes as well as the noncoding regions of fish mitogenomes are highly conserved, but gene intervals and lengths can vary between species. Satoh et al. (2016) in a study on fish mitogenomes reported that these 37 genes were found in 249 fish species, out of 250, with the exception of *Limnichthys fasciatus*, which was missing ND6 gene. Such information is useful in the identification of functions of particular genes and alternate genes present thereon. Fish mitogenomes have been widely used for species identification and evolutionary studies due to the various advantages they confer, like maternal inheritance, small size, high substitution rate, rapid evolution, limited recombination, and large copy number (Zhu et al. 2023; Alvarenga et al. 2024).

Advancements in NGS sequencing have immensely helped mitogenome sequencing projects by enabling rapid generation of high-quality data. Since the mitogenome is smaller, as compared to the nuclear genome, elucidating mitogenomes can be undertaken by students and researchers using modest computing resources. There are several popular fish mitogenome analysis tools and software available nowadays, like NOVOPlasty (Dierckxsens et al. 2017), MitoHiFi (Uliano-Silva et al. 2023), DeGeCI (Fiedler et al. 2023), and MitoFish (Zhu et al. 2023).

There are specialized databases dedicated to store, compare, and analyse fish mitogenomes. MitoFish (Mitochondrial Genome Database of Fish) database version 4.00 maintained by the University of Tokyo, Japan, reports a total of complete 4210 fish genomes and 44,825 complete + partial (accessed as of 08-05-2024, https://mitofish.aori.u-tokyo.ac.jp/). MitoFish enables mitogenome annotation, primer designing, and data downloading (Zhu et al. 2023). Another database, FMiR (Fish Mitogenome Resource) database developed and maintained by ICAR-National

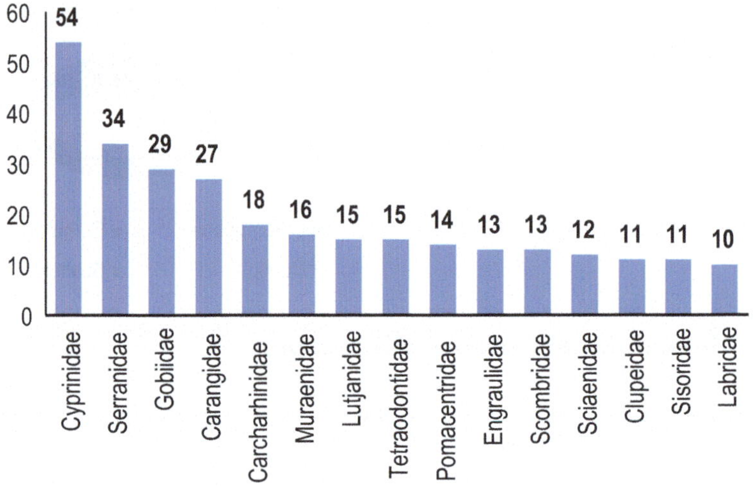

Fig. 4.3 Top 15 families of Indian fish mitogenome assembly data

Bureau of Fish Genetic Resources (NBFGR), Lucknow, India, has 3727 fish mitogenome collection (accessed 04-03-2024) representing about 459 families under 67 orders with tools for data analysis and retrieval of information on mitogenome, taxonomy, habitat, distribution, conservation status of fishes, SSR analysis, primer designing, and similarity search (https://mail.nbfgr.res.in/fmir/) (Nagpure et al. 2015). Data compiled from the FMiR database revealed that the mitogenomes of 659 Indian fishes are available. Of these, the majority of the sequenced mitogenomes is from marine habitats (479) followed by fresh water (146) and brackish water (34). Maximum species belonged to Cyprinidae (54) followed by Serranidae (34) (Fig. 4.3). IUCN classification of threatened species revealed that most of the species targeted for mitogenome studies are of the Not Reported category (393) followed by Least concern (161) (Fig. 4.4).

4.2.5 Chromosome Mapping in Fishes

Majority of genetic material is located in the nucleus, in the form of chromosomes, whose numbers present in the cell varies among species. The chromosomes transmit specific information to the next generation, which are necessary for growth, development, and functioning of an organism. DNA within the chromosomes holds the genetic code and the genes that provide instructions, and ultimately the chromosomes package and organize this information within our cells. The genes are the important hereditary units and are organized in a specific arrangement on each chromosome, where it occupies a particular location, known as its locus. The forms (one inherited from the mother and one from the father) of the gene at the same locus on each chromosome pair is called an allele.

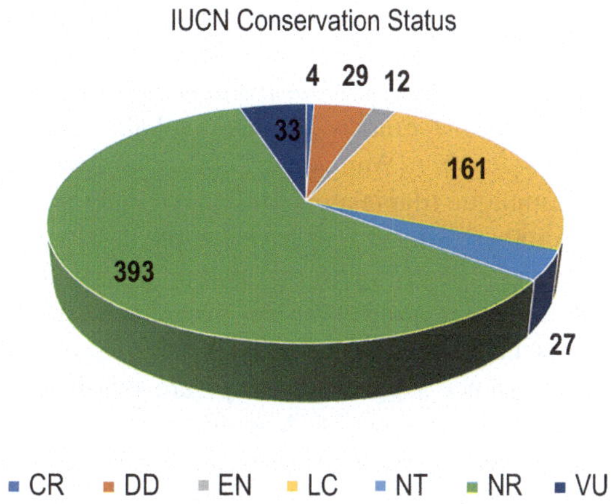

Fig. 4.4 IUCN conservation status of Indian fish species for which mitogenome is available

The advancement in molecular biology has made possible the wide-scale portrayal of genomes, including defining chromosomes from cytological level down to their direct alignment at individual nucleotide base pair level. The chromosome mapping is an important genetic tool that portrays the relative positions and spatial arrangement of genes and other marker sequences on the chromosomes. These maps not only depict the positions of genes but also the distances between them, scaled to a specific unit. Thus, the mapping of chromosomes is instrumental in understanding the intricate architecture of the genome, which in turn gives the information on genetic composition of the traits, including genetic diseases, and their inheritance patterns.

Chromosome mapping studies in fishes have majorly progressed towards the organizational level and subsequently towards the addition of genetic/molecular markers to define linkage groups within the species. The chromosome mapping may broadly be categorized in two groups, viz. linkage/genetic mapping and physical mapping (Brown 2002). The linkage mapping uses classical genetic tools, like breeding experiments, pedigree information, etc., to the chromosome construct map, while the physical mapping uses molecular biology techniques, like fluorescence in situ hybridization (FISH), to localize gene/DNA region on the chromosome.

The ultimate goal of creation of a linkage/genetic map is to generate a chromosome map of a species that depicts, in graphical form, the linear order and relative locations between the linked genes/DNA regions, by plotting their recombination frequency (i.e. frequency at which recombination happens between the two loci). The map gives relative locations along each chromosome rather than the distance between the chromosome and provides a first level of mapping, generally measured in centimorgan (cM) unit, where 1 cM of a genetic map is equal to one percentage of recombination frequency. The unit was named after the developer of the first linkage map, Thomas Hunt Morgan, who developed it with his student, Alfred Sturtevant, in *Drosophila* in 1913. Since then, extensive linkage groups have been developed in important fish species, such as salmonids (a review on salmonids' genomics by Dysin et al. 2022). Other species where genetic markers, such as microsatellites, used in the creation of genetic maps in aquaculture species, in rainbow trout (Guyomard et al. 2012), in Atlantic salmon Lien et al. 2011), Japanese flounder (Song et al. 2012), *Salmo trutta* (Gharbi et al. 2006), *Oreochromis niloticus* (Lee et al. 2005), *Ictalurus punctatus* (Waldbieser et al. 2001; Kucuktas et al. 2009), *Lates calcarifer* (Wang et al. 2007), *Ctenopharyngodon idella* (Xia et al. 2010), and Mozambique tilapia and red tilapia (Liu et al. 2013).

Another form of mapping is a physical map of a chromosome, where physical distances (ideally measured and/or represented in nucleotide bases) between the chromosomal landmarks (like genes, markers, or other DNA region of interest) are represented. Various techniques (such as restriction map, restriction hybrid map, and sequence tagged site map) are employed to construct physical map in the absence of complete DNA sequence. FISH is another powerful molecular biology tool, generally used to visualize the position and distribution of a target DNA regions, on the chromosome within a genome. Physical maps have been reported in a few fish species, such as *Ictalurus punctatus* (Quiniou et al. 2007), *Oreochromis niloticus* (Katagiri et al. 2005), and *Dicentrarchus labrax* (Kuhl et al. 2010). Such physical chromosome maps were constructed using BAC (bacterial artificial chromosome) clones.

In addition to the above, there are extensive works on chromosomal mapping of rDNA sites, especially in fish cytogenetics (Symonová and Howell 2018). The works of physical mapping of major ribosomal RNA genes across the teleost species has been extensively analysed and reviewed by Gornung (2013) in 330 species belonging to 77 families and 22 orders of bony fishes and 11 species of basal ray-finned fishes. He found that around 87% of 45S rDNA sites are mapped on terminal position of the chromosomes. The cytogenetic mapping of rRNA genes may reflect the evolutionary lineages and pathways of certain taxa (Martins and Wasko 2004).

4.3 Functional Genomics

4.3.1 Transcriptomics

Transcriptome has been defined as one of the measures of cell status quantified by the full set of all RNA molecules, both protein-coding RNAs (mRNA) and noncoding RNAs (rRNA, tRNA, lncRNA, pri-miRNA), present at various physiological conditions within a cell type or tissue (Thompson et al. 2016). Transcriptome being the dynamic part of a cell, its sequencing can facilitate understanding the functional structure and large-scale expression profiles of the trait under study (Kumar et al. 2023). To understand the evolution of fish and effectively harness transcriptome sequencing technology, the Beijing Institute of Genomics (BGI) and China National Genebank (CNGB) established the Fish-T1K project in 2013 to generate transcriptome sequences for 1000 diverse fish species (Sun et al. 2016). NCBI SRA database (https://www.ncbi.nlm.nih.gov/sra/) provides results over 120,000 result1168 SRA records for the search terms 'transcriptome OR RNA-Seq AND Fishes' (accessed on 08-04-2024). FishGET database contains 1362 RNA-Seq paired-end data of 8 fish species, collected from 97 different studies, from which a user can obtain and study the gene expression patterns (Guo et al. 2023). In India, to accelerate genomics and transcriptomic research, ICAR launched Consortium Research Platform (CRP) on Genomics, led by ICAR-NBFGR, Lucknow. One of its aims is to generate genomes and transcriptome resources for prioritized fish and shellfish species. Transcriptome studies in fishes have resulted in a greater understanding of biotic and abiotic factors affecting biological processes such as growth, immunity, adaptation, reproduction, disease resistance, etc. In India, transcriptome studies have been undertaken, primarily for the identification of genes associated with economic traits like growth: *Tenualosa ilisha* (Divya et al. 2019), *Labeo catla* (Mohindra et al. 2022), *Labeo rohita* (Nandanpawar et al. 2023)); nutrition (*Rachycentron canadum* (Ebeneezar et al. 2023)); stress response (thermal stress tolerance in *Clarias magur* (Kumar et al. 2022a, b); ammonia tolerance in *C. magur* (Banerjee et al. 2019); salt tolerance in *Etroplus suratensis* (Arun Kumar et al. 2020), *L. rohita* (Harshini et al. 2023), and *Heteropneustes fossilis* (Chutia et al. 2023); hypoxia tolerance in *C. magur* (Mohindra et al. 2016); reproduction in *L. rohita* (Sahu et al. 2015), *Clarias magur* (Agarwal et al. 2020; Kushwaha et al. 2023), *Eleutheronema tetradactylum* (Vinodha et al. 2022), and *Tor putitora* (Shahi et al. 2023); etc. There is still potential for exploring the transcriptome repertoire of Indian fish species, which would need a targeted large-scale data generation and analysis efforts.

4.4 Use of Genomic Data for Genetic Variability Assessment

Fresh water biodiversity is more susceptible to decline than its counterparts (Birnie-Gauvin et al. 2023), for which reasons may be anthropogenic destruction, degradation of habitats, over-utilization/exploitation of natural resources, pollution, and to some extent by climate change. However, the loss of variation at the intraspecific level is more than that of species level (Roches et al. 2021). Thus, for planning and management of conservation programmes, genetic variability and diversity, both among species and within species levels, form fundamental material, and the populations, wild as well as domesticated, have to be characterized. It would identify evolutionarily significant units, enabling the recognition of threatened and endangered stocks/species, and on the basis of this information, strategies can be formulated for sustainable harvesting of populations.

Genetic approaches have analytical power to identify and validate genetic variation, genetic diversity (Singh et al. 2022) and uniqueness of individuals, populations, or species, even when with similar morphological characters. In addition to morphological features, DNA barcoding/meta-barcoding and recently environmental DNA (e-DNA; Manandhar et al. 2023) and genome skimming (Literman et al. 2023) methods have emerged as the effective tools to identify, reconfirm, or describe species, along with traditional taxonomy. These tools have advantage of characterizing the species at any developmental stage, sex, or body part or from DNA in aquatic environment.

The literature available on genome assemblies released and published indicate the importance of genomic research for understanding the biology of living beings, by developing trait-specific markers and identifying the genetic variations and probable causes of abnormalities or reduced performances; diseases concerning various kinds of traits and mitigation of these issues arise due to defects in the DNA sequences. Zebrafish, being a model species for studying the gene functions of vertebrates, is also a successful model for human genetic diseases as 'approximately 70% of human genes have at least one obvious zebrafish orthologue' (Howe et al. 2013). Besides zebrafish, many other fishes are being used as models in genetic, evolutionary, biotechnology, aquaculture improvements, etc. (Silveira et al. 2023). The information generated through genomic studies, particularly gene knockdown, has been proven to be useful for understanding the gene functions in fish (Haffter et al. 1996). In combination with the transcriptomes, gene discovery and understanding of biological processes, genes, and markers associated with important commercial traits, as well as adaptive evolution of fish species with reference to their environment, have been the major applications. The majority of the transcriptome studies in fish is undertaken to understand the molecular mechanism underpinning commercial traits, including growth (Mohindra et al. 2022), reproduction (Kushwaha et al. 2023), stress tolerance (Mohindra et al. 2016), disease resistance (Sudhagar et al. 2018), nutrition and toxicology (Chandhini and Rejish 2019), and osmoregulation (Mohindra et al. 2023). Discovery of novel transcripts and gene-associated markers have been a major output with large transcriptomic data, which may be responsible for the underlying traits (Cao et al. 2020; Xiang et al. 2010).

The fish genome contains several important reference points in understanding the evolution of land vertebrates (Gross 2013) and several unique genes for terrestrial and environmental adaptation like amphibians (Kushwaha et al. 2021). The genomic and transcriptome information provides unique information and insights into unique features of a species about its evolution and adaptation in a particular environment and many more. This information is also useful for comparative genomic studies, for gaining insights into evolutionary rate in teleosts (Zhang et al. 2020), as well as genic evolution since the divergence (Yang et al. 2012), and to find out signatures of local adaption (Fé-Gonçalves et al. 2020).

The genetic/linkage and physical/sequence maps have greatly been used in comparative genomic studies (Zhu et al. 2015) to map candidate genes or DNA markers of non-model fishes (Guyomard et al. 2012) and achieve functional annotation of homologous genes (Woods et al. 2000). Genetic maps may also be used to identify differences in recombination levels among the individuals and between the sexes when multiple mapping families are utilized in studies, pointing out to the varying genetic architecture (Danzmann and Gharbi 2001). In addition, comparative genomics among multiple genomes of single/multiple species helps to infer evolutionary processes and helps to discover lineages/ancestry within/among species (Nicolini et al. 2023).

Fish mitogenomes have been used to resolve higher-level relationships among teleosts and concluded that fish mitogenomes may be decisive in resolving persistent controversies over higher-level relationships, especially in teleosts (Miya et al. 2001). Several other researchers have successfully employed mitogenome studies to resolve taxonomic ambiguities and resurrect taxonomic families (Miya and Nishida 2000; Pasco-Viel et al. 2012; Alò et al. 2013; Roul et al. 2021; Pavan-Kumar et al. 2016; Radhakrishnan et al. 2019; Chowdhury et al. 2024). Mitogenome data can also have forensic applications in trade and conservation, especially for regulated species through accurate species identification and population genetic assessments. Some protected species, like sharks, are more vulnerable to over-exploitation due to their conservative life history traits and sensitivity to overfishing. Hence, such species are accorded legal protection in the form of local and international laws restricting their trade, but there are many situations where the morphological characters are difficult to interpret (i.e. eggs, early life stages, processed or damaged and other traded forms, etc.) for correct species identification. Mitochondrial gene, cytochrome-c-oxidase subunit I has been successfully employed in developing species-specific DNA barcoding sequence information for species identification, as it contain genetic differences at the species level (Kathirvelpandian et al. 2022). ICAR-NBFGR, Lucknow, has facilitated in the forensic identification of fish species and development of species-specific molecular codes, etc., based on molecular markers and bioinformatic analysis (Kathirvelpandian et al. 2022). DNA analyses based on mtDNA have also been used to distinguish threatened whales from minke whales in 'whale' meat, from Japan and Korea (Baker et al. 2010). Further, whale meat on sale in sushi restaurants in Los Angeles and Seoul, South Korea, matched whales from Japan, indicating illegal international trade in whales under protection (Baker et al. 2010).

With the availability of reference or multiple reference genomes and transcriptomes, for a commercially important species, genome re-sequencing or generation of pangenomes (Mwaura et al. 2023) would provide understanding into basic structure and organization of genes, identification of core gene sets, as well as discovering the genomic variations within the species over the spatial distribution. This, in turn, would facilitate development of genome-wide markers, to estimate admixture, genomic variability, and their associations with a particular trait. Transcriptome sequencing has been reported as an effective tool for the identification of gene-associated SNPs because it focuses on protein-coding RNAs (Qian et al. 2014) and for the development of SNP arrays (Liu et al. 2013). Gene-associated SNPs are important as they may be present in large numbers with genome-wide distribution. Additionally, SNPs associated with genes could be causative SNPs for traits themselves. In this direction, FishSNP database has been published, which provides information on SNPs from 15 fish species (Zhang et al. 2024). Such large-scale genomic and transcriptomic data aid to define species, strains and genomic variations, divergence, and differentiation.

In addition, with the availability of high-quality and chromosome-level genome assemblies in fishes, introduction of genetic variability by targeted mutations and gene knockouts through CRISPR-Cas technology may become possible in important commercial species (Coogan et al. 2022). This would enable researchers to study genotype-phenotype relationships, elucidate molecular mechanisms underlying complex traits, and identify potential therapeutic targets. This approach provides valuable insights into genetic variability and its impact on biological processes, disease susceptibility, and drug response.

4.5 Future Prospective and Action Points

Prospects for the sustainable management of fish genetic resources and future strategy in India have to be drawn up based on information on its large and diverse fish germplasm. Integrative multi-omics technologies would facilitate a deeper fundamental biological understanding of their genomic architecture, enabling effective genome mining for bioprospecting of novel genes and alleles from unexplored genetic diversity, for important production traits, stress (biotic and abiotic) tolerance, and their biomarkers. This would assist in genome-wide data generation, collection, and investigation for improved understanding of the intricate associations between genotypes and phenotypes, as well as hidden and cryptic variability.

By taking into consideration the vast germplasm involved, these studies need to be emphasized for commercially and evolutionary important fresh water and brackish/marine species. In this direction, omics data-intensive projects need to be planned and executed in consortium mode research, through which various expertise from different genomics and bioinformatic domains are brought together for effective execution and successful outcome.

In the process, systematic collection and compilation of the numerous genomic information on our biodiversity, which is now vast and scattered in nature, is need of the hour, into databases, and can be achieved though broad-based collaborative efforts, for meaningful decisions to be derived from them.

Large-scale genomic and transcriptomic data aid in defining species, strains, and genomic variations, divergence, and differentiation. Thus, arose the concept of population genomics, which combines genomic concepts and technologies with the population genetics objectives of understanding evolution and adaptation (Peng et al. 2023). Thus, genomics offers effective tools for characterization of biodiversity, and in combination with population genomic data and phenomics, it would facilitate genomic diversity to be incorporated into management and conservation of biodiversity.

In the face of climate change scenario, improved varieties would be needed for sustainable fish production, for nutritional security of the ever-increasing population; availability of a wide range of genetic diversity within the species would be a basic requirement. The use of genomic technologies to characterize our endemic fish diversity, including improved varieties developed through selection, will enable us to meet the changing global scenario, following conservation biodiversity and their sustainable use. It is envisaged that with the rapid advancements of genomic research in India, in the near future, genomic tools will also be utilized for characterizing uniqueness of individuals, populations, or species and documenting genetic variation and genetic diversity of fish genetic resources of India, empowering its conservation strategies.

Acknowledgement The authors are grateful to the director of ICAR-NBFGR, Lucknow, India, for providing facilities for this work.

References

Agarwal D, Gireesh-Babu P, Pavan-Kumar A et al (2020) Transcriptome analysis of *Clarias magur* brain and gonads suggests neuro-endocrine inhibition of milt release from captive GnRH-induced males. Genomics 112(6):4041–4052

Ahmad SF, Jehangir M, Srikulnath K et al (2022) Fish genomics and its impact on fundamental and applied research of vertebrate biology. Rev Fish Biol Fisheries 32: 357–385

Alò D, Correa C, Arias C et al (2013) Diversity of Aplochiton fishes (Galaxiidea) and the taxonomic resurrection of *A. marinus*. PLoS One 8(8):e71577

Alvarenga M, D'Elia AKP, Rocha G et al (2024) Mitochondrial genome structure and composition in 70 fishes: a key resource for fisheries management in the South Atlantic. BMC Genomics 25(1):215

Anderson S, Bankier AT, Barrell BG et al (1981) Sequence and organization of the human mitochondrial genome. Nature 290(5806):457–465

Aparicio S, Chapman J, Stupka E et al (2002) Whole-genome shotgun assembly and analysis of the genome of *Fugu rubripes*. Science 297(5585):1301–1310

Arun Kumar TV, Pradeep MA, Arshad E (2020) Transcriptomic approach to study salinity tolerance in euryhaline cichlid, *Etroplus suratensis* (Bloch, 1790). J Mar Biol Assoc India 62(1):24–32

Baker CS, Steel D, Choi Y et al (2010) Genetic evidence of illegal trade in protected whale's links Japan with the US and South Korea. Biol Lett 22:58–59

Banerjee B, Koner D, Hasan R et al (2019) Transcriptome analysis reveals novel insights in air-breathing Magur catfish (*Clarias magur*) in response to high environmental ammonia. Gene 703:35–49

Belton JM, McCord RP, Gibcus JH et al (2012) A comprehensive technique to capture the conformation of genomes. Methods 58(3):268–276

Birnie-Gauvin K, Lynch AJ, Franklin PA et al (2023) The RACE for freshwater biodiversity: essential actions to create the social context for meaningful conservation. Conserv Sci Pract 5(4):e12911

Brown TA, Genomes (2002) Chapter 5, Mapping Genomes. 2nd edition. Oxford: Wiley-Liss. Available from: https://www.ncbi.nlm.nih.gov/books/NBK21116/

Cao M, Zhang M, Yang N et al (2020) Full length transcriptome profiling reveals novel immune-related genes in black rockfish (*Sebastes schlegelii*). Fish Shellfish Immunol 106:1078–1086

Chandhini S, Rejish Kumar VJ (2019) Transcriptomics in aquaculture: current status and applications. Rev Aquac 11:1379–1397

Chen-Shan C, David HA, Patrick M et al (2013) Nonhybrid, finished microbial genome assemblies from long-read SMRT sequencing data. Nat Methods 10(6):563–569

Chowdhury LM, Divya PR, Mandal S (2024) Complete mitochondrial genome of critically endangered catfish *Hemibagrus punctatus* (Jerdon, 1849) and comparative analysis for insights into the phylogeny of hemibagrids through mitogenomic approach. Mol Biol Rep 51:601

Chutia P, Das M, Saha N (2023) Transcriptome analysis of gills reveals novel insights into the molecular response of stinging catfish (*Heteropneustes fossilis*) to environmental hypertonicity. Gene 851:147044

Coogan M, Alston V, Su B et al (2022) CRISPR/Cas-9 induced knockout of myostatin gene improves growth and disease resistance in channel catfish (*Ictalurus punctatus*). Aquaculture 557:738290

Danzmann RG, Gharbi K (2001) Gene mapping in fishes: a means to an end. Genetica 111:3–23
Dierckxsens N, Mardulyn P, Smits G (2017) NOVOPlasty: de novo assembly of organelle genomes from whole genome data. Nucleic Acids Res 45(4):e18
Divya BK, Mohindra V, Singh RK et al (2019) Muscle transcriptome resource for growth, lipid metabolism and immune system in Hilsa shad, *Tenualosa ilisha*. Genes Genomics 41(1):1–15
Dysin AP, Shcherbakov YS, Nikolaeva OA et al (2022) Salmonidae genome: features, evolutionary and phylogenetic characteristics. Genes (Basel) 13(12):2221
Ebeneezar S, Krupesha Sharma SR, Vijayagopal P et al (2023) Full-length transcriptome from different life stages of cobia (*Rachycentron canadum*, Rachycentridae). Sci Data 10:97
Eddy SR (2012) The C-value paradox, junk DNA and ENCODE. Curr Biol 22R:898
FAO (2021) Thematic background study—genome-based biotechnologies in aquaculture. FAO, Rome. https://doi.org/10.4060/cb7955en
FAO (2022) The state of world fisheries and aquaculture 2022. Towards blue transformation. FAO, Rome. https://doi.org/10.4060/cc0461en
Fé-Gonçalves LM, Araújo JDA, Santos CHDAD et al (2020) Transcriptomic evidences of local thermal adaptation for the native fish *Colossoma macropomum* (Cuvier, 1818). Genet Mol Biol 43(3):e20190377
Fiedler L, Middendorf M, Bernt M (2023) Fully automated annotation of mitochondrial genomes using a cluster-based approach with de Bruijn graphs. Front Genet 14:1250907
Gharbi K, Gautier A, Danzmann RG et al (2006) A linkage map for brown trout (*Salmo trutta*): chromosome homeologies and comparative genome organization with other salmonid fish. Genetics 172(4):2405–2419
Gornung E (2013) Twenty years of physical mapping of major ribosomal RNA genes across the teleosts: a review of research. Cytogenet Genome Res 141:90–102
Gross M (2013) What fish genomes can tell us about life on land. Curr Biol 23(10):R419–R421
Guo C, Duan Y, Ye W et al (2023) FishGET: a fish gene expression and transcriptome database with improved accuracy and visualization. iScience 26(4):106539
Guyomard R, Boussaha M, Krieg F et al (2012) A synthetic rainbow trout linkage map provides new insights into the salmonid whole genome duplication and the conservation of synteny among teleosts. BMC Genet 13:15
Haffter P, Granato M, Brand M et al (1996) The identification of genes with unique and essential functions in the development of the zebrafish, *Danio rerio*. Development 123:1–36
Harshini V, Shukla N, Raval I et al (2023) Interplay of gene expression and regulators under salinity stress in gill of Labeo rohita. BMC Genomics 24(1):336
Heather JM, Chain B (2016) The sequence of sequencers: the history of sequencing DNA. Genomics 107(1):1–8
Howe K, Clark M, Torroja C et al (2013) The zebrafish reference genome sequence and its relationship to the human genome. Nature 496:498–503
Katagiri T, Kidd C, Tomasino E et al (2005) A BAC-based physical map of the Nile tilapia genome. BMC Genomics 6:89
Kathirvelpandian A, Chowdhury LM, Kumar MS (2022) Species-specific molecular signatures for the commercially important scombrids using mitochondrial gene analysis; a tool for fisheries management. J Asia-Pac Biodivers 15(4):481–487
Kucuktas H, Wang S, Li P et al (2009) Construction of genetic linkage maps and comparative genome analysis of catfish using gene-associated markers. Genetics 181(4):1649–1660
Kuhl H, Beck A, Wozniak G et al (2010) The European sea bass Dicentrarchus labrax genome puzzle: comparative BAC-mapping and low coverage shotgun sequencing. BMC Genomics 11:68
Kumar MS, Baisvar VS, Kushwaha B et al (2022a) Thermal stress induces expression of nuclear protein and Parkin genes in endangered catfish, *Clarias magur*. Genes 825:146388
Kumar MS, Kumar R, Kushwaha B et al (2022b) Genomic research in fishery sector: current status and future prospects. Indian J Plant Gen Res 35(3):305–307
Kushwaha B, Pandey M, Das P et al (2021) The genome of walking catfish *Clarias Magur* (Hamilton, 1822) unveils the genetic basis that may have facilitated the development of environmental and terrestrial adaptation systems in air-breathing catfishes. DNA Res 28(1):dsaa031

Kushwaha B, Srivastava N, Kumar MS et al (2023) Protein-protein networks analysis of differentially expressed genes unveils the key phenomenon of biological process with respect to reproduction in endangered catfish, *C magur*. Gene 860:147235

Lee BY, Lee WJ, Streelman JT et al (2005) A second-generation genetic linkage map of tilapia (Oreochromis spp.). Genetics 170(1):237–244

Lien S, Gidskehaug L, Moen T et al (2011) A dense SNP-based linkage map for Atlantic salmon (*Salmo salar*) reveals extended chromosome homeologies and striking differences in sex-specific recombination patterns. BMC Genomics 12:615

Literman R, Windsor AM, Bart HL Jr et al (2023) Using low-coverage whole genome sequencing (genome skimming) to delineate three introgressed species of buffalofish (Ictiobus). Mol Phylogenet Evol 182:107715

Liu F, Sun F, Li J et al (2013) A microsatellite-based linkage map of salt tolerant tilapia (*Oreochromis mossambicus* x *Oreochromis* spp.) and mapping of sex-determining loci. BMC Genomics 14:58

Manandhar P, Manandhar S, Sherchan AM et al (2023) Environmental DNA (eDNA) based fish biodiversity assessment of two Himalayan rivers of Nepal reveals diversity differences and highlights new species distribution records. PLOS Water 2(6):e0000099

Martins C, Wasko AP (2004) Organization and evolution of 5S ribosomal DNA in the fish genome. In: Williams CR (ed) Focus on genome research. Nova Science Publishers, Hauppauge, NY, pp 335–363

Miya M, Nishida M (2000) Use of mitogenomic information in teleostean molecular phylogenetics: a tree-based exploration under the maximum-parsimony optimality criterion. Mol Phylogenet Evol 17(3):437–455

Miya M, Kawaguchi A, Nishida M (2001) Mitogenomic exploration of higher teleostean phylogenies: a case study for moderate-scale evolutionary genomics with 38 newly determined complete mitochondrial DNA sequences. Mol Biol Evol 18(11):1993–2009

Mohindra V, Tripathi RK, Singh A et al (2016) Hypoxic stress -responsive genes in air breathing catfish, *Clarias magur* (Hamilton 1822) and their possible physiological adaptive function. Fish Shellfish Immunol 59:46–56

Mohindra V, Chowdhury LM, Chauhan N et al (2022) Transcriptome analysis revealed hub genes for muscle growth in Indian major carp, *Catla catla* (Hamilton, 1822). Genomics 114(4):110393

Mohindra V, Chowdhury LM, Chauhan N et al (2023) Transcriptome analysis revealed osmoregulation related regulatory networks and hub genes in the gills of Hilsa shad, *Tenualosa ilisha*, during the migratory osmotic stress. Mar Biotechnol 25(1):161–173

Mwaura JG, Wekesa C, Kelvin K et al (2023) Pangenomics of the cichlid species (*Oreochromis niloticus*) reveals genetic admixture ancestry with potential for aquaculture improvement in Kenya. JoBAZ 84:28

Nagpure NS, Rashid I, Pathak AK et al (2015) FMiR: a curated resource of mitochondrial DNA information for fish. PLoS One 10(8):e0136711

Nandanpawar P, Sahoo L, Sahoo B et al (2023) Identification of differentially expressed genes and SNPs linked to harvest body weight of genetically improved rohu carp, *Labeo rohita*. Front Genet 14:1153911

Nicolini F, Martelossi J, Forni G et al (2023) Comparative genomics of Hox and ParaHox genes among major lineages of Branchiopoda with emphasis on tadpole shrimps. Front Ecol Evol 11:1046960

Pasco-Viel E, Veran M, Viriot L (2012) Bleeker was right: revision of the genus Cyclocheilichthys (Bleeker 1859) and resurrection of the genus *Anematichthys* (Bleeker 1859), based on morphological and molecular data of southeast Asian Cyprininae (Teleostei, Cypriniformes). Zootaxa 3586:41–54

Pavan-Kumar A, Raman S, Koringa PG et al (2016) Complete mitochondrial genome of threatened mahseer tor tor (Hamilton 1822) and its phylogenetic relationship within Cyprinidae family. J Genet 95(4):853–863

Peng Y, Jin M, Li Z et al (2023) Population genomics provide insights into the evolution and adaptation of the Asia corn borer. Mol Biol Evol 40(5):msad112

Qian X, Ba Y, Zhuang Q et al (2014) RNA-Seq technology and its application in fish transcriptomics. OMICS 18:1–13

Quiniou SM, Waldbieser GC, Duke MV (2007) A first generation BAC-based physical map of the channel catfish genome. BMC Genome 8:40

Radhakrishnan DP, Kumar RG, Mohitha C et al (2019) Resurrection and re-description of *Pampus candidus* (Cuvier), silver Pomfret from the northern Indian Ocean. Zool Stud 58:e7

Roches DS, Pendleton LH, Shapiro B et al (2021) Conserving intraspecific variation for nature's contributions to people. Nat Ecol Evol 5:574–582

Roul SK, Jeena NS, Kumar R et al (2021) Postulating the modality of integrative taxonomy in describing the cryptic congener *Pampus griseus* (cuvier) and systematics of the genus *Pampus* (perciformes: Stromateidae). Front Mar Sci 8:778422

Sahu DK, Panda SP, Meher PK et al (2015) Construction, De-novo assembly and analysis of transcriptome for identification of reproduction-related genes and pathways from Rohu, *Labeo rohita* (Hamilton). PLoS One 10(7):e0132450

Satoh TP, Miya M, Mabuchi K et al (2016) Structure and variation of the mitochondrial genome of fishes. BMC Genomics 17:719

Shahi N, Singh B, Mallik SK et al (2023) RNA-Seq reveals differential gene expression patterns related to reproduction in the Golden Mahseer. Aust Fish 8(7):352

Silveira T, Moreno Abril SI, Lucas CG et al (2023) Editorial: application of fishes as biological models in genetic studies. Front Genet 13:1092160

Singh RK, Divya PR, Mohindra V et al (2022) Diversity of fish genetic resources below the species level, characterization and applications in resource management. Indian J Plant Gen Res 35(3):301–304

Song W, Pang R, Niu Y et al (2012) Construction of high-density genetic linkage maps and mapping of growth-related quantitative trail loci in the Japanese flounder (*Paralichthys olivaceus*). PLoS One 7(11):e50404

Sudhagar A, Kumar G, El-Matbouli M (2018) Transcriptome analysis based on RNA-seq in understanding pathogenic mechanisms of diseases and the immune system of fish: a comprehensive review. Int J Mol Sci 19(1):245

Sun Y, Huang Y, Li X et al (2016) Fish-T1K (transcriptomes of 1,000 fishes) project: large-scale transcriptome data for fish evolution studies. Gigascience 5:18

Sun Z, Li S, Liu Y et al (2024) Telomere-to-telomere gapless genome assembly of the Chinese sea bass (*Lateolabrax maculatus*). Sci Data 11:175. https://doi.org/10.1038/s41597-024-02988-9

Symonová R, Howell MW (2018) Vertebrate genome evolution in the light of fish cytogenomics and rDNAomics. Genes 18(9):96

Thompson SD, Prahalad S, Colbert RA (2016) Integrative genomics. In: Textbook of pediatric rheumatology. Elsevier, Amsterdam, pp 43–53

Tyler AD, Mataseje L, Urfano CJ et al (2018) Evaluation of oxford nanopore's MinION sequencing device for microbial whole genome sequencing applications. Sci Rep 8:10931

Uliano-Silva M, Ferreira JGRN, Krasheninnikova K et al (2023) MitoHiFi: a python pipeline for mitochondrial genome assembly from PacBio high-fidelity reads. BMC Bioinfor 4(1):288

Vinodha V, Kalarani K, Inbaraj RM (2022) Transcriptome sequencing and analysis of male Fourfinger threadfin, *Eleutheronema tetradactylum* (Shaw, 1804). Asian Fish Sci 35:168–178

Waldbieser GC, Bosworth BG, Nonneman DJ et al (2001) A microsatellite based genetic linkage map for channel catfish, *Ictalurus punctatus*. Genetics 158(2):727–734

Wang CM, Zhu ZY, Lo LC et al (2007) A microsatellite linkage map of barramundi, *Lates calcarifer*. Genetics 175(2):907–915

Woods IG, Kelly PD, Chu F et al (2000) A comparative map of the zebrafish genome. Genome Res 10(12):1903–1914

Xia JH, Liu F, Zhu ZY et al (2010) A consensus linkage map of the grass carp (*Ctenopharyngodon idella*) based on microsatellites and SNPs. BMC Genomics 11:135

Xiang LX, He D, Dong WR et al (2010) Deep sequencing-based transcriptome profiling analysis of bacteria-challenged *Lateolabrax japonicus* reveals insight into the immune-relevant genes in marine fish. BMC Genomics 11:472

Yang E, Hulse AM, Cai JJ (2012) Evolutionary analysis of sequence divergence and diversity of duplicate genes in *Aspergillus fumigatus*. Evol Bioinforma 8:623–644

Zhang YH, Ravi V, Qin G et al (2020) Comparative genomics reveal shared genomic changes in syngnathid fishes and signatures of genetic convergence with placental mammals. Natl Sci Rev 7(6):964–977

Zhang L, Li H, Shi M et al (2024) FishSNP: a high-quality cross-species SNP database of fishes. Sci Data 11:286

Zhu C, Tong J, Yu X et al (2015) Comparative mapping for bighead carp (*Aristichthys nobilis*) against model and non-model fishes provides insights into the genomic evolution of cyprinids. Mol Gen Genomics 290:1313–1326

Zhu T, Sato Y, Sado T et al (2023) MitoFish, MitoAnnotator, and MiFish pipeline: updates in 10 years. Mol Biol Evol 40(3):msad035

In Situ Conservation: Tools, Strategies, and Challenges

5

Uttam Kumar Sarkar, Rejani Chandran,
T. K. Teena Jayakumar, and Charan Ravi

Abstract

India is recognized as a mega biodiversity hotspot, contributing 11.72% of the world's fish biodiversity. The loss of biodiversity is one of the most pressing global crises, raising concerns about the status of biological resources vital to human life. Following the signing of the Convention on Biological Diversity, there has been global focus on the conservation and sustainable use of natural resources. In India, conserving its extensive and varied aquatic resources is crucial, not only because a significant portion of food resources still comes from the wild but also to maintain ecological and socioeconomic balance. Conservation of fish diversity is essential to sustain ecological, nutritional, and socioeconomic stability. Species diversity and genetic variability are key for the long-term stability of complex ecosystems and species. Addressing the challenge of conserving the country's fish diversity and aquatic resources requires coordinated efforts, integrating in situ programs with the latest technological innovations and strict enforcement of laws. Therefore, conservation policies should encourage management practices that uphold the integrity of aquatic ecosystems, prevent endangerment, and enhance the recovery of threatened species. This chapter aims to review the relevant approaches and strategies of in situ conservation and to outline future research opportunities to promote conservation and sustainable fish production.

Keywords

Biodiversity · In situ conservation · Protected area · Sustainability, Management

U. K. Sarkar (✉) · R. Chandran · T. K. Teena Jayakumar · C. Ravi
ICAR-National Bureau of Fish Genetic Resources, Lucknow, Uttar Pradesh, India
e-mail: Uttam.Sarkar@icar.gov.in; rejani.chandran@icar.gov.in;
Teena.Jayakumar@icar.gov.in; Ravi.Charan@icar.gov.in

U. K. Sarkar et al. (eds.), *Sustainable Management of Fish Genetic Resources*,
https://doi.org/10.1007/978-981-97-5250-8_5

5.1 In Situ Conservation

Fish exhibits the greatest biodiversity of the vertebrates (animals with backbones) with over 36,775 species of which 18,688 species belong to freshwater habitats (Fricke et al. 2024). India has extensive and diverse natural water resources, like coastline (8118 km), ponds and tanks (2.45 m ha), beels, oxbow lakes and swamps (1.2 m ha), reservoirs (3.15 m ha), brackish water (1.24 m ha), and 14 major rivers which share 83% of total drainage (Department of Fisheries, India Annual Report 2023). The country has 3227 fish species, which belong to 1040 genera, 256 families, and 57 orders, which make up to 11.72% of the global finfish diversity (NBFGR 2024). Four out of 34 global biodiversity hotspots are located in India, with Western Ghats and Northeast India harboring the most endemic freshwater fishes. A total of 306 finfish species are found in the Western Ghats, with nearly 69% being endemic (Gopalakrishnan and Ponniah 2000). Similarly, Northeast India harbors about 240 finfishes, with nearly 40% being endemic to the region. Fish is an important resource as they offer ecosystem stability, food security, economic value, ecotourism, recreation, ecosystem services, and cultural importance among many others. In terms of habitat, fishes reside in almost all imaginable aquatic habitats, ranging from icecap to hot springs. A multitude of stressors and drivers including habitat destruction, overexploitation, increasing population, pollution, and invasive species are affecting the ichthyodiversity negatively. Ecosystems and species important to human life and the health of the environment are disappearing at an alarming rate. Deliberate efforts to protect and conserve a variety of fish species along with their habitat are being implemented across the globe to sustainably manage fisheries. As per the FAO, sustainable fishery development envisages an eco-friendly, equitable mode of development that can sustain livelihood over generations. Effective conservation and management of at least 30% of the world's lands, inland waters, coastal areas, and oceans is 1 of the 23 targets for achievement by 2030 under UN Biodiversity Agreement. The target gives special emphasis to areas of particular importance for biodiversity and ecosystem functioning and services which prioritize ecologically representative, well-connected, and equitably governed systems of protected areas and other effective area-based conservation.

In situ conservation is defined as "the conservation of ecosystems and natural habitats and the maintenance and recovery of a viable population of species in their natural surroundings and, in the case of domesticated or cultivated species, in the surroundings where they have developed their distinctive properties" (Convention on Biological Diversity, Article 2). In situ conservation programs integrate knowledge on fish and habitat diversity, habitat utilization, biology including genetic structure, life history traits, as well as human interference and other socioeconomic issues. The method allows continued coevolution of wild species with other forms and maintains genetic diversity of species and evolutionary adaptations. This technique helps in the preservation and protection of large populations and thus maintaining the natural balance of the ecosystem which eventually has higher potential of population recovery (Fig. 5.1). The in situ conservation strategy holds significant importance in safeguarding fish germplasm resources, encompassing diverse

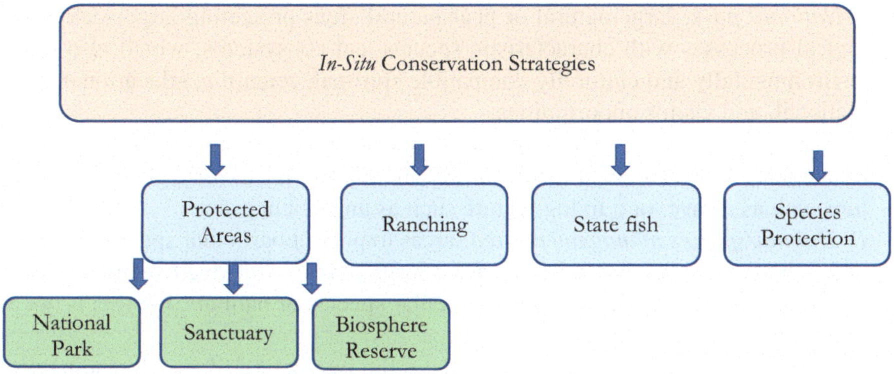

Fig. 5.1 In situ conservation strategies

methods such as marine reserves, freshwater sanctuaries, temple tanks, and protected water bodies within wildlife areas.

5.2 Methods of In Situ Conservation Techniques

5.2.1 Protected Sites or Reserves

In situ conservation methods entail the designation of specific zones as protected areas, ranging from complete prohibition of access to permitting various levels of human activity. IUCN defines a protected area as "A clearly defined geographical space, recognized, dedicated and managed, through legal or other effective means, to achieve the long-term conservation of nature with associated ecosystem services and cultural values." This conservation approach not only preserves the genetic diversity of species but also maintains their evolutionary adaptations, enabling them to continuously adjust to changing environmental conditions such as shifts in pest populations or climate. There is an urgent need of protected areas targeting freshwater habitats harboring rare or endangered species (Sarkar et al. 2008). The authority and responsibility in these protected areas can be governed by the government, private organizations, community, or shared.

The six management categories of protected areas as per IUCN, namely (source: IUCN, Dudley 2008):

- *Ia Strict nature reserve:* strictly protected for biodiversity and also possibly geological/geomorphological features, where human visitation, use, and impacts are controlled and limited to ensure the protection of the conservation values.
- *Ib Wilderness area:* usually large unmodified or slightly modified areas, retaining their natural character and influence, without permanent or significant human habitation, protected and managed to preserve their natural condition.

- *II National park:* large natural or near-natural areas protecting large-scale ecological processes with characteristic species and ecosystems, which also have environmentally and culturally compatible spiritual, scientific, educational, recreational, and visitor opportunities.
- *III Natural monument or feature:* areas set aside to protect a specific natural monument, which can be a landform, sea mount, marine cavern, geological feature such as a cave, or a living feature such as an ancient grove.
- *IV Habitat/species management area:* areas to protect particular species or habitats, where management reflects this priority. Many will need regular, active interventions to meet the needs of particular species or habitats, but this is not a requirement of the category.
- *V Protected landscape or seascape:* where the interaction of people and nature over time has produced a distinct character with significant ecological, biological, cultural, and scenic value and where safeguarding the integrity of this interaction is vital to protecting and sustaining the area and its associated nature conservation and other values.
- *VI Protected areas with sustainable use of natural resources:* areas which conserve ecosystems, together with associated cultural values and traditional natural resource management systems.

There are 106 existing national parks in India covering an area of 44,403 km^2, which is about 1.35% of the geographical area of the country (National Wildlife Database Centre, Nov. 2023). Madhya Pradesh has the largest number of national parks (11) which covers almost 1.41% of the state area. There are 573 existing wildlife sanctuaries in India covering an area of 123,762.56 km^2, which is 3.76% of the geographical area of the country, 115 existing conservation reserves in India covering an area of 5548.75 km^2, and 220 existing community reserves covering an area of 1455.16 km2 (National Wildlife Database Centre, Nov. 2023). India has several significant marine parks, sanctuaries, and biosphere reserves aimed at conserving marine ecosystems and resources. Studies indicate that protected areas harbor greater species diversity, population densities, and biological traits, underscoring their effectiveness in conserving aquatic biodiversity. The establishment of targeted fish protected areas is crucial, with policies required to create freshwater sanctuaries for optimal protection and in situ conservation. Peninsular India hosts 24 marine protected areas (MPAs), while the country's islands have over 100 MPAs. The 24 mainland MPAs cover approximately 8214 km^2, constituting about 5% of India's total protected area network and representing 0.25% of the nation's geographic area (Table 5.1). The Andaman and Nicobar Islands encompass a total area of 4947 km^2, with 21,510 km^2 protected under India's Wild Life Protection Act, 1972. There are 105 protected areas (PAs) in the Andaman and Nicobar Islands, all part of India's MPA network, covering about 60% of the islands' terrestrial area and safeguarding over 40% of coastal habitats. Notable MPAs in this region include Mahatma Gandhi Marine National Park and Rani Jhansi Marine National Park. In the Lakshadweep group of islands, Pitti Island is the sole island designated as an MPA. India has identified 12 protected areas as transboundary protected areas under the IUCN

Table 5.1 Major marine protected areas in India

Sl. no	Name of marine protected area	State/UTs	Year of establishment/ creation	Area (km^2)
Marine national park				
1.	Gulf of Kachchh	Gujarat	1995	162.89
2.	Sundarbans	West Bengal	1984	1330.1
3.	Gulf of Mannar	Tamil Nadu	1986	6.23
4.	Mahatma Gandhi marine National Park	Andaman and Nicobar Islands	1983	281.5
5.	Rani Jhansi marine National Park	Andaman	1996	256.14
Marine sanctuaries				
1.	Bhitarkanika	Odisha	1975	672
2.	Gahirmatha		1997	1435
3.	Gulf of Kachchh	Gujarat	1980	295.03
4.	Malvan marine sanctuary	Maharashtra	1987	29.12
5.	West Sundarbans	West Bengal	2013	556.45
6.	Chilika (Nalabana)	Odisha	1987	15.53
7.	Pulicat Lake	Andhra Pradesh/Tamil Nadu	1980	153.67
8.	Coringa	Andhra Pradesh	1978	235.7
9	Pitti	Lakshadweep	2002	0.01
Biosphere reserves				
1.	Sundarbans	West Bengal	1989	9630
2.	Gulf of Mannar	Tamil Nadu	1989	10,500
3.	Great Nicobar	Andaman and Nicobar Islands	1989	885
4.	Great Rann of Kutch	Gujarat	2008	12,454

Source: Lakra et al. 2007; Sivakumar et al. 2013

Transboundary Protected Area program, with two of these sites being MPAs (Sundarbans National Park and Gulf of Mannar Biosphere Reserve). Additionally, India has designated six UNESCO World Heritage natural sites, with Sundarbans National Park being one of them. There is a pressing need for focused endeavors to enhance and fortify community involvement in the management of India's marine protected area network.

The Ramsar Convention, an international treaty focused on conserving and sustainably utilizing wetlands, aims to prevent the ongoing encroachment on and loss of wetlands, while acknowledging their crucial ecological functions and diverse values including economic, cultural, scientific, and recreational aspects. Since India joined the Ramsar Convention in 1981, 75 wetlands across the country, representing various habitats, have been designated as Ramsar sites of international importance. These wetlands harbor significant fish diversity, but urgent conservation measures are required. The Ministry of Environment and Forests (MoEF), in collaboration with other agencies, has initiated several eco-restoration efforts in these identified wetlands, involving local community participation.

5.2.2 Stock Enhancement Through Ranching

Captive breeding programs not only compensate for the declining fish populations but also supplement and enhance the yields of wild fisheries. Implementation of captive breeding and stocking programs in the wild, alongside stock enhancement programs through ranching, is crucial to revive declining fish populations. Juveniles collected from their natural habitat are reared up to sexual maturity and bred within the confines of a culture unit or lab environment, and the young ones reared are released back (ranching) to the natural environment. Though the technology has been standardized for major carps, many indigenous freshwater fish species having enormous commercial value are yet to achieve the status. ICAR-NBFGR through its various in-house and collaborative efforts has captively bred nonconventional food fish species, viz., *Clarias dussumieri*, *Anabas testudineus*, *Cirrhinus reba*, *Chitala chitala*, *Puntius sarana*, *Semiplotus semiplotus*, *Hypselobarbus carnaticus*, *Ompok pabo*, *O. pabda*, and *Labeo dussumieri.* Additionally, the methodology has also been standardized for more than 25 species of indigenous ornamental species having export potentials such as *Pethia setnai*, *P. nigripinnis*, *Dawkinsia arulius*, *D. tambraparniei*, *D. rubrotinctus*, *Hemibagrus punctatus*, *Ompok malabaricus*, *Pristolepis marginata*, *Horabagrus nigricollaris*, *Chela fasciata*, *Danio malabaricus*, *D. filamentosus*, *Sahyadria denisonii*, *D. fasciatus*, *Mesonoemacheilus triangularis*, *Labeo dussumieri*, *Clarias dussumieri*, and *H. brachysoma*. The technique shoulders immense importance, as these can be sources of seed for ranching into natural waters which is one of the major components of in situ conservation program. A holistic collective research involving scientists and stakeholders is necessary for proper implementation and success. Stock-identified rehabilitation is required for a successful stocking program. Identifying suitable stocks for rehabilitation is essential for effective stocking programs. Successful mass-scale seed production and ranching need to be done for in situ conservation of these species in natural waters. In Kerala, successful stock-specific, breeding-assisted river ranching has been observed for three food fishes, *Horabagrus brachysoma*, *H. nigricollaris*, and *Labeo dussumieri*.

Sea ranching is another method used to boost fisheries' productivity by releasing animals into their natural habitat to be harvested when they reach a larger size. This technique, also known as artificial recruitment of aquatic organisms, aims to improve stock and enhance production while conserving the species. The process of sea ranching includes establishing brood stocks, breeding and rearing larvae, nurturing them in nurseries, releasing them at appropriate locations, and monitoring both released and natural stocks. Ranching is typically conducted in bays, lagoons, shallow water bodies, and protected ecosystems for optimal results. In India, natural stock enhancement marine environment has been carried out for a few species like Asian sea bass, *Lates calcarifer*; mud crab, *Scylla serrata*; penaeid prawns; and marine snail species along the coast of Tamil Nadu and Kerala (Serajuddin et al. 2018).

5.2.3 State Fish Concept

National Bureau of Fish Genetic Resources (NBFGR) in 2006 introduced an innovative approach to fish conservation by establishing the concept of "state fishes of India." This initiative aimed to recognize and prioritize certain fish species across different states in India, involving collaboration to develop conservation strategies for their designated state fish. The initiative, similar to designating state animals or flowers, engages local communities and stakeholders in fish germplasm conservation, fostering sustainable fishery development (Lakra et al. 2010; Sarkar et al. 2024). Currently, 22 Indian states have adopted 17 fish species (Table 5.2) under this program. However, detailed documentation of field-level conservation measures for managing state fishes is lacking. The selection criteria for state fishes are based on regional significance and stakeholder preferences, involving various government processes and multiple entities. The Department of Fisheries, Government of India, has urged remaining states to declare their state fish to further aquatic conservation and fishery development. State fish species are categorized based on themes such as conservation, local importance, ecological significance, and aquaculture potential. Examples include fish like *Etroplus suratensis*, *Tenualosa ilisha*, and *Pampus*

Table 5.2 State fishes of India (source: Sarkar et al. 2024)

Sl No.	Scientific name	Name of the state	IUCN conservation status
1.	*Channa striata* (Bloch,1793)	Andhra Pradesh; Telangana	Least concern
2.	*Clarias magur* (Linnaeus,1758)	Bihar	Least concern
3.	*Labeo calbasu* (Hamilton,1822)	Haryana	Least concern
4.	*Tor putitora* (Hamilton,1822)	Arunachal Pradesh; Himachal Pradesh; Jammu and Kashmir; Uttarakhand	Endangered
5.	*Puntius carnaticus* (Jerdon,1849)	Karnataka	Least concern
6.	*Etroplus suratensis* (Bloch,1790)	Kerala	Least concern
7.	*Osteobrama belangeri* (Valenciennes, 1844)	Manipur	Near threatened
8.	*Semiplotus modestus* (Day, 1870)	Mizoram	Data deficient
9.	*Neolissochilus hexagonolepis* (McClelland,1839)	Nagaland	Near threatened
10.	*Tor mahanadicus* (David,1953)	Odisha	–
11.	*Ompok bimaculatus* (Bloch,1794)	Tripura	Near threatened
12.	*Chitala chitala* (Hamilton,1822)	Uttar Pradesh	Near threatened
13.	*Tenualosa ilisha* (Hamilton,1822)	West Bengal	Least concern
14.	*Tor tor* (Hamilton, 1822)	Madhya Pradesh	Data deficient
15.	*Mugil cephalus* Linnaeus,1758	Goa	Least concern
16.	*Pampus candidus* (Cuvier,1829)	Maharashtra	–
17.	*Protonibea diacanthus* (Lacepède,1802)	Gujarat	Near threatened

candidus, which hold economic, cultural, and culinary significance in Kerala, West Bengal, and Maharashtra, respectively. Hilsa, revered in West Bengal, carries socio-religious importance during festivals, while *Channa striata* is valued for its medicinal properties in Andhra Pradesh and Telangana. *Protonibea diacanthus* is prized for its culinary and industrial uses. Additionally, mahseer fishes are culturally significant and indicative of a healthy riverine ecosystem. Following the overwhelming response and acceptance by the public, a few other states have also come forward to declare their state fish on the recommendation by ICAR–NBFGR for the conservation and promotion of sustainable fisheries management.

5.2.4 Species Protection

Numerous species receive legal protection either directly or indirectly through regulations on trade. Safeguarding species and their habitats legally necessitates understanding which species are most vulnerable, a process led globally by the World Conservation Union or International Union for the Conservation of Nature and Natural Resources (IUCN) and on a national level by various government bodies, NGOs, and researchers. Monitoring plays a crucial role in pinpointing the species and habitats most at risk and in gauging trends in the severity of specific threats. Various methods have been developed for the conservation assessment of fish. The major classification system used internationally for assessing the status of the threat to each species is that adopted and developed by IUCN. The IUCN Red List is a catalogue of taxa that are facing the risk of extinction or the threat in their wild habitats. The fish species restricted to limited geographic area and facing anthropogenic threats because of alteration in their habitats are considered to be more liable for becoming threatened species (IUCN 2023). The uses of Red List data are to develop awareness about the importance of threatened biodiversity, identification of endangered species, providing a global index of the decline of biodiversity, and defining conservation priorities at the local level and guiding conservation action. To conserve aquatic resources, the Ministry of Environment and Forests, government of India, has listed many marine finfish and shellfish species in schedules I–IV of the Wild Life Protection Act, 1972, through gazette notification no. 1–4/95 WL-1 dated 05.12.2001. Regulations remain pivotal in addressing the risks facing biodiversity. Legal safeguards for species and their habitats, as well as restrictions on activities endangering them, are essential, although they may not always be entirely effective, particularly when human livelihoods are also at stake (Serajuddin et al. 2018). Regulatory measures targeting risks rather than specific species or habitats offer a significant advantage by potentially safeguarding numerous species simultaneously. Nonetheless, many significant threats to biodiversity, such as climate change, are challenging to address through regulation.

5.3 Challenges and Threats

In situ fish conservation faces a multitude of challenges and threats that impede its efficacy (Fig. 5.2). Understanding these obstacles is paramount for devising effective conservation strategies to address them. By comprehensively understanding these challenges, conservationists can develop targeted interventions to mitigate their impacts and ensure long-term survival of fish species in their natural environments. In situ fish conservation in freshwater ecosystems poses unique challenges, increasing the risk of species extinction compared to marine or terrestrial environments. Just 3% of Earth's water is freshwater, with the majority locked in polar ice caps or deep underground (Gleick 1996). Of this freshwater, only 0.1% exists in a liquid form suitable for fish habitats, with rivers holding a mere 2% of surface freshwater (Dudgeon 2011). Human activities such as extraction, consumption, and pollution significantly diminish both the quantity and quality of fish habitats (Mestanza-Ramón et al. 2020). Human needs intensify competition with ecosystem demands, leading to diminished downstream flows and exacerbating freshwater scarcity for aquatic life. Changes in land use and habitat modification further impact freshwater bodies, tightly linking water and habitat quality to terrestrial activities. Fragmentation of freshwater habitats by barriers complicates conservation efforts, creating isolated ecosystems with distinct evolutionary histories. Additionally, rivers act as both recipients and transmitters of pollutants downstream, heightening

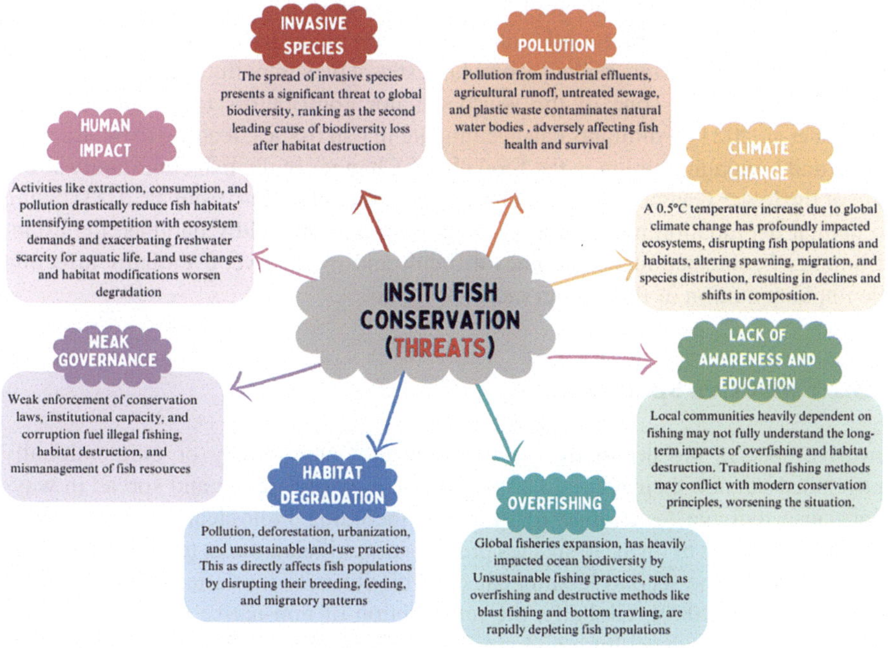

Fig. 5.2 Challenges for in situ conservation

vulnerability. Targeted conservation actions are crucial to safeguarding in situ fish populations in freshwater ecosystems. The conservation of fish species within their natural habitats faces significant challenges stemming from environmental, socio-economic, and policy-related factors.

5.3.1 Habitat Degradation and Pollution

Pollution, deforestation, urbanization, and unsustainable land-use practices are leading to the destruction and degradation of habitats such as wetlands, rivers, estuaries, and marine ecosystems. This degradation directly affects fish populations by disrupting their breeding, feeding, and migratory patterns. India's extensive irrigation system and hydropower projects further exacerbate the problem by significantly altering fish habitats (Ghosh and Ponniah 2008). Habitat degradation has led to the disappearance of species like *Hypselobarbus dubius* in Stanley Reservoir of Tamil Nadu due to silt smothering breeding grounds and decline of humpback mahseers, *Tor remadevii*, and *Neolissochilus hexagonolepis* in the Cauvery River system due to reservoir construction, and construction of Farakka Barrage on the Ganga has adversely affected the hilsa fishery and anadromous migration of hilsa (Ghosh and Ponniah 2008). Despite documentation of these impacts, research on early life stages, intraspecific variation, and habitat preferences of Indian fishes remains sparse. Moreover, there is insufficient research documenting the effects of sand mining, civic pollution, and hydroelectric projects on habitat, sediment dynamics, and aquatic life. Fish habitats are further deteriorating due to pollution from industrial effluents, agricultural runoff, untreated sewage, and plastic waste contaminants. The Ganges River system ranks second largest as a plastic pollution source globally and is recognized as one of the world's major contributors to plastic pollution (Nelms et al. 2021). Pollutants can accumulate in fish tissues, leading to bioaccumulation and biomagnification in the food chain. Changes in the fish community structure of the lower Ganga basin have also been documented (Parvez et al. 2023) owing to habitat degradation and pollution. With poor and unhealthy habitats, the establishment of protected areas becomes difficult and ineffective, eventually affecting the implementation of this conservation program.

5.3.2 Invasive Species

Invasive species are non-native species that are intentionally or unintentionally introduced outside their natural habitats where they proliferate and spread in ways that cause damage to a range of interests, including biological diversity (Shine 2007). The spread of invasive species presents a significant threat to global biodiversity, ranking as the second leading cause of biodiversity loss after habitat destruction (Sala et al. 2000). Globally, the introduction of non-native freshwater fishes, largely driven by aquaculture and stock enhancement activities, has doubled over the past three decades (Gozlan et al. 2010). For instance, Asian production of

African cichlid tilapia exceeds that of Africa by sevenfold, while Chile contributes approximately 30% of global farmed salmon production, both relying on introduced species (Gozlan et al. 2010). Despite meeting societal demands, these introductions pose ecological risks, impacting native fish species and ecosystem dynamics through predation, competition, hybridization, habitat alteration, and disease transmission. Fishes are among the most threatened aquatic animals due to ongoing introductions driven by socioeconomic factors. This highlights the inherent conflict between societal development and conservation priorities in fish conservation efforts. In India, most inland water bodies have established populations of invasive fishes, such as *Clarias gariepinus*, *Oreochromis mossambicus*, *Oreochromis niloticus*, *Piaractus brachypomus*, and *Pterygoplichthys* spp., which have high phenotypic plasticity and survive well even in adverse environmental conditions. The presence of invasive species in the water body affects the effectiveness of the in situ conservation measure, as the ranched seed may be predated by the exotic species present. Hence, success of conservation strategy is dependent of the spread and distribution of invasive species.

5.3.3 Overfishing

Unsustainable fishing practices, such as overfishing and destructive methods like explosive fishing, poisoning, and bottom trawling, are depleting fish populations leading to the decline of commercially important and ecologically significant species (Yan et al. 2021). Small meshed gears that capture juveniles and fingerlings of fishes are used by fishermen in water bodies which is ultimately affecting the abundance. Fishes are being captured before they are mature and have successfully bred in nature eventually affecting the biological performance and population structure. The absence of all sizes of individuals in a population and the skewed size population are indicative of an unhealthy and unstable population which is prevalent in most of the water bodies of the country. Scattered and unsystematic fishing activities along the freshwater bodies make it difficult to oversee and monitor appropriately. Overfishing not only threatens aquatic biodiversity and the livelihoods of fisherfolk but also disrupts entire ecosystems, leading to the loss of critical feeding and breeding grounds for fish fauna (Medhi et al. 2018). Through the establishment of in situ conservation measures, overexploitation practices may be curbed to a certain extent if properly implemented. If suitable measures to control the fishing activity are not executed, the ranching program will not yield any success as the seeds will be caught again by fishermen, before they reach maturity.

5.3.4 Climate Change

Global climate change presents a significant threat to conservation, with a 0.99 °C temperature increase over the last two decades leading to various impacts on ecosystems, including altered precipitation patterns, sea-level rise, and ocean acidification (Sarkar et al. 2022). These changes disrupt fish populations and their habitats, affecting spawning, migration, disease occurrence, behavioral changes, and distribution patterns leading to population declines and shifts in species composition (McCarty 2001). Increased temperature and reduced rainfall have altered the water quality of water bodies with reduced dissolved oxygen, increased salinity, high ammonia levels, eutrophication, and associated fish mortalities. Fluctuations in these parameters affect the conservation strategies negatively and may alter the success ratio. As maintaining the water quality parameters in ideal conditions for the conserved water body in a holistic manner is not practical, climate change poses an immediate threat to in situ conservation methods.

5.3.5 Weak Governance and Enforcement

The success and failure of any conservation and management strategy are heavily dependent on the enforcement of the suggested measures. Wide coverage of water bodies and shortage of officials to inspect and monitor the entire stretch of water body reduce the enforcement. Overfishing, destructive fishing practices, poaching, and indiscriminate release of effluents and waste can be regulated and controlled legally by proper governance. Inadequate enforcement of existing conservation laws and regulations, weak institutional capacity, and corruption contribute to illegal fishing activities, habitat destruction, and ineffective management of fish resources in India.

5.3.6 Lack of Awareness and Education

Creating awareness among the community and stakeholders is extremely important for the successful implementation of conservation strategies. Making them understand the long-term benefits associated with their actions is crucial. It is also important that every stakeholder associated with the water body follows the instructions as even one outlier will create doubt or uncertainty in everyone's mind. Limited awareness among local communities, fishermen, and policymakers about the importance of fish conservation and sustainable fishing practices hinders conservation efforts in India.

5.4 Strategies

5.4.1 More Representation of Hotspot Areas under PAs

Despite the development of various guidelines, approaches, and strategies for biodiversity conservation, there is a notable lack of representation of aquatic areas within India's protected area network (Stolton et al. 2013). While significant attention has been paid to terrestrial biodiversity, the richness of freshwater fish species within many protected areas remains poorly understood. Only a small proportion of protected areas are exclusively designated for wetland protection, and around 10% of protected areas lack specific aquatic habitats. Although the Wild Life Protection Act of India offers legal protection to terrestrial and some aquatic animals, there is less emphasis on freshwater fish biodiversity. There is poor representation and coverage of biodiversity in protected areas as neither do the protected areas adequately represent important species, communities, or regions nor reflect the ecological, cultural, or socioeconomic significance (Sarkar et al. 2008). Establishing a comprehensive network of freshwater aquatic sanctuaries within the protected area network should be a priority component of conservation policies for protected areas. A comprehensive examination of India's protected area network reveals a notable deficiency in adequately representing and covering aquatic biodiversity within these designated areas (Sarkar et al. 2008). This oversight fails to recognize the ecological, socioeconomic, and cultural importance of aquatic ecosystems. Historically, India's protected area network has not fully encompassed several biologically significant regions, communities, and species. More recently, a classification proposed by Rodgers et al. (2000) identifies 10 biogeographical zones divided into 26 provinces, including Transhimalaya, Himalaya, desert, semiarid, Western Ghats, Deccan, Gangetic Plains, coasts, Northeast India, and islands. This classification highlights that the proportion of protected area coverage is highest in islands (18.54%), followed by Western Ghats (10.12%), and both Himalayas and Transhimalayas (19.12%). Several potential areas rich in aquatic biodiversity could be considered (Fig. 5.3) for inclusion in protected areas in India, namely:

- *Hotspots:* India is blessed with naturally rich and diverse ecosystems under the four designated hotspots of the country (Himalayas, Indo-Burma, Western Ghats and Sri Lanka, and Sundaland). The protected area in these hotspots covers approximately 10% which is less than the ideal value set as per Aichi Target. Considering the anthropogenic activities, effects of climate change, habitat degradation, and their impact on diversity and endemism, in situ conservation strategies should be enhanced significantly. Efforts to identify and design the network of "Protected Areas Resilient to Climate Change (PARCC)" should also be initiated bearing in mind the adaptive changes and behavioral changes observed in fishes.
- *Estuaries and mangrove forests:* estuarine ecosystems, where rivers meet the sea, are vital breeding grounds and nurseries for numerous aquatic species. Mangrove forests along the coasts of states like West Bengal, Odisha, Andhra Pradesh, and Gujarat support a wide variety of marine life and provide important ecosystem services. Protecting these areas as wildlife sanctuaries or marine reserves can safeguard their biodiversity.
- *Coral reefs:* India's coral reefs, found along the coastal states like Andhra Pradesh, Odisha, Lakshadweep Islands, and Andaman and Nicobar Islands, are hotspots of marine biodiversity. Establishing marine protected areas or marine parks around coral reef ecosystems can help mitigate threats such as overfishing, habitat destruction, and coral bleaching.

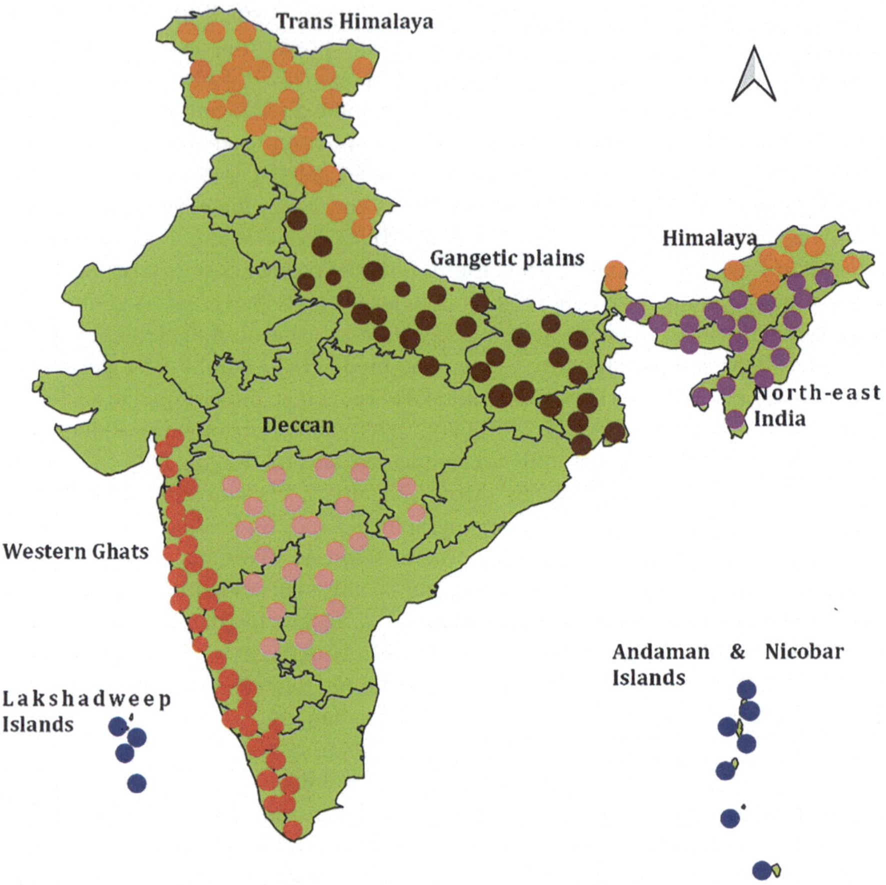

Fig. 5.3 Potential biodiverse rich area for more representation under aquatic PA

- *Rivers and streams:* India's rivers and streams are home to numerous freshwater species, including fish, amphibians, and invertebrates. Protecting key riverine habitats like deep pools through riverine sanctuaries or riparian reserves can conserve the unique biodiversity found in these ecosystems.
- *Floodplain wetlands:* wetland ecosystems are highly productive and ideal natural habitats for in situ conservation of threatened species or species with universal value from a conservation of science point of view.
- *Special habitats:* offshore areas beyond coastal boundaries, including seagrass meadows, deep-sea habitats, and pelagic zones, also harbor diverse aquatic life. Establishing marine protected areas in these offshore regions can help safeguard vulnerable species and habitats from threats such as pollution, overfishing, and habitat destruction.

5.4.2 Artificial Intelligence

Automated monitoring facilitated by artificial intelligence (AI) can provide a cost-effective solution to provide tools for monitoring restored ecosystems (Ditria et al. 2022). These techniques can reduce the long-term cost and bottlenecks experienced in monitoring. The technology could provide automated collection, transfer, and processing of information which enable timely and effective management. Such techniques have been efficiently used in coral reef monitoring (González-Rivero et al. 2020). Similarly, the technology also has the potential to accurately estimate the abundance of fish species in enclosed water bodies (Rourke et al. 2022). Automated monitoring of biodiverse areas to obtain data will help in broadening our understanding of these inaccessible ecosystems. Coupled with algorithms, this technology shall provide accurate information to researchers and resource managers which will eventually aid in making agile and accurate predictions.

5.5 Recommendations

As in situ conservation strategies protect the species in their natural habitat/environment, their applicability and efficacy are greater than ex situ conservation. The challenge lies in implementing the strategy in a large open water body, unlike the controlled conditions of ex situ conservation. Given the odds, the successful execution of in situ conservation strategies may be achieved by considering the following aspects:

- Strengthen governance mechanisms, enhance institutional capacity, and improve enforcement of conservation laws (Indian Fisheries Act, Convention on Biological Diversity, Ramsar Convention, the Environment (Protection) Act, Wild Life Protection Act, River Boards Act, Water Cess Act, Water Act) to combat illegal fishing, habitat destruction, and other threats to fish populations.
- Strict regulations on fishing quotas, gear dimensions, and closed seasons must be implemented. Rigorous management of fish spawning aggregation (FSA) sites is necessary to aid recovery, including establishing protected areas, imposing fishing restrictions, and managing human activities nearby.
- Linkage and collaborative effort of organizations and departments can effectively increase the success ratio of in situ conservation. Challenges of finance and manpower can be easily overcome, and with proper monitoring, successful implementation is further guaranteed. More organizations should partner together toward this combined prospective approach.
- For the successful effective management of stocked populations, improvements in stocking programs considering institutional arrangements involving all stakeholders, the social and legal framework to integrate stocking plans with harvest regulations, cost-recovery, and fisheries management plans for the possible achievements should be developed.

- The proposed concept by ICAR-NBFGR to establish freshwater aquatic sanctuaries (FAS) within the protected area network presents a promising solution, emphasizing the importance of targeted initiatives to tackle the specific challenges confronting freshwater habitats and species (Sarkar et al. 2007).
- Aquatic Diversity Management Areas (ADMAs): Creating Aquatic Diversity Management Areas (ADMAs) is a systematic approach to watershed management, primarily aimed at protecting local aquatic biodiversity (Pawde et al. 2023). ADMAs can vary from small-scale initiatives targeting biodiversity to extensive laws safeguarding particular species. The most effective strategy for managing ADMAs involves halting or substantially reducing any human activities that degrade habitat within those areas.
- Advancement needs to be made for selecting ranching sites, microhabitat, and the magnitude of stocking; density dependence and carrying capacity are key considerations. Adaptive management or better-improved methods for evaluating stocking effectiveness, species interactions, and environmental influences are key to understanding the uncertainties about stocking success realistically and objectively in light of the specific objectives of a stocking program.
- Majority of organisms are exploited from the wild, often involving the adoption of unsustainable fishing practices increasing the pressure on the wild stock leading to the collapse of the natural habitats. The captive propagation of organisms can bring long-term sustainability to the resources, as the natural population is not disturbed. In this context, the ICAR-NBFGR introduced the concept of "Marine Ornamental Aquaculture with Community Participation" at Lakshadweep Islands for local livelihood and marine biodiversity conservation. This innovative approach to community aquaculture of ornamental organisms will open new avenues, where native people access their heritage biodiversity to derive benefits while conserving these invaluable resources.
- Implementing habitat restoration projects and regulating land-use practices are crucial strategies to mitigate degradation. The National Commission on Integrated Water Resource Development facilitates these efforts, using the river continuum concept and eco-restoration principles for sustainable fish habitat management.
- Implement early detection and rapid response programs for invasive exotic species, conduct risk assessments for species introductions, and regulate trade and transportation of exotic species to prevent their establishment and spread. Environmental DNA technology (E-DNA) for early detection of invasive species in lakes and larger water bodies should be popularized. Additionally, ICAR-NBFGR has developed a national plan and quarantine for aquaculturists and other stakeholders culturing alien fishes in India.
- Improve wastewater treatment facilities, promote eco-friendly agricultural practices, and enforce regulations on industrial discharge and waste management to reduce pollution in aquatic environments.
- Integrating climate change deliberations into ecological restoration and conservation efforts is essential which safeguards biodiversity and helps mitigate climate change's adverse effects on social and economic systems.

- Conduct outreach and education programs to raise awareness about the value of fish conservation, promote community participation in conservation initiatives, and integrate fish conservation into school curriculum and public awareness campaigns. Initiatives such as educational programs, incentives, and volunteer monitoring can be employed to accomplish this goal. For instance, the state of Delaware, USA, maintains an Adopt-a-Wetland Program intended to raise public awareness of the need for conservation and the value of wetlands. This strategy promotes the efforts to improve Delaware's wetland resources through increased agency coordination, data availability, education, monitoring, and restoration efforts. Similar programs can be taken up in India.
- Efforts at restoration and mitigation: aquatic areas that have undergone habitat degradation or declining species populations might be the focus of restoration efforts. In riverine environments, several management techniques such as the creation of riparian buffer zones and restoration of natural flow patterns and discharge regimes are being implemented.

5.6 Conclusion

In situ fish conservation in India faces multifaceted challenges, including habitat degradation, overfishing, invasive species, pollution, climate change, lack of awareness, weak governance, and conflicting interests. Addressing these challenges requires coordinated efforts from governments, conservation organizations, researchers, and local communities. By implementing sustainable management practices, promoting habitat restoration, enhancing monitoring and enforcement, raising public awareness, and addressing underlying socioeconomic factors, it is possible to effectively conserve fish species and safeguard aquatic ecosystems for future generations. The lack of species-specific data on spawning grounds and habitat needs of endangered fishes hinders targeted recovery efforts. Establishing aquatic sanctuaries is a crucial yet overlooked aspect. An integrated approach to sustainable resource management, with community involvement in planning, is essential, particularly focusing on living aquatic resources.

References

Ditria EM, Buelow CA, Gonzalez-Rivero M et al (2022) Artificial intelligence and automated monitoring for assisting conservation of marine ecosystems: a perspective. Front Mar Sci 9:918104

Dudgeon D (2011) Asian river fishes in the Anthropocene: threats and conservation challenges in an era of rapid environmental change. J Fish Biol 79(6):1487–1524

Dudley N (ed) (2008) Guidelines for applying protected area management categories. IUCN, Gland, p 86

Fricke R, Eschmeyer WN, Van der Laan R (eds) (2024) Eschmeyer's catalog of fishes: genera, species, references. (http://researcharchive.calacademy.org/research/ichthyology/catalog/fishcatmain.asp). Electronic version accessed 27 04 2024

Ghosh SK, Ponniah AG (2008) Freshwater fish habitat science and management in India. Aquat Ecosys Health Manag 11(3):272–288

Gleick PH (1996) Basic water requirements for human activities: meeting basic needs. Water International 21(2):83–92. https://doi.org/10.1080/02508069608686494

González-Rivero M, Beijbom O, Rodriguez-Ramirez A et al (2020) Monitoring of coral reefs using artificial intelligence: a feasible and cost-effective approach. Remote Sens 12:489

Gopalakrishnan A, Ponniah AG (2000) An overview of "Endemic fish diversity of western Ghats". pp 1–12. In: Ponniah AG, Gopalakrishnan A (eds) Endemic fish diversity of western ghats. NBFGR-NATP Publication. National Bureau of Fish Genetic Resources, Lucknow, U.P., India. 1,347 p

Gozlan RE, Britton JR, Cowx I, Copp GH (2010) Current knowledge on non-native freshwater fish introductions. J Fish Biol 76(4):751–786

IUCN (2023) The IUCN red list of threatened species. Version 2023-1. https://www.iucnredlist.org. Accessed 25 Apr 2024

Lakra WS, Mohindra V, Lal KK (2007) Fish genetics and conservation research in India: status and perspectives. Fish Physiol Biochem 33:475–487

Lakra WS, Sarkar UK, Gopalakrishnan A et al (2010) Threatened freshwater fishes of India. ICAR-National Bureau of Fish Genetic Resources, Lucknow

McCarty JP (2001) Ecological consequences of recent climate change. Conserv Biol 15(2):320–331

Medhi K, Paul T, Borah AK et al (2018) Fish stock depletion in Indian inland open waters: present status and approaches for sustainable use, conservation and management. Aquac Times 4(5):26–32

Mestanza-Ramón C, Henkanaththegedara SM, Vásconez Duchicela P et al (2020) In-situ and ex-situ biodiversity conservation in Ecuador: a review of policies, actions and challenges. Diversity 12(8):315

Nelms SE, Duncan EM, Patel S et al (2021) Riverine plastic pollution from fisheries: insights from the Ganges River system. Sci Total Environ 756:143305

Parvez M, Mohsin A, Arnob S et al (2023) Fish diversity decline in the lower Gangetic plains: a victim of multiple stressors. Biodivers Conserv 32(1):341–362

Pawde ED, Thaware VH, Paul K (2023) Conservation strategies for fish biodiversity to maintain healthy ecosystem. Int J Fish Aquat Stud 11(1):147–149

Rourke ML, Fowler AM, Hughes JM et al (2022) Environmental DNA (eDNA) as a tool for assessing fish biomass: a review of approaches and future considerations for resource surveys. Environ DNA 4:9–33

Rodgers WA, Panwar HS, Mathur VB (2000) Wildlife protected area network in India: a review. Wildlife Institute of India, Dehradun, India, p 44

Sala OE, Chapin FS III, Armesto JJ et al (2000) Global biodiversity scenarios for the year 2100. Science 287:1770–1774

Sarkar UK, Kapoor D, Paul SK et al (2007) Fish biodiversity in the water bodies of Samaspur bird sanctuary, Uttar Pradesh: towards developing a freshwater aquatic sanctuary. J Bomb Nat Hist Soc 104(1):51–54

Sarkar UK, Pathak AK, Lakra WS (2008) Conservation of freshwater fish resources of India: new approaches, assessment and challenges. Biodivers Conserv 17:2495–2511

Sarkar UK, Puthiyottil M, Das A et al (2022) Impact of climate change and technologies for climate resilient inland open water fisheries. Indian Farming 72(7):11–14

Sarkar UK, Kathirvelpandian A, Kantharajan G et al (2024) The concept of "state fishes of India": current status, knowledge gaps, and strategic plans for conservation and sustainable utilization. J Fish Biol 104:1675. https://doi.org/10.1111/jfb.15729

Serajuddin M, Bano F, Awasthi M et al (2018) Marine stock enhancement in India: current status and future prospects. In: Marine ecology-biotic and abiotic interactions. IntechOpen, London, pp 175–195

Shine C (2007) Invasive species in an international context: IPPC, CBD, European strategy on invasive alien species and other legal instruments. EPPO Bull 37(1):103–113

Sivakumar K, Mathur VB, Pande A et al (2013) Coastal and marine protected areas in India: challenges and way forward. ENVIS Bull 15:292–298
Stolton WS, Shadie P, Dudley N (2013) IUCN WCPA best practice guidance on recognising protected areas and assigning management categories and governance types, best practice protected area guidelines series no. 21. IUCN, Gland
Yan HF, Kyne PM, Jabado RW et al (2021) Overfishing and habitat loss drive range contraction of iconic marine fishes to near extinction. Sci Adv 7(7):eabb6026

Ex Situ Conservation: Status, Strategies and Challenges

6

Ravindra Kumar, Rajeev Kumar Singh, Murali S. Kumar, and Santosh Kumar

Abstract

Sustainable use of genetic resources is a pressing priority for ensuring food security, nutrition, and livelihood to a large section of the population. Ex situ conservation tools, which mainly focus on "off-site preservation of biological material", have become popular for preserving genetic resources from harmful anthropogenic activities as well as natural disasters and conserving genetic diversity through sustainable utilization of the resources. Genetic variation is crucial for the long-term survival of the species. The aim of ex situ conservation is not only to preserve the species in captivity but also to maintain its genetic makeup for eventual release back into the wild. The development of genomics and associated molecular markers has boosted genetic conservation efforts by enabling us to compute and determine the genetic variability, population structure, genetic bottlenecks, associated historical events, and phylogeographic relationships. Captive breeding-assisted conservation programs are one of the most important ex situ conservation methods. Other methods include conservation programs in aquariums, live gene banks, milt cryopreservation, cell lines, etc. Ex situ approach makes biological material available for conservation education and display, generating awareness as well as long-term interest in the general public about the need for germplasm conservation. Such conservation requires a multi-stakeholder approach with the participation of all, like local communities, subject matter specialists, policymakers, government institutions, NGOs, and the general public. India, with its vast biological diversity, is well dignified to lead ex situ conservation plans for sustainable management of aquatic genetic resources.

R. Kumar (✉) · R. K. Singh · M. S. Kumar · S. Kumar
ICAR-National Bureau of Fish Genetic Resources, Lucknow, Uttar Pradesh, India
e-mail: Ravindra.Kumar1@icar.gov.in; Rajeev.Singh1@icar.gov.in; S.Murali@icar.gov.in; Santosh.Kumar5@icar.gov.in

U. K. Sarkar et al. (eds.), *Sustainable Management of Fish Genetic Resources*,
https://doi.org/10.1007/978-981-97-5250-8_6

Keywords

Aquaria · Captive breeding · Cell lines · Ex situ · Sperm cryopreservation

6.1 Ex Situ Conservation?

Conservation usually refers to the cautious management and protection of natural resources, biodiversity and ultimately the ecosystems to ensure their sustainable use for contemporary and future generations. According to FAO, the sustainable use of genetic resources is pivotal for food security and nutrition. Therefore, conserving and utilizing the diversity (among and within species) are important for meeting future challenges. The subject encompasses different facets, including the conservation of land, water, forest, wildlife, minerals, energy, as well as climate change mitigation. All components (viz., individuals, communities and governments) play crucial roles in these conservation efforts. Basically, there are two strategies of conservation, viz., in situ and ex situ, that is practiced globally for safeguarding a variety of living species. Under in situ approach, the resource is conserved within their natural habitats. This method is convenient and cost-effective and allows species to grow naturally and acclimate to the ecological conditions. On the contrary, the ex situ conservation refers to the "off-site preservation" of biological material for protection and efficient use of resources, or the conservation of various quantities under physical laws. Under the basic ex situ approach (Fig. 6.1), the organisms are bred and maintained in artificial or non-native ecosystems under captivity/controlled conditions, such as botanical gardens, zoos, live gene banks, etc. It is usually a human-controlled environment with less competition for resources and also allows longer lifetime and breeding activity, often utilizing genetic tools. Thus, the ex situ conservation plays a vital role in safeguarding endangered species; however, it requires careful management and consideration of both advantages and disadvantages.

6.2 Methods/Modes of Ex Situ Conservation

6.2.1 Captive Breeding

Theoretically, the aim of ex situ conservation is not only to preserve the species in captivity but also to maintain its genetic makeup for eventual release back into the wild. Captive breeding is among the most popular ex situ conservation approach. As a result, many programs, aiming at conservation along with research and education, are being executed. Froehlich et al. (2017) defined conservation aquaculture as a tool, which harmonizes both natural resources and the people. In this aligned relationship, natural populations are supported well, and socioeconomic benefits to human beings are also maintained. There are good examples, where such programs have been useful in the recovery of endangered species, such as white sturgeon (Ridlon et al. 2021) and white abalone (Rogers-Bennett et al. 2016). Other species

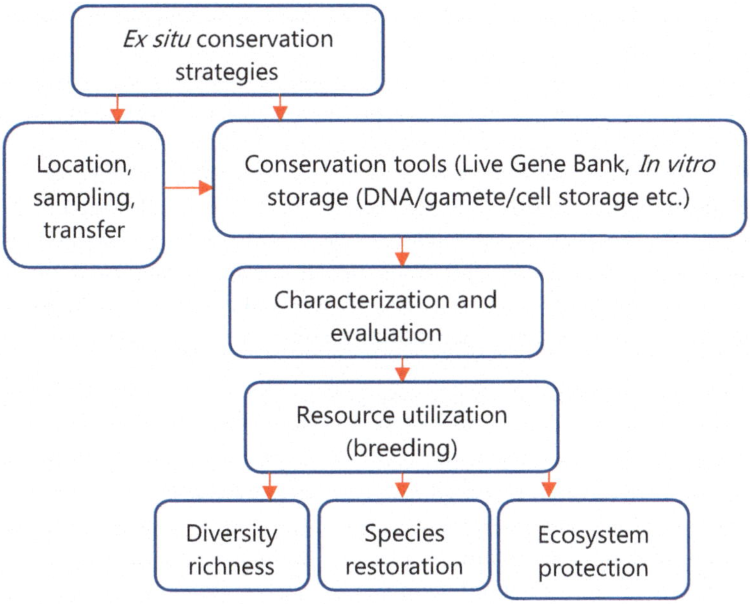

Fig. 6.1 Basic strategy of ex situ conservation

being supported for declining populations include eastern oyster *Crassostrea virginica* (Brumbaugh et al. 2000) and European flat oyster *Ostrea edulis* (Pogoda et al. 2019).

The optimized breeding and larval rearing protocols are crucial for successful propagation. With demonstration and training to stakeholders, the production may be enhanced, which is likely to reduce the dependency on natural resources for food and trade. With its eight germplasm resource centers, the ICAR-NBFGR has prioritized important species for conservation and propagation. The captive breeding of marine ornamentals are being executed, which is expected to alleviate the current fishing pressure on corals, shrimps, etc. In this context, germplasm resource centers are established for marine ornamental invertebrates (at Lakshadweep, India) and marine ornamental fishes (at Airoli, Maharashtra, India) are working to harmonize biodiversity conservation with livelihood enhancement of the local stakeholders.

6.2.2 Aquariums

An aquarium is a man-made habitat that mimics the natural counterparts of aquatic animals (fish, amphibians, mammals) and plants living in water bodies. Historically, the first type of aquaria was the artificial fish ponds in the ancient Mesopotamia and Egypt, where the fish was considered as a source of food along with amusement. It is likely that the Chinese started domestication of the carp, presently known as goldfish for their ornamental value. The evidence from archaeological studies indicated the usage of glass in indoor aquaria by the Romans. They might also have used ships to transport live fish.

Over the time, these aquaria have become a source of attraction for the people, in general. Being an integrated part of the tourism industry, the modern public aquaria made a start within the London Zoo as early as in 1853. Currently, apart from being associated with ornamental and sports values, several aquaria are considered important for educational, scientific and aesthetic perspectives. Such organizations are now more vigilant for their expansion into conservation efforts and applied research. Educational programs play an important role in aquatic conservation by creating awareness and participatory approaches. In these, the target clients are usually school children, academicians and families. The World Association of Zoos and Aquariums (WAZA) defines conservation education as the source of interpreting living collections, mainly to attract and inspire people to make conservation a priority.

Several advancements have been made in developing interpretive methodologies, which use live cams, recorded videos and graphics. For long-term impact, animal exhibits, along with interpretive commentaries and signages, are good alternatives. Moreover, the incorporation of abiotic and biotic stressors, which the fish might be experiencing in a natural habitat, can be even better option. Live views of fish/animal/insect in their natural habitat, in real time, is a wonderful experience. A website, named Explore (www.explore.org), provides a live view of considerable number of animals/fishes in their natural habitat. Viewing these live shots is the best source of awakening conservation concerns in all age groups, more prominently in kids, who have been away from nature due to rising urbanization. Aquaria serve to be an important source of take-home messages for the general public and more importantly school-going children, who can be the ambassadors for biodiversity conservation. They can be sensitized to inculcate and adopt day-to-day habits, which are environmentally conducive for other creatures on this planet toward sustainable future.

The Chimelong Ocean Kingdom in China is the world's largest aquarium having interactive exhibits and allowing visitors to learn about marine life in fun and engaging ways. The aquarium also allows visitors to have interactive experiences such as feeding sessions or swimming with dolphins. In India, VGP Marine Kingdom in Chennai is the largest walk-through aquarium, which houses a large variety of aquatic animals, including marine. This is home to the aquatic species, which are rare and near extinct. Marine World in Thrissur, Kerala, has been reported as the largest public aquarium in India with displays marine and freshwater species. ICAR-National Bureau of Fish Genetic Resources, Lucknow, maintains Ganga Aquarium, which maintains fishes from freshwater and marine ecosystems. Generally, there is a huge footfall of school children, who visit as part of their educational trips. This attracts several tourists too, giving them a chance to watch live fish, not available usually. Other popular aquaria in India include Taraporewala Aquarium in Mumbai, Lalbagh Aquarium in Bengaluru, Bagh-e-bahu in Jammu, Jagdish Chandra Bose Aquarium at Surat, etc.

6.2.3 Gamete/Zygote/Embryo Cryopreservation

Cryopreservation of gametes, zygotes and embryos has become an increasingly important tool of assisted reproduction. It is one of the important ex situ methods of germplasm conservation. According to FAO, this can be considered as a major strategy for the conservation of fishery resources (Khoshoo 1997). Gamete cryopreservation is done by lowering the temperature and, thereby, reducing their metabolic rate, which increases their longevity without any drastic change in the fertilizing capacity. It has wide-ranging applications in aquaculture and fisheries management. Though the need is felt for cryopreservation of fish eggs and embryos, the attempts have met with no or limited success so far (Hagedorn et al. 1997; Ahammad et al. 2002). However, studies on the cryopreservation of eggs, embryos and larvae in invertebrate species have met with some success (Diwan and Kandasami 1997).

6.2.3.1 Sperm Cryopreservation

It is a process where sperms are preserved by cooling to ultralow temperatures. At these low temperatures, any biological activity, including the biochemical reactions that would lead to cell death, is effectively stopped. This technology got a breakthrough when the cryoprotectant property of glycerol was discovered (Polge et al. 1949). All this led to the development of the first cryopreservation protocol in herrings 4 years later, by Blaxter. Though, sperm cryopreservation has been successful in fish and over 200 species-specific protocols have been developed (Tsai and Lin 2012), the same for eggs and embryos still remains far from reality. The working strategy for milt cryopreservation is depicted in Fig. 6.2. Moreover, the adoption of this technique at commercial level for fish seed production has been very limited.

A high degree of procedural variability and success, not only between the species but at times within the species too, is considered among the limiting factors in its utilization. Species-specific protocols are developed through experimental standardization of various parameters after the captive breeding protocol is developed and optimized. However, there is a bottleneck due to short breeding season and low domestication in most of the fish species. In all such cases, the time available in a year for conducting the experiments is short and determined by the breeding cycle of the species in question. Therefore, it is essential to prioritize the candidate species for sperm cryopreservation. In India, ICAR-NBFGR has been extensively working on sperm cryopreservation since 1989, and around 30 species-specific protocols have been developed, so far (*Source: Handbook of Fisheries and Aquaculture, 2011*).

6.2.3.2 Eggs and Embryo Cryopreservation

In order to enhance fish stocking through aquaculture technologies and reduce the dependency on wild stock for fish production, cryobanking of eggs and embryos becomes very important. Successful cryobanking of eggs and embryos will support not only the enhanced fish production but also help in the restoration of threatened genetic resources. Though, cryopreservation of embryos has largely been established in the mammalian systems, but unsuccessful in finfishes, and if it was

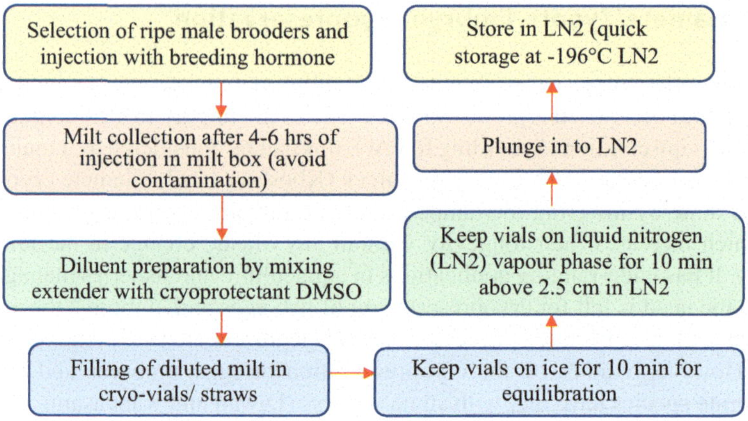

Fig. 6.2 Working strategy for cryopreservation

successful repeatability was less. Two main reasons for this unsuccessful result may be the low membrane permeability of cryoprotective agents in the egg as well as embryos and the presence of a large yolk mass, which results in ice crystal formation during the freezing process. In addition, the oocytes and embryos are prone to chilling injuries under extremely low temperatures. In this regard, several attempts were carried out to overcome the pointed challenges, such as microinjection of cryoprotectant directly into the cytoplasm (Leung and Jamieson 1991), use of negative pressure to increase the permeability of cryoprotectants (Hagedorn et al. 1997), chorion removal (Routray et al. 2002), microinjection of antifreeze proteins (Robles et al. 2006) and application of hydrostatic pressure (Zhang and Rawso 1995). The use of ultrasound technique has been reported to enhance the embryo permeability in zebrafish (Bart 2000). However, methanol demonstrated a limited degree of penetration into prim-6 stage of zebrafish embryos, but it did not penetrate in the subsequent stages (Liu et al. 2001). However, Aril et al. (1987) observed DMSO to be a good cryoprotectant for medaka (*Oryzias latipes*) fish embryos. In common carp, the morulae-stage embryos were found to be partially protected against chilling in DMSO and sucrose mixture; half-epiboly in DMSO, sucrose and methanol solution; and heartbeat stage in methanol and glycerol solution (Dinnyes et al. 1998). Using isotope-labeled DMSO and glycerol, it was found that these solutes penetrated into both dechorionated and intact zebrafish embryos after a duration of 5 h (Harvey et al. 1983). While working with hatching of *Cyprinus carpio* embryos stored at 4 °C and −2 °C at different concentrations of methanol and sucrose, Ahammad et al. (2003a) found that the hatching performance was maximum (41%) in sucrose at 4 °C. No survival was observed at −2 °C in any concentration of sucrose solutions. Further, it was reported that the combination of methanol and sucrose produced the best results among all concentrations tested at both temperatures. While studying the hatching responses of rohu embryos at different concentrations of cryoprotectants and temperatures, the same researchers (Ahammad et al.

2002, 2003a, b, 2004) reported the highest hatching percentage (72%) when embryos were stored at −4 °C temperature in combination with methanol and trehalose. Furthermore, cryopreservation of pluripotent blastomere and the grafting of thawed cells into recipient embryos opened an alternate pathway for the production of improved strains of fish and shellfish (Tsai and Lin 2012). The cryopreservation of an isolated embryonic cell is, thus, another option to preserve both maternal and paternal genomes.

The attempts to cryopreserve eggs, embryos and larvae of invertebrates, especially crustaceans, have been relatively more successful than the finfish (Diwan et al. 2010). In penaeid shrimp, the successful survival of thawed eggs and larvae has been reported up to a freezing temperature of −10 °C (Diwan and Kandasami 1997; Diwan 2000). Asahina and Takahashi (1978) were successful in the cryopreservation of sea urchin eggs and embryos. During the same period, some efforts were also made to cryopreserve the embryos and nauplii of shrimp, *Penaeus indicus* (Subramoniam and Newton 1993; Simon et al. 1994; Subramoniam 1994). Subramoniam and Newton (1993) successfully preserved the nauplii of *P. indicus*. They held them at −30 °C using liquid nitrogen and reported 82% survival for nauplii frozen at −30 °C and 63% at −40 °C. Diwan and Kandasami (1997) also reported the success of freezing viable embryos and larvae of shrimp *P. semisulcatus*. Cryopreservation of Pacific oyster, *Crassostrea gigas*, larvae (10 dph) got limited success and resulted in less than 10% survival (Labbe et al. 2018).

6.2.4 Cell Lines

Cell banking is considered an important option for ex situ conservation. Ryder and Onuma (2018) highlighted the characterization and conservation of biodiversity using cell banking. The cell banking could help to mitigate otherwise irretrievable loss of living material from species facing extinction. They also reported that the application of induced pluripotent stem cells (iPSCs) and in vitro gametogenesis in the case of cryobanked somatic cells may provide new options for maintaining sustainable populations and genetic rescue. The frozen zoo concept was initiated in 1975 on similar thinking by the San Diego Zoo at the Beckman Center for Conservation, USA, which now contains 10,000 cell cultures, oocytes, sperm and embryos representing nearly 1000 taxa that are a valuable resource for conservation and assisted reproduction (https://science.sandiegozoo.org/resources/frozen-zoo%C2%AE). Further, de-extinction concept is also propagating, which includes approaches such as somatic cell nuclear transfer (SCNT), where the nucleus of a somatic cell is transferred to an egg cell, whose own nucleus is removed. Interspecies somatic cell nuclear transfer (iSCNT) may have utilities in the genetic restoration of endangered species (Wisely et al. 2015). There are other examples also where the cell banks, along with iPSCs, have been reported as important tools for the conservation of endangered species, like Anatolian gray cattle (Arat et al. 2011), mandrill, white rhinoceros, and Somali wild ass (Ben-Nun et al. 2011; Verma and Sumer

2015). Additionally, the usage of cell lines from endangered species aids in its conservation efforts by the fact that no specimen has to be disturbed or sacrificed.

Fish cell lines, as a resource, have been used as an in vitro model in disease diagnosis and virus isolation, pharmacology, toxicity, drug screening etc., for the last three decades and their use has been increasing every year. Schirmer et al. (2021) developed and validated a toxicity assay as a latest guideline in the field of environmental toxicology, approved by the Organisation for Economic Co-operation and Development (OECD), based on rainbow trout gill cell line (RTgill-W1) to predict fish acute toxicity. This assay was the first validated alternative to live-animal testing saving other valuable resources, like water, human resource etc. and can pave way for many such assays from other fish cell lines.

Cellosaurus, an online resource lists 945 fish cell lines developed globally (https://www.cellosaurus.org/ release 49 of May 2024) with over 280 new fish cell lines were established from 44 fish families in 21 countries during the last 10 years alone (Thangaraj et al. 2021). Kumar et al. (2024) reported a detailed list of fish cell line depositories and their cell line availability. A total of 81 fish cell lines have been reported in the National Repository for Fish Cell Lines (NRFC at ICAR-National Bureau of Fish Genetic Resources, India); 25 in European Collections of Authenticated Cell Cultures (ECACC), UK; 13 in American Type Culture Collection (ATCC), USA; 11 in IZSLER, Italy; 8 in RIKEN, Japan; and 4 in German Collection of Microorganisms and Cell Cultures (DSMZ), Germany. The NRFC (https://mail.nbfgr.res.in/nrfc/) at Lucknow, India, currently houses the world's largest collection of fish cell line accessions, developed from 16 tissues of 39 fish species belonging to 32 genera, 16 families, and 8 orders (29 freshwater, 3 brackish water, and 7 marine species) (Kumar et al. 2023). This includes 44 cell lines deposited by ICAR-NBFGR, 19 cell lines were contributed by C. Abdul Hakeem College, Vellore, Tamil Nadu; 12 by ICAR-CMFRI, Kochi, Kerala; 4 by FCRI, Tamil Nadu; and 2 by ICAR-DCFR, Bhimtal, Uttarakhand. These cell lines are available for research and development. These are authenticated cell lines and are free from mycoplasma.

6.2.5 Cellular Agriculture

Cellular agriculture has been defined as a set of technologies to produce or manufacture products in a lab, typically obtained from livestock, poultry, or fishery using a combination of cell culture, molecular as well as synthetic biology approaches (Stephens et al. 2018; Kumar et al. 2023). This is assumed to be a viable option to cater to the increasing need for food and nutritional security. Lab-grown meat has the potential to reduce pressure on natural resources and the environment. Dr. Marcus Johannes "Mark" Post, a Dutch pharmacologist and a professor of vascular physiology at Maastricht University and of angiogenesis in tissue engineering at the Eindhoven University of Technology, was the first in the world to present the concept for cultured meat and also popularize cell-based meat by developing the world's first cell-based hamburger in August, 2013 (Halpern et al. 2021). Of late, cell-based meat products became popular after being approved for sale in Singapore (Eat Just

Inc. for lab-grown chicken in 2020) and the United States (Upside Foods and Good Meat for lab-grown chicken in 2023) (Smith et al. 2022).

The fisheries sector, in particular, has good potential for cell-based meat production, as the fish cells are considered to be more suited for lab-grown meat due to their ability to be subcultured more frequently before reaching senescence (Potter et al. 2020). Cell-based fish meat techniques may also be developed for high-priced fishes, such as mahi-mahi, crustaceans, etc., so that the fishing pressure on natural stocks could be reduced (Kumar et al. 2023). Bluu Seafood Company claims that they have developed hybrid lab-based seafood (e.g., fish balls and fish fingers) from Atlantic salmon and rainbow trout mixed with plant-based proteins like soy. They expect to secure permission to sell in Singapore and the United States by the year 2025 (https://www.bluu.bio/product). In India, ICAR-Central Marine Fisheries Research Institute (CMFRI), Kochi, is working with Neat Meatt Biotech, a start-up company for developing cultivated meat. The ICAR-CMFRI envisaged to develop marine cell lines, while Neat Meatt will develop scaffolds and scale production through bioreactors. Similarly, ICAR-Central Institute of Fisheries Education, Mumbai, has partnered with Good Food Institute to develop lab-based fish meat. ICAR-NBFGR, Lucknow, has a fish cell line repository that can be accessed by researchers for evaluating cell-based meat techniques. The development of cell-based fish meat may help in aquaculture diversification and reduce the fishing pressure. This technology may soon see a great surge in the fisheries sector.

6.2.6 DNA/Gene Construct/BAC Banking

DNA/gene/BAC bank is a biorepository that preserves genetic material to ensure their availability for future use. This is also known as genetic material banking and involves the systematic collection, storage and preservation of genetic materials, in terms of DNA, gene construct, BAC (bacterial artificial chromosome), YAC (yeast artificial chromosome) etc. from various organisms. These banks serve as repositories to safeguard genetic diversity and facilitate research so as to support conservation efforts. DNA banking involves storing an individual's genetic material for a long term, where the total DNA is isolated from the cells of the animal. In genome resource banking (GRB), the focus is given to collecting, storing and distributing gene constructs from threatened and endangered species. Such genetic materials are preserved indefinitely in liquid nitrogen. In India, the Banaras Hindu University at Varanasi is undertaking pioneering efforts by creating the human DNA bank and nearly completing the wildlife DNA bank (https://currentaffairs.adda247.com/).

BAC library is another widely applied tool of genomic resource, which has further been used both in scientific research and conservation efforts. BAC is usually a DNA molecule cloned in DNA sequences of bacterial plasmids, typically *E. coli*. The BAC may accommodate DNA segments of 100 kb to 300 kb size of an organism. They are valuable tools for DNA sequencing, positional cloning, physical genome mapping, etc. In several fish species, the BAC library has been constructed with high quality in terms of average insert size, good genome coverage, less quantity of blank clones as well as fewer missing wells. Some of the good examples of BAC clones, with average insert size and genome coverage are: 80 kb size, 9× coverage in pufferfish (Liu et al. 1998); 188 kb, 18.8× in salmon (Thorsen et al. 2005); 161 kb, 10.6× in channel catfish (Quiniou et al. 2003); 65 kb, 6× in Nile tilapia (Katagiri et al. 2001); and 145 kb, 6× in medaka (Khorasani et al. 2004). Additional considerations for good-quality libraries include a low proportion of chimeric clones and finally a large number of independent clones. BAC libraries were also constructed in several other commercially important species, like barramundi, rainbow trout, common carp, prawn, etc. The BAC library has also been developed for many economically important animals such as cattle, sheep, frog, chicken, porcine, etc.

6.3 Models/Examples of Ex Situ Conservation

The fish preservation program CARES (Conservation, Awareness, Recognition, Encouragement and Support), founded in 2004 and consisting of 30 aquarium societies and international organizations, is one of the largest conservation hobbyist organizations. This is devoted to maintain populations of endangered or extinct in the wild species through captive breeding and rearing protocols for conservation and reintroductions. They have prepared a priority list to focus on conservation of species, where some species are not even in the IUCN Red List (Valdez and Mandrekar 2019). The European Union of Aquarium Curators (EUAC; https://www.euac.org/conservation-programmes), consisting of aquarium curators, has funded 42 conservation projects in 21 countries targeting more than 30 species during the last two decades in different areas of biological/ecological research, veterinary/conservation medicine, animal welfare, captive breeding, reintroduction/restocking/translocation etc.

To conserve native gene pools of brown trout in wild populations of the eastern Pyrenees River basins, a law on biodiversity conservation (Article 52.2, Law 42/2007 of Natural Heritage and Biodiversity) was enacted in Spain, resulting in the creation of genetic refuges, where the release of hatchery-bred stock was banned. The effect of such refuges showed mixed results, wherein in some river basins, the genetic integrity of native trout was preserved during 24 years of monitoring, but in some places, significant amounts of introgression with hatchery-bred individuals were reported (Araguas et al. 2009). In another example, the US Fish and Wildlife Service, in 1988, successfully translocated 200 individuals of watercress darter, *Etheostoma nuchale*, an endangered fish native, to 4 springs in Alabama, USA, outside the species' native range, to establish a new population in Tapawingo Spring, USA, where the population of fish subsequently increased (Olden et al. 2011).

Some of the live fish gene banks across the globe are presented in Table 6.1. The Research Center for Fisheries and Aquaculture (HAKI) at the Hungarian University of Agriculture and Life Sciences, Hungary (https://aquaculture.uni-mate.hu/general-introduction-haki) maintains and continuously develops 16 domestic, European and Asian carp lines in their live carp gene bank. Four genetically selected

Table 6.1 Live fish gene banks across the world

Sl. No.	Country	Location	Species	More information
1.	India	Lucknow, Kochi, Airoli, Guwahati, Manipur, Telangana, and Lakshadweep	IMCs, endemic species	ICAR-NBFGR annual reports; https://www.nbfgr.res.in/
2.	Hungary	Szarvas at HAKI	16 domestic, European, and Asian carp lines and 5 native and 3 non-native sturgeon species	https://aquaculture.uni-mate.hu/gene-bank-collection
3.	Philippines	BFAR-NFTC, Manilla	Tilapia	https://www.bfar.da.gov.ph/national-freshwater-fisheries-technology-center-nfftc/
4.	Canada	Mactaquac facility, New Brunswick	Atlantic salmon	https://www.dfo-mpo.gc.ca/science/organization/facilities-installations/maritimes/ index-eng.html
5.	Norway	Eidfjord, Haukvik, and Bjerka	Atlantic salmon	https://conbio.onlinelibrary.wiley.com/doi/full/10.1111/csp2.552 https://nasco.int/wp-content/ uploads/2020/02/AquacultureFAR_Norway.pdf
6.	Russia	Federal living bank, Krasnodar Institute of Fisheries, Krasnodar	Sturgeons	http://krasniirh.kuban.ru/eindex.html

native sturgeon species in the gene bank provide a chance as ex situ conservation measure for the restoration of endangered Danube sturgeon stocks in Europe. A tilapia live gene bank has been established at the National Inland Fisheries Technology Center (https://www.bfar.da.gov.ph/national-inland-fisheries-technology-center-niftc/) of the Bureau of Fisheries and Aquatic Resources within the Central Luzon State University, Manilla, Philippines. It has a collection of both cryopreserved tilapia sperm cells from various tilapia stocks as well as live specimens of various strains of Nile tilapia from Egypt, Israel, Singapore, Taiwan, Thailand, Kenya, Senegal, Ghana and the Philippines. Live gene bank for wild Atlantic salmon was established at Mactaquac Facility, New Brunswick, Canada (https://www.dfo-mpo.gc.ca/science/organization/facilities-installations/maritimes/index-eng.html) in 1998 and in Norway as part of Norwegian Gene Bank (NGB) program, housing 68 Atlantic salmon populations (Boe et al. 2021). The National Fish Museum and Repository Facility at ICAR-NBFGR, Lucknow, India, currently holds over 1200 finfish species specimens and > 250 or mollusk voucher species. It also houses over 19,000 tissue and DNA accessions belonging to commercial/prioritized fish species of India. Sperm repository with cryopreservation protocols for several prioritized species is also available. It also holds models of the crafts, nets, and other indigenous gears used by fishermen across the country. With eight live gene banks (Lucknow in Uttar Pradesh, Kochi in Kerala, Airoli in Maharashtra, Guwahati in Assam, Imphal in Manipur, Nagarjuna Sagar in Telangana, Pichavaram in Tamil Nadu and Agatti Island in Lakshadweep) across the country, the ICAR-NBFGR, Lucknow, has emphasized on conservation and management of indigenous and threatened fish species, particularly used in food and trade.

6.4 Benefits/Importance of Ex Situ Conservation

In the present human development scenario, the climate change is affecting our ecosystems at several levels, from the populations to the services they provide to economies, communities, biodiversity, etc. Climate controls the growth and behavior of the plants and animals, including fish, and which organism has to succeed depends on its interactions with the physical environment. As a result of climate change, many species are either migrating to more conducive environments or being extinct in nature, thereby, changing their population dynamics. Further, the habitat deterioration due to natural catastrophes and anthropogenic activities have led to many species getting displaced and/or vulnerable. In such situations, ex situ conservation has served several purposes, such as:

- Rescue of threatened germplasm from the changed and/or challenged environment. The organisms in ex situ facilities (such as live gene or seed banks) are entirely protected from predation, poaching and/or habitat loss. As these are human-controlled environments, some of the adverse impacts of climate change can also be minimized. Populations can be more effectively managed and divided if a disaster strikes. This flexibility allows for better conservation planning.

- Ex situ approach facilitates reproduction-assisted rehabilitation. The chances of reproductive success can be enhanced by the conservationists by mating genetically diverse wild type. Germplasm is produced off-site for reintroduction in their native tracts, translocation to other suitable ecosystems, reinforcement and habitat/landscape restoration as well as management. Under captivity, the researchers can study the reproduction, physiology, lifestyle and ecology of the endangered species more easily.
- Produce germplasm/material from wild type for conservation biology research.
- Large-scale storage of germplasm in various forms (e.g., gamete, embryo, DNA, etc.) in ex situ facility.
- Supply material from culture systems for various purposes, such as to remove or reduce pressure on natural collection.
- The health of the individuals can closely be monitored and the necessary medical assistance is provided, which would ultimately help in maintaining the well-being of the population.
- Generate skills and knowledge to support wider conservation aims. Conservation sites can serve as attraction for educational centers and may be used for fundraising for further conservation efforts.
- Genetic variation is crucial for the long-term survival of the species/population/ecosystem. The ex situ approach makes material available for conservation education and display. The researcher can measure the genetic diversity at the population level, which is important for the long-term survival of the population/stock.

6.5 How Genomics Can Help Biodiversity Conservation?

Habitat restoration has primarily been considered as the foremost way of conservation; however, the advanced genomic approaches have arbitrated substantially to support the conservation efforts. With the use of available genomic marker panels, the issues of conservation (including identification and preservation of novel genetic variants/alleles, minimizing events of inbreeding and increasing adaptive potential of threatened species) may be addressed (Ghildiyal et al. 2023). The integration of genomics in conservation efforts has been discussed at several platforms and its need was felt worldwide for various taxa. The affordable cost of sequencing has further advocated this deliberation. Whole genome information is surely required for conducting gene editing research. Some researchers have linked epigenetics to biological conservation (Rey et al. 2020). These epigenetic modifications are important events during major ecological transitions and, therefore, can potentially identify vulnerability to environmental changes.

The development of molecular tools and computational analysis can precisely determine the genetic variability, population structure, genetic bottlenecks, associated historical events and phylogeographic relationships (Frankham et al. 2002). The allele frequency variations within fish populations have been used to identify genetic stocks. These stocks are considered important for conservation, as these differences might be of adaptive value. Such assessments have made important contributions to the conservation genetics of a species by deciphering intraspecies genetic variability. Attard et al. (2022) used genomics, especially SNPs, to detect the admixture populations and differentiate hatchery-produced golden perch from wild-threatened endemic stocks. They recommended the use of genomics to check the broodstock ancestry for better hatchery management. With increased genomics data on Indian fish stocks, it is now possible to evaluate our hatchery stocks and recommend management policies. Through a flagship program on Outreach Activity on Fish Genetic Stocks, the ICAR-NBFGR has characterized 28 freshwater and marine finfish as well as shellfish species for their conservation/aquaculture significance. This multi-institutional program was aimed at identifying genetic stocks of cultivable species, along with their descriptors for valuation, registration and ex situ conservation. Such information is important for initiating the genetic improvement programs, requiring broader genetic diversity in the base populations.

With the advent of modern ways of high-throughput sequencing, i.e., next-generation sequencing (NGS), the genomic resources are developed in massive quantities. Several genomes and transcriptomes have become available, while many are in the pipeline. These resources will lead to the bioprospecting of useful genes, marker-assisted selection and genetic improvement to enhance productivity. The ICAR-NBFGR has been executing a program on genomics in a consortium mode to generate genomic resources for the characterization and conservation of fish germplasm resources, wherein nine institutes across the commodities (animals, crops, fish, insects, and microbes), are involved.

6.6 Role of Ex Situ Conservation in Biodiversity Restoration

According to Whitmore (1990), in situ conservation is the most preferred approach. However, there are several reasons that make ex situ an important alternative. Despite being relatively expensive and risk-prone, the ex situ approach provides scope for propagation-assisted rehabilitation. There are several examples, where captive breeding has resulted in the conservation of several important species, including blue whale, the mountain gorilla, and the European bison (www.newscientist.com), among others. In fishes, the iconic Indian sport fish mahseers, a high-priced fish were reported to be declining in size and abundance in the Indrayani River of Western Ghats, owing to prevalent anthropogenic activities. Here, the mahseer had been facing threats due to hydropower projects, blocked migration routes and loss of breeding grounds. The ex situ approach has been helpful in restoring the germplasm in the wild. The seeds obtained from Tata Electric Company, Lonavala, were reared, raised and bred using the protocol developed by Kulkarni and Ogale (1978). The seeds were subsequently ranched in the river. An investigation, conducted by Nagalgaon and Patil (2021), concluded that the riverine experimental fishing witnessed the presence of adult fish in the stretch, where mahseers had not been observed during the last several years. Recently, after Maharashtra, Himachal Pradesh has also taken up captive breeding of mahseers for restoration in to rivers. There are several such instances, where captive breeding has successfully been achieved in threatened species, such as *Pethia setnai*, an ornamental fish with high market demand. An endangered priced fish, *Dawkinsia denisonii*, also known as red-line torpedo barb and having high demand in aquarium trade was also bred in captivity (Anna Mercy et al. 2015) to provide relief to natural resources. Other species include *Ompok pabda* (Purkayastha et al. 2012), *O. bimaculatus* (Malla and Banik 2015), *Chitala chitala* (Sarkar et al. 2006), *Channa barca* (Choudhury and Biswas 2003), and an endemic species to peninsular India, *Clarias dussumieri*. There are also a few successful models, which depict the involvement of stakeholders in the captive breeding of important species. An endangered species Pyrenean sculpin, *Cottus hispaniolensis*, was captively bred in 2006 by a nonprofit association, named ADEFFA, with the support from public administration.

6.7 Challenges of Ex Situ Conservation

- *Genetic diversity loss:* In ex situ approach, the captive populations usually have limited genetic diversity due to the limited space and small number of individuals, which may lead to fitness issues and compromised adaptability. Animals may be exposed to a wider range of diseases as compared to the wild ones. Araguas et al. (2009) collated the case of brown trout, *Salmo trutta*, population enhancement through hatchery-bred stock and found a loss of genetic diversity and reduced fitness. They advocated that fish refuges must be built as genetic reservoirs containing native individuals, where hatchery-bred individuals are banned to offset this problem. These genetic consequences depend on several factors, including the number and genetic relatedness of founders.

Population size is an important attribute for conservation priorities for a species. There is always a greater risk for smaller populations due to genetic drift and demographic stochasticity. As the accumulation of inbreeding is through generations, it also depends on the period for which breeding has been coordinated (Ruiz-Lopez et al. 2009). Moreover, these issues can be resolved by appropriate genetic management of the focal taxa. Several studies have been conducted to determine the factors to contain inbreeding. Franklin (1980) found in his investigation that inbreeding can be minimized by keeping an effective population size of 50. Later on, Hochkirch (2011) advocated at least 100 individuals are required to effectively address the inbreeding and genetic diversity loss. Lande (1995) highlighted the significance of larger effective population sizes for retaining the evolutionary potential. It is believed that the species with small ranges are somewhat less affected by inbreeding depression, as the genetic load may have been reduced due to the act of natural purging.

Further, proper management is also essential; thus, maintaining ex situ facilities is expensive, including food, veterinary care and infrastructure. According to Hochkirch (2011), a captive breeding program should focus on the following points: (1) delineation of the natural genetic structure of the focal species, (2) identification of management units for prioritization, (3) baseline information on the captive population, (4) analysis of the genetic structure of the captive population, (5) selection of appropriate individuals for the breeding program, (6) the aim should be to maximize the genetic variability, and (7) determining the genetic structure after some generations.

- Ex situ *conservation sites*: Due to the very nature of ex situ conservation, protected species are reared outside their natural native habitat which may have evolutionary implications or, some cases, disturb the local ecosystem. Living outside the natural habitat can affect behavior and reproduction in the artificial environment. Further, creating suitable environmental conditions, similar to the natural ones, can also be challenging. Moreover, live germplasm centers can be expensive and difficult to maintain without sustained funding efforts. Further, fish translocation and modifying the environmental conditions may be difficult. Involving stakeholders and addressing the issue with co-management approach are important in such endeavors. Public aquariums are key to raising awareness and support for fish conservation efforts. However, they are mainly seen housing fish species without much regard for conservation. In such a scenario, plans must be drawn up for aquariums to set up a sustainable breeding program, especially for threatened species so that the fishing pressure in the wild might be reduced. Additionally, they must be incentivized to aid the recovery of threatened species via collaborative reintroduction or translocation efforts (Kasso and Balakrishnan 2013).
- *Reintroduction challenges*: Animals bred in captivity may struggle to survive when reintroduced into the wild, on account of their behavioral differences and lack of natural skills. Reintroduced animals may also face difficulties in their acceptance by the existing wild relatives. In the case of aquaria and live gene banks, there can be escapees which may be invasive or exotic to the new habitat. Olden et al. (2011) listed a few cases where regional or intercontinental introductions of aquatic species led to a deleterious consequence affecting the whole ecosystem. Hence, proper planning and execution are required.
- *Policy and governance issues*: Ex situ conservation usually requires a multi-stakeholder approach with the participation of local communities, subject matter specialists, policymakers, government institutions, NGOs and the general public. Usually, species with high commercial or socioeconomic values are considered ignoring many endemic small species which may require urgent conservation efforts. Even though their importance may not be evident to us, they may play an important role in the ecosystem as a whole. Hence, experts from taxonomy, genetics, fisheries resource management, ecology, policy management, and extension subjects need to be involved in decision-making.

6.8 Ethical Issues of Ex Situ Conservation

The ex situ approach of conservation of natural resources, biodiversity and ecosystems raises several ethical concerns, as it involves breeding and preserving vulnerable/threatened species outside their natural habitats. There are several ethical concerns for ex situ conservation, like animal welfare, relocation issues, cost, success rate of relocation of the germplasm etc. There are many a time conflicts between the ethical duties and the research and management practices being used to safeguard the welfare of the animals in situ approaches. Harmonizing animal welfare, species viability, and ecological integrity becomes vital, so as to safeguard fish diversity from decline and extinction.

There are always challenges in translocating populations from native habitats to new environments that invite ethical, philosophical and technical queries. Due to relocation from the native to ex situ approach, there are rapid loss in biodiversity which demands careful consideration of ethics and responsibilities in conservation investigation and preparation. Further, captive propagation and reintroduction programs also face challenges in establishing self-sustaining captive populations and many times, these captive breeding programs are costing high. Moreover, the species reintroduction attempts may not always work. Additionally, as per genetic considerations, the ex situ approach affects the reproductive rates, genetic diversity, inbreeding depression, and genetic adaptations. In summary, the developing scenario of conservation beliefs, the urgency of global ecological changes and the need for realistic trade-off shapes the ethical considerations for ex situ conservation efforts. Looking at these details, our aim is to protect the species, simultaneously respecting their well-being in the ecological background.

6.9 Conclusion

The genetic material banking plays a crucial role in preserving biodiversity, advancing scientific knowledge and safeguarding our planet's genetic heritage. The aim of ex situ conservation is not only to preserve the species in captivity but also to maintain its genetic makeup for eventual release back into the wild. Ex situ strategies like captive breeding, cell culture, gamete cryopreservation, aquaria, and live gene banks are widely used as conservation tools. While breeding fish in captivity to restock habitats is a good conservation strategy, it may not achieve the intended benefit, if genetic aspects of conservation are not given due importance. Overlooking genetic concerns may result in a different sets of problems, like inbreeding, genetic bottleneck, founder population effect, genetic drift etc., resulting in a reduction of fitness of the species. Cryopreservation of gametes, zygotes and embryos is routinely applied in mammalian systems and is being used for artificial insemination and embryo transfer in the livestock industry. Sperm and embryonic cell cryopreservation have been successful in many teleosts and invertebrates; however, cryopreservation of embryos and oocytes remains a major challenge. Larvae of oysters, shrimps and sea urchins have been cryopreserved, and the technique is ready for adoption. Moreover, cryopreservation of stem cells will be helpful in the revival of animals. Establishment a cryobank will help in utilizing genetic resources worldwide for various purposes. Thus, the efforts made through ex situ conservation may play vital role in complementing in situ conservation approach toward safeguarding biodiversity and ensuring restoration of the threatened species.

References

Ahammad MM, Bhattacharyya D, Jana BB (2002) The hatching of common carp (*Cyprinus carpio* L.) embryos in response to exposure to different concentrations of cryoprotectant at low temperatures. Cryobiology 44(2):114–121

Ahammad MM, Bhattacharyya D, Jana BB (2003a) Hatching of common carp (*Cyprinus carpio* L.) embryos stored at 4 and– 2° C in different concentrations of methanol and sucrose. Theriogenology 60(8):1409–1422

Ahammad MM, Bhattacharyya D, Jana BB (2003b) Stage-dependent hatching responses of rohu (*Labeo rohita*) embryos to different concentrations of cryoprotectants and temperatures. Cryobiology 46(1):2–16

Ahammad MM, Bhattacharyya M, Banerjee D et al (2004) Toxicity of protectants to embryos of silver carp (*Hypophthalmichthys molitrix*) after cold storage at different storage time periods. Cell Preserv Technol 2:227–233

Araguas RM, Sanz N, Fernández R et al (2009) Role of genetic refuges in the restoration of native gene pools of brown trout. Conserv Biol 23(4):871–878

Arat S, Caputcu AT, Akkoc T et al (2011) Using cell banks as a tool in conservation programmes of native domestic breeds: the production of the first cloned Anatolian grey cattle. Reprod Fertil Dev 23(8):1012–1023

Aril N, Namai K, Gomi F et al (1987) Cryopreservation of medaka embryos during development. Zoo Sci 4:813–818

Asahina E, Takahashi T (1978) Freezing tolerance in embryos and spermatozoa of the sea urchin. Cryobiology 15(1):122–127

Attard CRM, Sandoval-Castillo J, Gilligan DM et al (2022) Genomics outperforms genetics to manage mistakes in fisheries stocking of threatened species. Biodivers Conserv 31:895–908

Bart AN (2000) New approaches in cryopreservation of fish embryos. The World Aquaculture Society: Cryopreservation in Aquatic Species, Baton Rouge, pp 179–187

Ben-Nun IF, Montague SC, Houck ML et al (2011) Induced pluripotent stem cells from highly endangered species. Nat Methods 8(10):829–831

Boe K, Bjøru B, Tangvold Bårdsen M et al (2021) Opportunities and challenges related to sperm cryopreservation in Atlantic salmon gene banks. Conserv Sci Pract 3(12):e552

Brumbaugh RD, Sorabella LA, Johnson C et al (2000) Small scale aquaculture as a tool for oyster restoration in Chesapeake Bay. Mar Technol Soc J 34(1):79–86

Choudhury M, Biswas SP (2003) Maturity and spawning habit of an ornamental snakehead fish *Channa barca* (Hamilton). J Ecophysiol Occup Health 3:49–152

Dinnyes A, Urbanyi B, Baranyai B et al (1998) Chilling sensitivity of carp (*Cyprinus carpio*) embryos at different developmental stages in the presence or absence of cryoprotectants: work in progress. Theriogenology 50(1):1–13

Diwan AD (2000) Shrimp embryo cryobanking is now possible. Curr Sci 78:1282–1212

Diwan AD, Kandasami D (1997) Freezing of viable embryos and larvae of marine shrimp, *Penaeus semisulcatus* de Haan. Aquac Res 28:947–950

Diwan AD, Ayyappan S, Lal KK et al (2010) Cryopreservation of fish gametes and embryos. Indian J Anim Sci 80(4):109–124

Frankham R, Ballou JD, Briscoe DA (2002) Introduction to conservation genetics. Cambridge University Press, Cambridge

Franklin IR (1980) Evolutionary change in small populations. In: Soule ME, Wilcox BA (eds) Conservation biology: an evolutionary—ecological perspective. Sinauer, Sunderland, MA, pp 135–150

Froehlich HE, Gentry RR, Halpern BS (2017) Conservation aquaculture: Shifting the narrative and paradigm of aquaculture's role in resource management. Biol Conserv 215:162–168

Ghildiyal K, Nayak SS, Rajawat D et al (2023) Genomic insights into the conservation of wild and domestic animal diversity: a review. Gene 886:147719

Hagedorn M, Kleinhans FW, Wildt DE et al (1997) Chill sensitivity and cryoprotectant permeability of dechorionated zebrafish embryos, Brachydanio rerio. Cryobiology 34(3):251–263

Halpern BS, Maier J, Lahr HJ et al (2021) The long and narrow path for novel cell-based seafood to reduce fishing pressure for marine ecosystem recovery. Fish Fish 22:652–664

Harvey B, Kelly RN, Ashwood-Smith MJ (1983) Cryobiology permeability of intact and dichorionated zebra fish embryos to glycerol and dimethyl sulfoxide. Aquaculture 20:432–439

Hochkirch A (2011) Ex situ conservation genetics: a review of molecular studies on the genetic consequences of captive breeding programmes for endangered animal species. Biodivers Conserv 20:1843–1861

Kasso M, Balakrishnan M (2013) Ex situ conservation of biodiversity with particular emphasis to Ethiopia. ISRN Biodiversity 13:985037

Katagiri T, Asakawa S, Minagawa S et al (2001) Construction and characterization of BAC libraries for three fish species; rainbow trout, carp and tilapia. Anim Genet 32:200–204

Khorasani MZ, Hennig S, Imre G et al (2004) A first-generation physical map of the medaka genome in BACs essential for positional cloning and clone-by-clone based genomic sequencing. Mech Dev 121:903–913

Khoshoo TN (1997) Conservation of India's endangered mega animals: Tiger and lion. Curr Sci 73(10):830–842

Kulkarni CV, Ogale SN (1978) The present status of mahseer (fish) and artificial propagation of *tor khudree* (Sykes). J Bombay Nat Hist Soc 75:651–660

Kumar MS, Kumar R, Kushwaha B et al (2023) National Repository of fish cell lines: a potential fish cell line depository for tomorrow's cellular agriculture. Indian Farming 73(12):34–37

Kumar MS, Singh VK, Mishra AK et al (2024) Fish cell line: depositories, web resources and future applications. Cytotechnology 76(1):1–25

Labbe C, Haffray P, Mingant C et al (2018) Cryopreservation of Pacific oyster (*Crassostrea gigas*) larvae: revisiting the practical limitations and scaling up the procedure for application to hatchery. Aquaculture 488:227–234

Lande R (1995) Mutation and conservation. Conserv Biol 9:782–791

Leung LKP, Jamieson BGM (1991) Live preservation of fish gametes. In: Fish evolution and systematics: evidence from spermatozoa. Cambridge University Press, Cambridge

Liu S, Liu Y, Zhang X et al (1998) Construction of a pufferfish bacterial artificial chromosome library. Zhongguo Yi Xue Ke Xue yuan Xue Bao. Acta Acad Med Sin 20:161–166

Liu XH, Zhang T, Rawson DM (2001) Differential scanning calorimetry studies of intraembryonic freezing and cryoprotectant penetration in zebrafish (*Danio rerio*) embryos. J Exp Zool 290(3):299–310

Malla S, Banik S (2015) Reproductive biology of an endangered catfish, Ompok bimaculatus (Bloch, 1794) in the lotic water bodies of Tripura, North-East India. Int J Fish Aquat 2:251–260

Mercy TVA, Sajan S, Malika V (2015) Captive breeding and developmental biology of *Sahyadria denisonii* (day 1865) (Cyprinidae), an endangered fish of the Western Ghats, India. Indian J Fish 62(2):19–28

Nagalgaon RK, Patil PY (2021) Ex-situ conservation of endangered mahseers (*tor* spp) through ranching in Indrayani River, Pune (MS), India. Vidyabharati Int Interdiscip Res J 12(2):727–734

Olden JD, Kennard MJ, Lawler JJ et al (2011) Challenges and opportunities in implementing managed relocation for conservation of freshwater species. Conserv Biol 25(1):40–47

Pogoda B, Brown J, Hancock B et al (2019) The native oyster restoration alliance (NORA) and the Berlin oyster recommendation: bringing back a key ecosystem engineer by developing and supporting best practice in Europe. Aquat Living Resour 32:13

Polge C, Smith AU, Parkes AS (1949) Revival of spermatozoa after vitrification and dehydration at low temperatures. Nature 164(4172):666–666

Potter G, Alec ST, Smith NTK et al (2020) A more open approach is needed to develop cell-based fish technology: it starts with zebrafish. One Earth 3:54–64

Purkayastha S, Sarma, S, Sarkar, UK, Lakra, WS, Gupta S, Biswas SP (2012) Captive breeding of endangered Ompok pabda with ovatide. J Appl Aquacult 24:42–48

Quiniou SM, Katagiri T, Miller NW et al (2003) Construction and characterization of a BAC library from a gynogenetic channel catfish *Ictalurus punctatus*. Genet Sel Evol 35:673
Rey O, Eizaguirre C, Angers B et al (2020) Linking epigenetics and biological conservation: towards a conservation epigenetics perspective. Funct Ecol 34:414–427
Ridlon AD, Wasson K, Waters T et al (2021) Conservation aquaculture as a tool for imperiled marine species: evaluation of opportunities and risks for Olympia oysters, *Ostrea lurida*. PLoS One 16(6):e0252810
Robles V, Cabrita E, Anel L et al (2006) Microinjection of the antifreeze protein type III (AFPIII) in turbot (*Scophthalmus maximus*) embryos: toxicity and protein distribution. Aquaculture 261(4):1299–1306
Rogers-Bennett L, Aquilino KM, Catton CA et al (2016) Implementing a restoration program for the endangered white abalone (*Haliotis sorenseni*) in California. J Shellfish Res 35(3):611–618
Routray P, Suzuki T, Strüssmann CA et al (2002) Factors affecting the uptake of DMSO by the eggs and embryos of medaka, *Oryzias latipes*. Theriogenology 58(8):1483–1496
Ruiz-Lopez MJ, Roldan ERS, Espeso G (2009) Pedigrees and microsatellites among endangered ungulates: what do they tell us? Mol Ecol 18:1352–1364
Ryder OA, Onuma M (2018) Viable cell culture banking for biodiversity characterization and conservation. Annu Rev Anim Biosci 6:83–98
Sarkar UK, Deepak PK, Negi RS et al (2006) Captive breeding of endangered fish *Chitala chitala* (Hamilton-Buchanan) for species conservation and sustainable utilization. Biodivers Conserv 15:3579–3589
Schirmer K, Fischer M, Eszter S et al (2021) Validation report for the test guideline 249 on fish cell line acute toxicity—the RTgill-W1 cell line assay (series on testing and assessment report no.: 334). OECD Publishing, Paris. https://doi.org/10.1787/c66d5190-en
Simon C, Dumont P, Cuende FX et al (1994) Determination of suitable freezing media for cryopreservation of *Penaeus indicus* embryos. Cryobiology 31(3):245–253
Smith DJ, Helmy M, Lindley ND et al (2022) The transformation of our food system using cellular agriculture: what lies ahead and who will lead it? Trends Food Sci Technol 127:368–376
Stephens N, Di Silvio L, Dunsford I et al (2018) Bringing cultured meat to market: technical, socio-political, and regulatory challenges in cellular agriculture. Trends Food Sci Tech 78:155–166
Subramoniam T (1994) Cryopreservation of crustacean gametes and embryos. Proc Natl Acad Sci India Sect B Biol Sci 60:229–236
Subramoniam T, Newton SS (1993) Cryopreservation of embryos of the penaeid prawn *Penaeus indicus*. Curr Sci 65:176–177
Thangaraj RS, Narendrakumar L, Geetha PP et al (2021) Comprehensive update on inventory of finfish cell lines developed during the last decade (2010–2020). Rev Aquac 13(4):2248–2288
Thorsen J, Zhu B, Frengen E et al (2005) A highly redundant BAC library of Atlantic salmon (*Salmo salar*): an important tool for salmon projects. BMC Genomics 6:50
Tsai S, Lin C (2012) Advantages and applications of cryopreservation in fisheries science. Braz Arch Biol Techn 55:425–434
Valdez JW, Mandrekar K (2019) Assessing the species in the CARES preservation program and the role of aquarium hobbyists in freshwater fish conservation. Aust Fish 4(4):49
Verma PJ, Sumer H (2015) Cell Reprogramming. In: Methods in molecular biology. Springer, Cham
Whitmore TC (1990) An introduction to tropical rain forests. Clarendon Press, Oxford
Wisely SM, Ryder OA, Santymire RM et al (2015) A road map for 21st century genetic restoration: gene pool enrichment of the black-footed ferret. J Hered 106(5):581–592
Zhang T, Rawso DM (1995) Studies on chilling sensitivity of zebrafish (*Brachydanio rerio*) embryos. Cryobiology 32(3):239–246

7 Habitat Degradation in Coral Reef Ecosystems and Mangroves: Current Status and Management Measures

T. K. Teena Jayakumar and Uttam Kumar Sarkar

Abstract

Marine ecosystems exhibit vast biodiversity and fulfil critical roles in supporting diverse species. Despite their distinct characteristics, these ecosystems are interconnected, as the health of one can significantly affect others. This interdependence underscores the importance of ecosystem-level conservation strategies to safeguard marine biodiversity and ecosystem services. Coral reefs and mangrove ecosystems are facing severe habitat degradation due to human activities and climate change impacts. As climate change impacts intensify, they can amplify the effects of existing stressors, necessitating the development of new or modified management approaches. This chapter explores the diversity of coral and mangrove ecosystems in India, major threats faced and various mitigation measures employed to address this degradation and promote ecosystem resilience. Protection strategies such as marine protected areas (MPAs) and zoning regulations have been effective in managing human activities and conserving critical habitats. Restoration efforts involving coral gardening, artificial reef deployment, and mangrove rehabilitation have shown promise in rebuilding degraded habitats. Additionally, community engagement and education programmes play a vital role in fostering local stewardship and reducing unsustainable practices. Innovative approaches, including nature-based solutions and sustainable tourism initiatives, are increasingly recognized for their potential to enhance ecosystem health and support biodiversity. Despite challenges, collaborative efforts involving governments, NGOs, scientists, and local communities offer hope for the long-term conservation of these vital coastal ecosystems. This chapter underscores the importance of proactive measures to combat habitat degradation and

T. K. Teena Jayakumar (✉) · U. K. Sarkar
ICAR-National Bureau of Fish Genetic Resources, Lucknow, Uttar Pradesh, India
e-mail: Teena.Jayakumar@icar.gov.in; Uttam.Sarkar@icar.gov.in

U. K. Sarkar et al. (eds.), *Sustainable Management of Fish Genetic Resources*,
https://doi.org/10.1007/978-981-97-5250-8_7

underscores the urgency of global action to protect coral reefs and mangroves for future generations.

Keywords

Coral reef · Mangrove · Restoration · Rehabilitation · Diversity

7.1 Introduction to Coral Reef Ecosystem

Coral reefs are shallow-water marine ecosystems found in tropical regions, known for their exceptionally high biomass production and diverse flora and fauna, akin to tropical rainforests in their richness and variety of species. The term 'coral' refers to a wide variety of exclusively marine invertebrate animals from the phylum Cnidaria of the class Anthozoa which includes hard corals, soft corals, precious corals, and hydrocorals. The 'corallum' refers to the entire coral structure, while the 'corallite' is the calcareous cup where coral polyps reside. These polyps secrete mucus to capture plankton and other food particles, which they then consume. A crucial symbiotic relationship exists between coral polyps and photosynthetic algae in the reef environment. The algae receive nutrients and carbon dioxide from the polyps, while the polyps obtain photosynthetic products from the algae. Additionally, an organism known as zooxanthellae regulates the growth and secretory functions of coral polyps. The coral structure consists of an outer layer of living tissue that secretes a lower layer of hard limestone skeleton (Turner et al. 2009). The attachment of free-swimming planulae (coral larvae) to the submerged edges of islands and subsequent growth eventually lead to reef formation.

The coral fauna mainly consists of hermatypic corals and ahermatypic corals. Hermatypic corals have symbiotic relationships with zooxanthellae and are responsible for building vast reef structures beneath the ocean's surface. These reefs thrive in the photic zone of tropical waters around the world. Ahermatypes are solitary or colonial forms without symbionts that gain nutrients by filtering planktons/detritus from seawater and they do not contribute to reef building. Corals possess a remarkable capacity to adapt their growth patterns according to the specific conditions of their habitat, including factors like depth, light levels, temperature, and water flow. As a result, a single reef can harbour a wide array of diverse habitats, each tailored to the unique needs of the organisms that depend them. Apart from the corals themselves, they support a broad array of marine creatures such as fish, molluscs, crustaceans, sharks, dolphins, and sea turtles within their habitats. They require specific conditions to thrive, including relatively shallow depths, direct sunlight exposure, optimal temperatures between 23 and 25 °C, and clear water free from suspended sediments. These reefs play a crucial role in supporting marine life and provide various economic and environmental benefits to millions of people worldwide (Fig. 7.1).

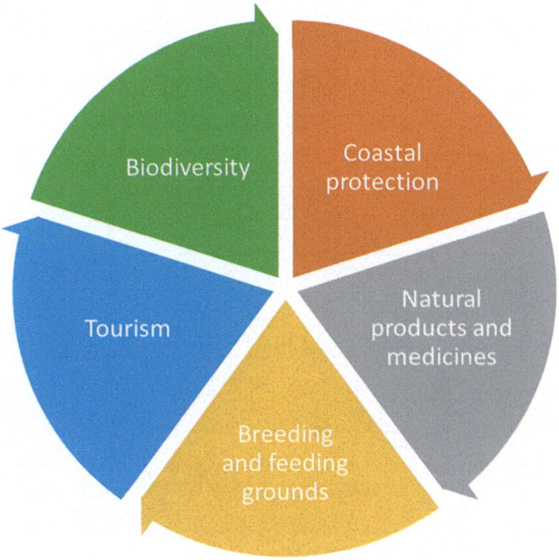

Fig. 7.1 Ecosystem services of coral reef

7.2 Coral Reef in India

There are mainly four types of coral reefs found in India:

(a) Platform reefs: these reefs are nearly flat and lack a lagoon. They are situated on the shallower parts of continental shelves and can be associated with atolls or positioned between a coastline and a barrier reef. These reef types are predominantly located in the Gulf of Kachchh in India.
(b) Fringing reefs: fringing reefs grow closely along the shores of mainland or islands, extending into the sea like a submerged platform. They are characterized by a narrow channel called a lagoon, separating them from the land. The most significant coral growth typically occurs at the edges of these reefs. Fringing reefs can be found in the Gulf of Mannar, Palk Bay, and the Andaman and Nicobar Islands of India.
(c) Barrier reefs: these reefs resemble fringing reefs but are separated from the mainland by a narrow stretch of deep sea, potentially up to 1000 feet deep, which is deeper than the lagoons they border. Barrier reefs are primarily found in the Andaman and Nicobar Islands of India. Outside of India, notable examples include the Great Barrier Reef in Australia. Barrier reefs are distinguished by their separation from the land by extensive bodies of water, running parallel to the coastline.
(d) Atolls: atolls are formations found in the ocean and are not connected to land. They are comprised of low reefs that rise no more than 28 feet above sea level and a lagoon. In India, atolls are primarily located in the Lakshadweep Islands: circular rings of reefs surrounding a lagoon, often accompanied by low-lying islands. Atolls are common in the Indian and South Pacific oceans.

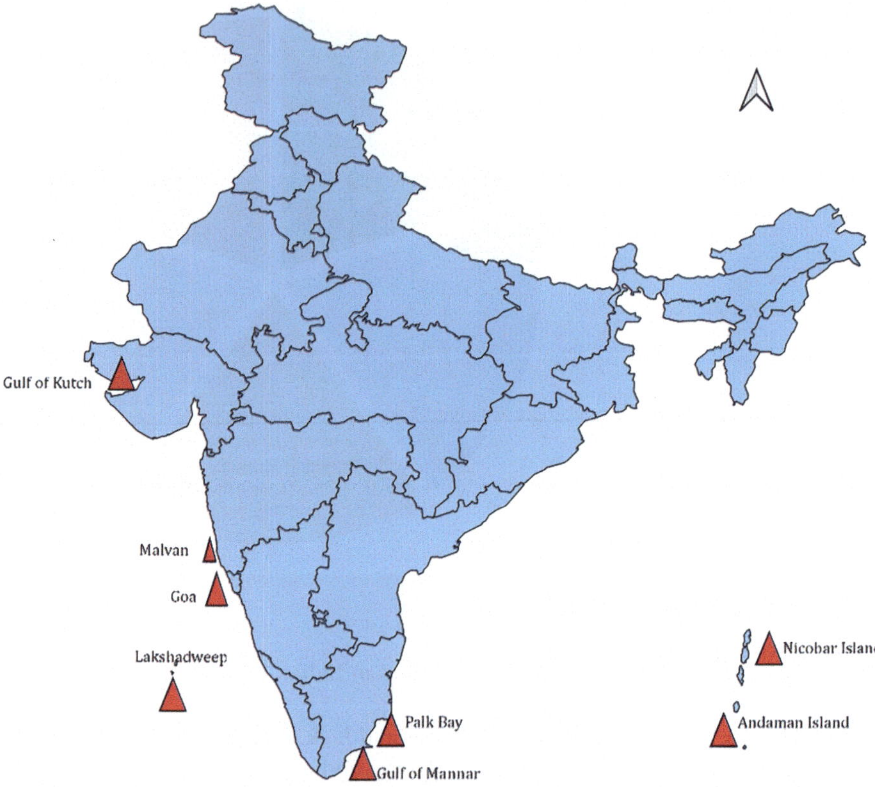

Fig. 7.2 Map depicting major coral reefs in India

India has a coastline of nearly 8118 km and possesses relatively few coral reef areas. Coral reefs in the Indian Ocean developed primarily during the tertiary and quaternary periods (Chanda 2022). In India, coral reefs are primarily found in seven regions: (1) the Palk Bay, (2) the Gulf of Kachchh, (3) the Gulf of Mannar, (4) the Lakshadweep Islands, (5) the Andaman and (6) Nicobar Islands, and (7) west coast (Fig. 7.2; Table 7.1). Corals also occur on the east coast between Parangipettai, south of Cuddalore (10°50′ N, 79°80′ E), and Pondicherry, but these communities are less surveyed (Ramaiyan et al. 1995). The lack of coral reefs in the Bay of Bengal is attributed to the substantial influx of fresh water and sediment transported by rivers such as the Ganga, Krishna, and Godavari. These factors pose a threat to coral polyps, making it challenging for them to thrive in this environment. Additionally, heavy monsoonal rainfall and significant human presence along the coastline further inhibit coral reef growth in this region.

Coral reefs in India are known for their remarkable biodiversity, serving as critical ecosystems that support a wide range of marine life. According to a comprehensive study by De et al. (2020), a total of 585 coral species, are reported distributed across 108 genera and 23 families. Among these, the Acroporidae family

Table 7.1 Major coral reefs in India, dominant coral genera, and conservation strategy

Reef location	Reef area (Km^2)	Dominant genera	Protection established (year)
Andaman and Nicobar Islands	1021.46	*Acropora*, *Favia*, *Montipora*, *Dipsastrea*, *Pocillopora* and *Porites*	Mahatma Gandhi (1983) and Rani Jhansi Marine National Parks (1996) and Great Nicobar Biosphere Reserve (2013)
Gulf of Kachchh	325.5	*Montipora*, *Goniopora*, and *Favia*	Marine National Park (1995)
Gulf of Mannar and Palk Bay	94.3	*Acropora*, *Montipora*, and *Porites*	Gulf of Mannar Biosphere Reserve (1980)
Lakshadweep Islands	933.7	*Acropora*, *Porites and Montipora*	Dr. K.K. Mohammed Koya Sea Cucumber Conservation Reserve (2020) and Attakoya Thangal Marine Conservation Reserve (2020)

Source: Muley et al. (2000), Sivakumar et al. (2013), Rajan et al. (2015), Saroj et al. (2016) and De et al. (2020)

contributed the most to species assemblage, with 184 species across six genera. The Merulinidae family followed, comprising 100 species within 19 genera. Additionally, the Poritidae family included 52 species from four genera, while the Fungiidae family was represented by 42 species spanning 15 genera. The dominant genera are *Acropora* (104 species), *Montipora* (54 species), *Porites* (30 species), *Dipsastrea* (20 species), *Goniopora* (20 species), *Favites* (18 species), *Lobophyllia* (16 species) and *Pavona* (16 species) (De et al. 2020, Table 7.1). This diversity is not only indicative of the health and complexity of coral reef ecosystems in India but also highlights the importance of these habitats for marine biodiversity conservation. The varied assemblages of coral species contribute to the structural complexity of the reefs, providing essential habitats for numerous marine organisms, including fish, invertebrates, and algae.

7.2.1 Andaman and Nicobar Islands

The islands are characterized by narrow, linear fringing reefs and a long barrier reef (329 km) on the west coast, emerged as a part of a mountain chain and lie on a ridge that extends southward from the Irrawaddy Delta area of Burma, continuing the trend of the Arakan Yoma range. Scientific understanding of the reefs in the area is limited, yet they have the potential to be the most diverse in India and those in the best condition. A deep oceanic ridge along 10° N separates the Andaman group and the Nicobar group of islands. The orientation of the chain of island groups is north to south. In these island groups, there are two marine national parks, viz. Mahatma Gandhi and Rani Jhansi Marine National Parks and Great Nicobar Biosphere Reserve. The coral fauna is highly diverse when compared to other parts of India.

7.2.1.1 Diversity

The coral reefs of the Andaman Islands exhibit a similar species diversity to the South-East Pacific Coral Triangle. They are among the most diverse and extensive coral reefs in the Indian Ocean, with relatively low levels of human disturbance (Wallace and Muir 2005). These reefs consist of fringing, channel, and patch types, hosting a total of 523 species of hard corals (Mondal et al. 2017; De et al. 2020). Specifically, the Andaman Islands alone harbour424 species, which accounts for 89% of India's coral diversity, while the Nicobar group of islands is home to 242 species, representing 51% of India's coral diversity (Raghuraman et al. 2012; Mondal et al. 2016). The reefs are mainly dominated by family Acroporidae (164 species), Merulinidae (100 species), Fungiidae (42 species), Lobophyliidae (32 species) and Dendrophylliidae (28 species) (De et al. 2020). Acroporidae alone contributes 30% of the total coral biodiversity of Andaman and Nicobar with 164 species which is about 83% of the Indian acroporids (Raghuraman et al. 2012; De et al. 2020). Areas with high wave action such as Car Nicobar, Little Andaman, and some open sea islands of Andaman have high diversity of live acroporids (Venkataraman et al 2003). The reefs of South Andaman are characterized by high diversity and abundance of species. Unlike the sandy and muddy bottom reefs found in North and Middle Andaman, the reef bottoms in South Andaman are predominantly covered with dead corals, rubbles, and live rock. This unique habitat composition results in high coral recruitment, substantial coral live cover, and a rich diversity of reef-associated fauna. The dominant coral genera in South Andaman reefs include *Acropora*, *Favia*, *Fungia*, *Pocillopora* (Fig. 7.3), and a few mussid genera, with acroporids and *Porites* being particularly prevalent. Despite the challenges, the Andaman and Nicobar Islands host diverse marine life, including 112 species of sponges, 1200 species of fish, and various other marine organisms (Venkataraman et al. 2003; Raghuraman et al. 2012). The genus *Acropora* is the predominant coral on the reefs followed by *Porites* in few islands (Venkataraman et al. 2003; De et al. 2020).

Fig. 7.3 *Porites* sp. in Andaman Island

7.2.1.2 Current Status

In the Andaman and Nicobar, many reef areas remain remarkably pristine, with significant live coral presence ranging from 56 to 65% (Muley et al. 2000). Coral mortality due to siltation possibly exacerbated by oil palm plantation was observed in many areas (Muley et al. 2000; Raghunathan et al. 2012). Siltation, likely exacerbated by effluents, has also been implicated in coral destruction around Port Blair. Neil Island near Ritchie's Archipelago faces substantial coral mortality, possibly due to agricultural run-off and pesticide use, although no formal studies have explored this link (Muley et al. 2000; Raghunathan et al. 2012). Construction activities, particularly sand mining for infrastructure projects, contribute significantly to coral reef degradation. Additionally, the collection of coral reef associates and invasive species like the crown-of-thorns starfish and white band disease poses further threats (Raghunathan et al. 2012). Coral bleaching events have been observed in this region, particularly in 1998, attributed to a 2 °C rise in sea temperature. This resulted in the loss of approximately 80% of live corals, but recent surveys indicate a mean live coral cover of 56%, dead coral cover of 22%, and coral rubble of 11% in Little Andaman and Great Nicobar Islands (Muley et al. 2000). Destructive tsunami disrupted the reefs of Andaman and Nicobar to a large extent. In shallower regions, recruited corals are more common due to high algal cover and the presence of hard bottoms, indicating positive signs of recovery (Majumdar et al. 2018). Deep-water corals in Car Nicobar reefs are thriving and are minimally affected, exerting dominance over the ecosystem (Raghunathan et al. 2012). A detailed investigation into the heat-induced bleaching of scleractinian corals during the El Niño event in the Andaman Sea and the Bay of Bengal, along with their resilience during the post-bleaching period, was conducted in the Andaman and Nicobar Islands in 2016 by Mondal and Raghunathan (2022). The study found that 83.6% of scleractinian corals in the Andaman Islands were bleached between April and May 2016. The highest bleaching rate, 91.5%, was observed in the South Andaman region of the Andaman Sea, while the lowest rate, 83.2%, was recorded in the North Andaman region. Coral recovery was noted from July to December 2016, predominantly among corals from the families Lobophylliidae, Merulinidae, and Dendrophylliidae. Conversely, corals from the families Acroporidae and Poritidae exhibited minimal recovery. Pre- and post-bleaching surveys indicated that the massive bleaching event resulted in a total loss of 23.58% of the live coral cover (Mondal and Raghunathan 2022).

7.2.2 Gulf of Kachchh

The Gulf of Kachchh, situated on the northern side of the Saurashtra Peninsula, lies within the coordinates of 22°15′–23°40′ N latitude and 68°20′–70°40′ E longitude. Spanning approximately 7350 km^2, it encompasses 42 islands and having rich diversity of flora and fauna. The Gulf of Kachchh hosts mostly fringing reefs, along with offshore platform reefs, patch reefs, and coral pinnacles. The tidal range in the gulf is reported to be as great as 12 m but may have seasonal changes with extreme low tides at certain times of the year (Brown 1997).

7.2.2.1 Diversity

De et al. (2020) have reported 76 species of coral consisting of 30 genera and 11 families in the area. Merulinidae is the most common family (25 species) followed by Acroporidae (13 species) and Poritidae (10 species) (Kumar et al. 2014; De et al. 2020). Three species of turtles and three species of marine mammals (dugongs, dolphins, and whales) are also distributed in the area. *Montipora venosa*, *Coscinaraea monile*, *Hydnophora exesa*, *Turbinaria peltata*, *Goniastrea pectinata*, *Platygyra sinensis*, *Cyphastrea serialia*, *Porites compressa*, and *Goniopora stutchburyi* are some of the common species found in the Gulf of Kachchh (Venkataraman 2011). Species such as *Siderastrea savignayana* and *Acanthastrea hillae* are reported only from the Gulf of Kachchh (Venkataraman 2011). Joshi et al. 2015 reported prevalence of the coral families Poritidae (*Goniopora* sp.) and Faviidae (*F. favus*) in the Gulf of Kachchh, which play a crucial role in the community structuring of these coral reefs. Ramose corals such as *Acropora*, *Pocillopora*, *Stylophora*, and *Seriatopora* are not found at present though semi-fossilised specimens of *Acropora* are found on some beaches (Pillai 1996). The coral fauna is comparatively less diverse when compared to other parts of India.

7.2.2.2 Current Status

The coral reefs in the area are highly degraded when compared with other three major coral reef regions of India, due to mud deposits, with living coral area rarely exceeding 20–30% (Wafar et al. 2000; Joshi et al. 2014). Efforts in marine national parks have contributed to significant restoration of coral reef areas. The coral reef ecosystem is under threat due to the extraction of corals and sands by cement industries, increased turbidity of water, oil refineries, chemical industries, and mechanized fishing boats. There were two catastrophic and localized bleaching events that happened in this region (Adhavan et al. 2014; Adhavan 2015). Coral bleaching and its resulting mortality not only harm coral communities but also affect fish populations and the human communities reliant on coral reefs and associated fisheries for their livelihoods and overall well-being. Bleached corals typically exhibit reduced growth rates, diminished reproductive capabilities, heightened vulnerability to diseases, and increased mortality rates. Wafar et al. (2000) reported that 70% of live corals were destroyed during the 1998 bleaching event. Coral colonies serve as vital sources of food and shelter for numerous reef-associated organisms, highlighting the fundamental significance of coral health to the ecological balance of the reef community. The corals in the Gulf of Kachchh survive through extreme environmental conditions such as high temperatures, salinity changes, and high-suspended particulate loads (Wafar et al. 2000; Joshi et al. 2014). In correlation with geographical and oceanographical changes such as transgression of the sea levels of the Late Pleistocene-Holocene period and upliftment of tectonic plates of the Gulf (Srivastava 1965), its arid climate and the semi-diurnal tidal amplitude fluctuations hamper the recovery of the reef. Soft corals may play the alternative key role in maintaining the ecological role in the degraded reef ecosystem in the absence of hard corals as they are comparatively more adapted in adverse hydrographic condition than the hard corals. The reef flats and intertidal pools of most of the islands of the gulf are covered with dead coral boulders. Species diversity of corals is low in the Gulf of Kachchh with only 37 species of hard corals recorded and a total absence of ramose growth forms (Wafar et al. 2000). Human activities such as mangrove deforestation, port development, offshore moorings, and pollution from urban centres have significantly impacted coral reef ecosystems. These anthropogenic influences have led to the degradation of coral reef habitats, resulting in a reduction of coral cover by over 50% on many reefs (Wafar et al. 2000). The 2010–2014–2016 El Nino Southern Oscillation (ENSO) caused prolonged marine heatwaves. Arora et al. (2019) have concluded that in the last decades, the coral cover in the Gulf of Kachchh has dramatically declined due to higher sea surface temperature (SST).

7.2.3 Lakshadweep Islands

The Lakshadweep is a group of the least studied coral atoll enclosing lagoons, submerged reefs, and banks lying between latitude 8°–12° N and longitude 71°–74° E, covering an area of only 32 Km^2, situated in the Arabian Sea. The Laccadive Islands are situated on the northern part of the Laccadive-Chagos Ridge in the western Indian Ocean extending over 2500 km. The ridge is divided into three main segments which are referred to the Laccadive Islands (Lakshadweep), the middle segment as the Maldives islands, and southern segment as Chagos Islands. This ridge is supposed to be a continuation of the Aravalli Mountain, and the islands are believed to be remnants of the submerged mountain cliffs. The Lakshadweep Islands are predominantly composed of atolls, with a single platform reef located at Androth. The reef flat spans an area of 136.5 km^2. The archipelago comprises 6 small islands, 12 atolls, 3 reefs, and 5 submerged banks, totalling an area of 32 km^2, while the lagoons cover approximately 4200 km^2.

7.2.3.1 Diversity

The coral fauna of Lakshadweep harbour 165 species, 54 genera and 17 families (Pillai 1996; Caeiro 1999; Kumar et al. 2011a, b; De et al. 2020). Major families of corals in the area includes *Acroporidae (51 species)*, Merulinidae (34 species), *Poritidae (17 species), Pocilloporidae, Favidae, Fungiidae, and Mussidae (De et al.* 2020*). Acropora* spp. are most dominant followed by *Porites and Montipora (De et al.* 2020*). Acropora* spp., *Pocillopora* spp., *Porites* spp., and favids dominate the lagoon and reef flat fauna (Hoon 1997; Venkataraman et al. 2003; Dhaneesh et al. 2009). In northern islands *Psammocora* spp. are common. In lagoons, blue coral *Heliopora coerulea* and *Millepora* spp. form the dominant species (Hoon 1997). *Halimeda* is the primary component of the lagoon sands and plays a crucial role in reef formation, along with other organisms such as foraminifers, echinoderms, bryozoans, annelids, and crustaceans. Pocillopora also occurs in great abundance, viz. *Pocillopora eydouxi* and *Pocillopora damicornis* are common (Venkataraman et al. 2003). *Fungia (Pleuractis) scutaria* (young form) of fungid group are also seen in reef areas. Some members of the *Montipora* group are also present. The most prolific growth of the corals occurs in the inner and outer sides of the reef areas. Eighty-six species of macrophytes, 81 brachyuran crabs, 10 anomuran crabs, 24 bivalves, 155 gastropods, 13 sea stars, 6 brittle stars, 15 sea urchins, 23 sea cucumbers, and 856 species of fish, along with 4 species of marine turtles, are so far reported from Lakshadweep (Hoon 1997; Ghosh 1991; Dhaneesh et al. 2009; Rajan et al. 2021). The green turtle *Chelonia mydas* and the hawksbill turtle *Eretmochelys imbricata* are also found in all the islands (Venkataraman et al. 2003).

7.2.3.2 Current Status

Coral reefs here have suffered damage from coral bleaching attributed to warmer sea temperatures, particularly on islands like Kavaratti and Kadmat. This has led to bacterial diseases. Another threat to the coral reefs is the presence of the crown-of-thorns starfish *Acanthaster planci* which was initially noted on Agatti Island in 1977, and since then, it has spread across numerous islands and reefs (Mallik 2017). Additionally, black and white band diseases have been observed in shallow coral areas, with reports also mentioning pink band disease. Besides coral bleaching, various factors contribute to the decline in reef biodiversity, including coral mining, dredging of navigational channels, unsustainable fishing practices, coastal development, souvenir collection, and other human population-related pressures. Regular nutrient monitoring has not revealed any sewage or oil pollution in the reef lagoons. Although the practice of blasting corals to create navigational channels has ceased, concerns arise from the construction of breakwaters on some islands, leading to potential coastal erosion issues. Recent observations indicate a decrease in fish catch within the reef lagoons, which could be attributed to coral loss from bleaching or increased harvesting due to population growth on the islands (Mallik 2017). Methods employed for catching live bait for the tuna fishery contribute to reef damage, impacting the local economy as the tuna fishery is vital to the islanders. The bleaching event in 1998 severely damaged coral cover around Lakshadweep, with varying reports estimating coral loss between 43–87% and 60–80% (Wafar 1999; Muley et al. 2000; Mallik 2017). Post-bleaching, live coral cover notably declined, reaching around 10% on Kadmat Island (Muley et al. 2000). Sporadic events of coral bleaching (Fig. 7.4) at local scales at Agatti Island has been reported by many authors (Vinoth et al. 2012; Kumar et al. 2012). The increase of SST has also resulted in the bleaching and mortality of coral-associated organisms like sea anemones (Fig. 7.5), giant clams, zoanthids, and foraminiferans (Kumar et al. 2012). The researchers have also documented the recovering of the bleached corals and anemones on decrease of SST (Vinoth et al. 2012; Kumar et al. 2012; Kumar et al. 2013). Saranya et al. (2022) has reported a significant increase in marine heatwaves, due to the rapid warming in the Indian Ocean and strong El Niños. Earlier, these heatwaves used to be rare in the tropical Indian Ocean but have now become an annual affair. The western Indian Ocean region experienced the largest increase in marine heatwaves at a rate of about 1.5 events per decade, followed by the north Bay of Bengal at a rate of 0.5 events per decade (Saranya et al. 2022). During 1982–2018, the

Fig. 7.4 Bleached coral at Lakshadweep (photo credit: Dr. T. T Ajith Kumar)

Fig. 7.5 Bleached sea anemone at Lakshadweep (photo credit: Dr. T. T Ajith Kumar)

western Indian Ocean had a total of 66 events, while the Bay of Bengal had 94 events (Saranya et al. 2022). Recently, works are being carried out on coral reefs in the Lakshadweep Sea which have undergone severe bleaching due to prolonged occurrences of marine heatwaves since October 2023. With shallow-water temperatures exceeding seasonal averages by 1.6 °C, researchers in Lakshadweep are observing widespread signs of coral bleaching and bleaching and mortality of sea anemones as well as giant clams.

7.2.4 Palk Bay

Coral reefs are situated in the Palk Bay near Rameswaram, centred at 9°17′ N and 79°15′ E. The fringing reef, stretching approximately 25–30 km in length, runs along the Pamban channel towards Rameshwaram Island.

7.2.4.1 Diversity

Sixty-five species of coral have been recorded from the area (Venkataraman and Rajan 2013; Edward et al. 2015). The dominant family is Acroporidae. Dominant species recorded were *Acropora cytherea*, *Porites solida*, *P. lichen*, *Favia pallida*, *Symphyllia radians*, *Platygyra* sp., *Leptoria* sp., *Favites* sp., *Goniastrea retiformis*, *Sarcophyton* sp., *Sinularia* sp., *Cladiella* sp., and *Lobophytum* sp. (Edward et al. 2015). *Dugong dugon* and marine turtles are found in the area.

7.2.4.2 Current Status

Coral bleaching was previously observed during 1998 and 2002 in Palk Bay. The recovery from bleaching in Palk Bay is slow. But the Indian Ocean tsunami during December 2004 reduced the live coral from 26.7% to 19.2% (Venkataraman and Rajan 2013). Sedimentation and siltation are high which have led to smothering of coral polyps and led to the growth of turf algae; there is filamentous, turf, and fleshy algae and the presence of excessive sediments over the coral surface (Ravindran et al. 2012). Visibility is poor due to siltation and it is influenced by the northeast monsoon. The recent increase in occurrences of heatwaves in the Bay of Bengal have resulted in the bleaching of corals in Palk Bay.

7.2.5 Gulf of Mannar

In the Gulf of Mannar, coral reefs have formed around a series of islands spanning approximately 140 km between Rameswaram and Tuticorin. A variety of reef structures, including shore reefs, platforms, patch reefs, coral pinnacles, and atolls, can be observed in this area. Fringing coral reefs and patch reefs are particularly prevalent around the islands.

7.2.5.1 Diversity

A comprehensive account of the coral fauna of this region is provided by Pillai (1986) consisting of 96 species of corals belonging to 36 genera. Later on, De et al. (2020) has reported 169 species consisting of 46 genera and 16 families. Dominant families are Acroporidae (58 species), Merulinidae (45 species) and Poritidae (19 species) and dominant genera of corals are *Acropora*, *Montipora*, and *Porites* (Venkataraman et al. 2003; Muley et al. 2000; De et al 2020). One hundred seventeen coral reef fish species under 79 genera and 42 families are reported from the area (Kumar et al. 2011a, b). In addition, green turtles *Chelonia mydas*, olive ridley turtles *Lepidochelys olivacea*, and *Dugong dugon* are also distributed in the area consisting of sea grasses.

7.2.5.2 Current Status

Though, bleaching and tsunami have damaged coral reefs of the area. It is reported that a few islands have submerged and vanished due to coral mining (Rajasuriya et al. 2005; Patterson Edward et al. 2005). Coral mining activities carried out before 1960s still have an impact on the ecosystem leading to siltation. Since the early 1960s, coral reefs in the region have been heavily exploited for industrial purposes, resulting in significant loss and submergence of islands like Vilanguchalli and Poovarasanpatti. Overfishing and the utilization of harmful fishing techniques persist as ongoing challenges in numerous areas within the region. Visibility in the area is impacted by monsoons, coral mining, and high sedimentation levels. Despite these challenges, the reefs in this region are more abundant and diverse compared to those in Palk Bay.

The reefs in the Gulf of Mannar and Palk Bay serve various purposes including reef fishing, chank and pearl fishing, ornamental shell trade, and illegal coral mining (Prakash et al. 2017). Local villagers in the vicinity of Palk Bay are involved in harvesting activities such as holothurians, seahorses, pipefish, chanks, and milkfish fry. Unfortunately, turtles and dugongs are also victims of poaching, with an estimated 1000 turtles being harvested annually (Hoon 1997). Extensive damage to the coral reefs in the area is attributed to over-exploitation of algae, shell collection, and boat-related damage during algae collection. Furthermore, the export of live crabs and lobsters and the use of fish traps contribute to the continued destruction of coral reefs. Problems such as siltation, agricultural run-off, sewage discharge, and faecal pollution exacerbate the challenges facing these ecosystems. The 1998 bleaching event led to a drastic increase in dead coral cover by 60–80%, particularly affecting branching corals like *Acropora* spp. and *Pocillopora* spp. (Venkataraman 2000). Massive corals now dominate the reefs, with branching corals nearly eradicated in certain island groups. The increase in heatwaves in the recent years has also resulted in coral bleaching events in the Gulf of Mannar. Since 1998, the reefs in both the Gulf of Mannar and Palk Bay have experienced multiple severe bleaching events, reflecting the ongoing and escalating challenges due to marine heatwaves. Major bleaching events have been recorded in 1998, 2010, 2016, and 2024, indicating a global pattern. Sadhukhan et al. (2022) carried out extensive underwater surveys from August 2018 to August 2020 in the Gulf of Mannar and observed severe coral bleaching in the Gulf of Mannar and Palk Bay during the summer of 2 consecutive years, 2019 and 2020, triggered by increased sea surface temperature (SST) of 4 °C–5 °C than the average temperature (28 °C–31 °C). The survey reported that 85% of the corals in the Gulf of Mannar got bleached after the marine heatwave in May 2020 (Sadhukhan et al. 2022). In the current year also, sporadic bleaching events are being reported from both Palk Bay and Gulf of Mannar.

7.2.6 West Coast

The west coast of India, stretching from Bombay to Goa, features submerged banks with isolated coral formations, as documented by Nair and Qasim (1978). Coral patches have been observed in the intertidal regions of Ratnagiri, Grande Islands in Goa, Malvan, and Gaveshani Bank, situated west of Mangalore, Angria bank off Malvan coast and Netrani Island in Karwar coast (Qasim and Wafar 1979; Nair and Qasim 1978) (Manikandan et al. 2016; Singarayan and Rethnaraj 2016; De et al. 2020; Viswambharan et al. 2021). In this chain, several islands exist including Vengurla Rocks at the southern tip and Sindhudurg Fort at the northern tip. Other small islets around Sindhudurg Fort are Mandel Rock, Malvan Rock, etc. There are numerous exposed rocky outcrops in this area. Pillai (1996) reported hermatypic corals along the coastline from Quilon in the Kerala coast to Enayam in Tamil Nadu.

7.2.6.1 Diversity

The major species reported are *Porites*, *Coscinaraea*, *Turbinaria*, some faviids, and *Pseudosiderastrea (Venkataraman et al. 2003)*. Coral patches are present from Quilon along the Kerala coast to Enayam in Tamil Nadu coast constituting mainly of hermatypic corals, dominated by *Pocillopora* spp. and *Acropora* spp. *Pseudosiderastrea* and *Porites* spp. (Fig. 7.3) are also common in the area. Pillai (1996) reported 29 species in 17 genera of scleractinians in this area. Of these, 13 species belonging to 6 genera are hermatypes, and the rest 16 species are ahermatypic (Jasmine et al. 2009). From Netrani waters, 14 species of corals belonging to 11 genera were observed (Zacharia et al. 2008) (Fig. 7.6). Along the coast of

Fig. 7.6 *Pocillopora* sp. in Netrani island

Bombay and Goa, Porites, *Coscinaraea*, *Turbinaria*, some favids, and *Pseudosiderastrea* are reported (Pillai 1996). All the genera recorded are massive or encrusting without any of representation of ramose forms (Pillai 1996). Anita Mary and Lazarus (2004) surveyed in 1997 along the southwest coast of India, from Kanyakumari to Vizhinjam, and identified 15 species of gorgoniids spanning 5 families and 11 genera. Particularly along the Vizhinjam and Enayam coast, the dominant species observed are *Pocillopora* sp. and *Montipora* sp. (Jasmine et al. 2009).

7.2.6.2 Current Status

The southwest coast of India, from Kanyakumari to Vizhinjam, is experiencing significant depletion of gorgoniid populations due to high demand for export and historical over-exploitation using bottom set nets (Anita Mary and Lazarus 2004). Along the west coast, siltation is of high rate and salinity may drop to 20 ppt during monsoon, restricting the growth of ecologically sensitive forms of ramose corals (Bakus et al. 1994). No comprehensive status survey has been conducted in the coral reef areas along the west coast of India. However, there are recent plans by the government of India to protect this region from anthropogenic disturbances. This initiative reflects growing recognition of the need to conserve and manage these marine ecosystems to mitigate human impacts and preserve biodiversity.

7.3 Threats to Coral Reefs

The health of corals, as deduced from the literature records since the 1960s, has been on a steady decline mainly due to anthropogenic interference and stress. Global warming is causing increased acidity in the oceans, which poses a threat to coral ecosystems. Additionally, the melting of glaciers worldwide is resulting in rising sea levels, leading to the submergence of coral reefs and inadequate sunlight for their growth, slowing their development. Coral reefs are extremely susceptible to thermal stress, which can lead to bleaching events characterized by the loss of symbiotic algae when sea surface temperatures (SSTs) surpass a certain threshold. Coral bleaching occurs when the tissue of corals turns white due to the partial or complete expulsion of zooxanthellae symbiotic algae. Factors such as increased irradiance, reduced salinity, and metal pollution also can contribute to bleaching, but the primary concern is elevated sea surface temperatures induced by rising global temperatures. The upsurge of stronger storms has led to oceanic waves that break and damage the coral reefs. They might even break colonies and make them unsafe to live. As the sea absorbs more and more CO_2, the pH levels continue to increase. High pH levels lead to weaker coral skeletons, and their vulnerability to diseases and destruction by storms. Various epidemic diseases such as white plague, white pox, white band, and black band pose significant threats to coral reefs, leading to extensive damage. In Indian reefs, recent reports indicate the emergence of diseases like white band, pink line, pink spot, white pox, yellow band, fungal blotch, white plague, narcotic patches, black band, and coralline lethal orange disease (CLOD) (Raghukumar and Raghukumar 1991; Muley et al. 2000).

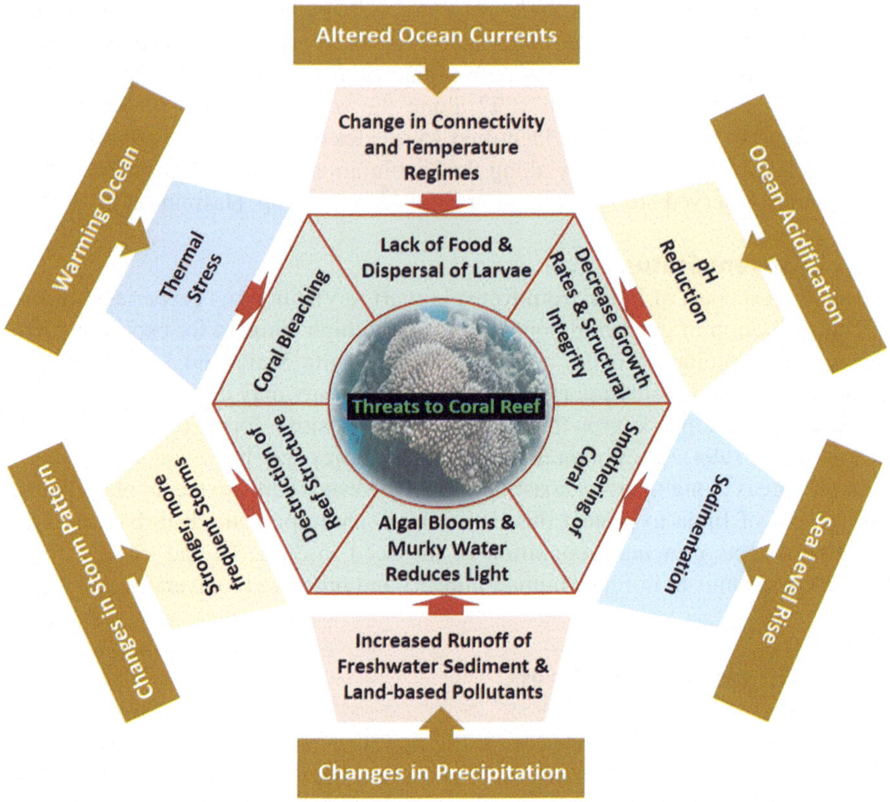

Fig. 7.7 Major threats of coral reefs in India

Illegal and destructive fishing practices, including blast fishing, contribute to reef degradation (Fig. 7.7). Overfishing removes herbivorous fish, leading to unchecked algal growth and a shift towards algal-dominated barren reefs devoid of fish and corals. The invasion of macroalgae such as *Kappaphycus alvarezii* and snowflake octocoral *Carijoa riisei* further exacerbates the damage to reef corals (Edward et al. 2012). Outbreaks of coral-eating crown-of-thorns starfish (*Acanthaster planci*) have also been reported in Indian coral reefs (Muley et al. 2000). Extreme natural events like cyclones, local tectonic upheavals, and tsunamis cause substantial destruction to corals. Besides, uncontrolled reef tourism contributes to degradation by inadvertently or intentionally damaging shallow-water corals through trampling and fragmenting (Kumara et al. 2008; Obura et al. 2008). Furthermore, anthropogenic activities such as illegal shell collection for souvenirs, unregulated harvesting, and smuggling of exotic marine organisms and reef-derived products remain major concerns for Indian reefs. Marine pollution, particularly from plastics and other contaminants, is suffocating coral reefs. Corals require ample space and clean water to thrive, but escalating pollution levels jeopardize their survival.

7.4 Conservation Strategies

In India, efforts have been made for the coral reef conservation through the establishment of marine parks and biosphere reserves. Properly managed MPAs and biosphere reserves protect habitat and contribute strongly to the conservation of biodiversity, as well as sustainable use. In the face of continued acceleration in the over-exploitation of marine ecosystems in many parts of the world, MPAs represent an essential part of any strategy for maintaining marine biodiversity. Examples are the Gulf of Kachchh Marine Park in Gujarat, Mahatma Gandhi Marine Park in Wandoor South Andamans, the Gulf of Mannar Biosphere Reserve, and Jhansi Rani Marine Park in Andaman and Nicobar. India has a total of 31 marine protected areas (MPAs), with five coral reef regions identified and surveyed for conservation purposes.

Eliciting local communities' participation towards conservation and sustainable use of marine resources has to be the focus area. Every effort should be made to ensure that local communities stand to gain economically and socially from the operation of protected areas. Attempts to exclude traditional human uses from protected areas may jeopardize the physical or economic survival of the people. Therefore, every effort should be made to ensure that local communities stand to gain economically and socially from the operation of MPA. There is also an urgent need for improvement in the capacities of local communities and of officials to manage human activities so that use of the marine environment is ecologically sustainable. Alternative or supplemental livelihoods are another approach to conserve the coral reef and strengthen community participation in reef management. NBFGR's initiative in this regard at Lakshadweep is a way forward in which the local communities are involved in the rearing of ornamental shrimps as an alternative livelihood. This helps for sustainable aquarium trade and conservation of coral reef ecosystem, besides promotion of livelihood to the islanders.

Legislation plays a crucial role in coral reef conservation in India. Various laws have been enacted by the government of India to protect coral reefs. For instance, all scleractinians and gorgoniids were brought under the Wild Life (Protection) Act of 1972 from July 2001. Conservation efforts are also enforced under the Marine Fisheries Regulation Act (MFRA) of 2000 and the Coastal Regulation Zone (CRZ) Notification of 2011, which legally protect Indian coral reefs. The Coastal Regulation Zone Notification, 1991, offers the only legal protection to all coral reefs. In this coral reef areas come under the CRZ1 category. A special category CRZ4 has been prepared for the islands of Andaman, Nicobar, and Lakshadweep. Norms for regulation of activities within the CRZ state that corals and sand from beaches and coastal water shall not be used for construction and other purposes. Dredging and underwater blasting in and around coral formations shall not be permitted. Section 7 (2) also states that construction of beach resorts/hotels shall not be permitted in ecologically sensitive areas such as marine parks and coral reefs (Notification 8.0114 (E) of 19 February 1991). The administration of Lakshadweep has already prohibited the collection of corals for construction purposes, but urgent action is needed to establish a national marine park in certain areas to safeguard the biodiversity of the atolls. To begin, designated zones where no interference is permitted should be established with the cooperation of local communities, respecting their traditional rights to utilize resources. Tourism in Lakshadweep, now permitted on a significant scale, must be strictly regulated as ecotourism. The coral reef ecosystem is destroyed while collecting marine ornamental shrimps from islands. As a part of fortification to the culture of marine ornamental invertebrates, the ICAR-National Bureau of Fish Genetic Resources established a germplasm resource centre for marine ornamental invertebrates at Agatti Island, Lakshadweep. The captive production of marine ornamental shrimps is carried out in the centre which would help to alleviate pressure on wild stocks. Community aquaculture units are established as a part of the project, where marine ornamental shrimps are reared by women islanders and marketed to earn profit (Jose et al. 2023). Marine ornamental shrimp rearing is a potential alternative livelihood option for coastal and island communities in India, and it will help conserve marine biodiversity.

Numerous studies have shown that damaged and degraded coral reefs can be restored using various ecological engineering techniques, such as coral translocation, transplantation of whole coral colonies, coral fragging or reef gardening, setting up nurseries, and deploying artificial substrates for reef development (Nanajkar et al. 2019). Coral reef restoration is a key conservation strategy aimed at rehabilitating degraded, damaged, or destroyed reef ecosystems to restore them to a healthy state and increase live coral coverage. Few studies on coral restoration activities were carried out in Tuticorin group of islands (Raj et al. 2015). Only a limited number of coral restoration experiments have been conducted in India, though some have achieved success (e.g. Kumar et al. 2017 in the Gulf of Kachchh). Recently, coral restoration project has been undertaken by the Ministry of Earth Science, and it is being implemented by the National Centre for Coastal Research in the dead reef areas in the 21 islands of GoM (Ramesh et al. 2020). Bleaching-resistant *Acropora formosa*, *A. millepora*, *A. muricata*, and *Favia favus* were selected for extensive transplantation, and a remarkable recruitment of corals has been observed thereafter (Ramesh et al. 2020). Recently, the Zoological Survey of India (ZSI), with help from Gujarat's forest department, is attempting for the first time a process to restore coral reefs using biorock or mineral accretion technology. Biorock method uses safe low-voltage electrical currents to grow natural limestone rock out of the sea on steel structures of any size or shape. They act as artificial reefs on which corals grow at very rapid rates (Goreau 2010).

Incorporating these approaches into coral reef conservation and management practices could yield tangible results beyond just monitoring coral decline due to local and global stressors. Considering the ecological benefits and economic reliance of coastal communities on coral reefs in India, the identification of appropriate locations for relocating and transplanting live corals from potential coastal development sites could contribute to the preservation and sustainability of coral reefs.

Despite some successful coral restoration efforts in India, such initiatives are still in their early stages. A successful large-scale coral restoration programme could positively impact coastal protection, fisheries, tourism, and overall ecosystem services. Such programmes require sound scientific understanding, careful planning, and long-term monitoring of restored reefs. Although costly, they are more economical than the consequences of inaction. However, cautious implementation is necessary to avoid failures due to insufficient scientific understanding and planning.

Initiatives and involvement from long-term beneficiaries and corporate entities are essential, rather than relying solely on government agencies. Enforcing the 'polluter pays' principle has the potential to generate financial resources for the conservation and restoration of coral reefs, thereby easing the burden on state finances. Restoration activities can be considered part of corporate social and environmental responsibility, promoting ecological justice. This should stimulate a conscious debate among experts, corporates, policymakers, and managers in India and other developing nations to establish appropriate reef restoration targets based on their resources and capabilities.

7.5 Introduction to Mangrove Ecosystem

Mangroves are highly productive ecosystems and serve as custodians of their juvenile stock (Odum, 1971). This term encompasses a group of salt-tolerant plant species forming forests in coastal areas, offering diverse ecological and economic benefits while supporting various coastal and marine ecosystems. They are the only 'blue carbon forests of the ocean' distributed in tropical and subtropical between 30° S to 38° N in the world covering 15.2 million hectares (Mathew et al. 2010; Kathiresan 2018). Mangroves serve multiple purposes, including providing wood for construction, fuel, fodder, bark for tannin extraction used by fishermen to dye fishing nets, and edible fruits and shoots as well as highly valued commercial products and fishery resources. They also have medicinal uses and serve as natural barriers against disasters like the 2004 tsunami in Tamil Nadu, where mangrove-rich areas played a protective role. Along extensive coastlines, mangrove vegetation plays a crucial role in preserving coastal biodiversity. These ecosystems teem with diverse flora and fauna and serve as vital nurseries and breeding grounds for fish, shellfish, crustaceans, molluscs, birds, reptiles, and mammals (Fig. 7.8). Mangroves function as structural engineers, equipped with features like pneumatophores, salt excretion glands, vivipary, and buoyant propagules that enhance their survival and recruitment in environments that are poorly aerated, saline, dynamic, and subject to tidal flooding. These plants are highly efficient in their use of nitrogen and nutrient resorption. Mangroves generate significant amounts of detritus, such as leaves, twigs, and bark, which are either buried in waterlogged sediments or consumed by crabs and gastropods. This detritus is further broken down by meiofauna, fungi, and bacteria, transforming it into a protein-rich food source for other organisms within the mangrove and estuarine ecosystems. They protect against erosion and also serve as sites for brackish water shrimp culture, aquaculture ponds, and developing burgeoning ecotourism. They help in carbon sequestration and thus in mitigating impacts of global warming and climate change. Mangroves are the best bioindicator of environmental pollution and the health of the coastal ecosystem.

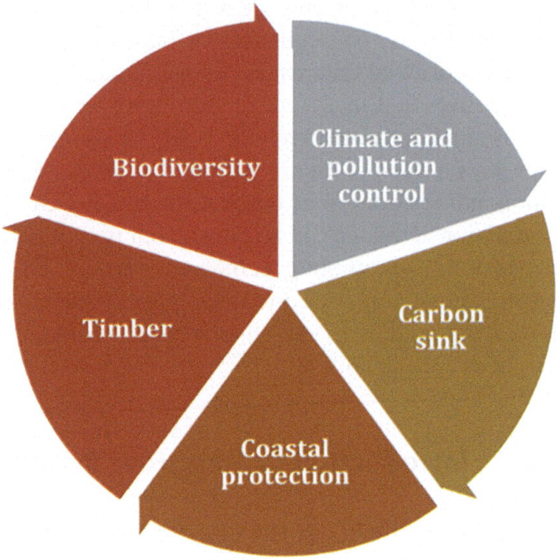

Fig. 7.8 Ecosystem services of mangroves

7.6 Indian Scenario: Diversity and Status

India has a vast coastline spanning approximately 8118 km, including its island territories, with a mangrove cover extending over about 4992 km^2, making it the world's fourth largest mangrove area (FSI 2021). These mangrove habitats, situated between 69°E–89.5°E longitude and 7°N–23°N latitude, consist of three distinct zones: the east coast facing the Bay of Bengal (with a coastline of around 2700 km), the west coast bordering the Arabian Sea (with a coastline of about 3000 km), and the island territories (with approximately 1816.6 km of coastline) (Ragavan et al. 2016). In India, states like West Bengal, Odisha, Andhra Pradesh, Tamil Nadu, Andaman and Nicobar Islands, Kerala, Goa, Maharashtra, and Gujarat hold significant mangrove areas (Fig. 7.9; Table 7.2). Approximately 58% of mangroves are found on the east coast along the Bay of Bengal, 29% on the west coast along the Arabian Sea, and 13% on the Andaman and Nicobar Islands (Ragavan et al. 2016). Sundarbans has the largest mangrove cover (23% of total cover in India) followed by Gujarat (23%) (Kathiresan 2018).

The total mangrove area is classified into very dense (canopy density over 70%), moderately dense (canopy density between 40 and 70%), and open mangrove cover (canopy density between 10 and 40%) (FSI 2021). The gradual topography along the east coast, particularly in deltaic regions, facilitates extensive intertidal expanses conducive to significant mangrove formation (Jagtap and Komarpant 2003). Mangrove coverage is more extensive on the east coast compared to the west coast due to terrain and slope variations, as well as the presence of nutrient-rich alluvial soil in river deltas like the Ganges, Brahmaputra, Mahanadi, Godavari, Krishna, and Cauvery.

Fig. 7.9 Map depicting mangrove areas in India

7.6.1 Diversity

India ranks third globally in terms of mangrove biodiversity richness, following Indonesia and Australia. Indian mangrove ecosystem consists of a total of 3985 biological species including 919 florae and 3066 faunae (Mathew et al. 2010). The Indian mangroves comprise approximately 44 species in 29 families (Untawale 1986; Mathew et al. 2010; Kathiresan 2018). Important mangrove species are *Avicennia officinalis*, *Rhizophora mucronata*, *Sonneratia alba*, *Avicennia alba*, *Bruguiera cylindrica*, *Heritiera littoralis*, *Phoenix paludosa*, *Morinda citrifolia*, and *Ceriops tagal.* Of these, 34 species belonging to 25 genera and 21 families are present along the west coast (Mathew et al. 2010). There are about 25 mangrove species which have restricted distribution along the east coast and are not found on the west coast (Mathew et al. 2010). Similarly, there are eight species of mangroves like *Sonneratia caseolaris*, *Suaeda fruticosa*, *Urochondra* setulosa, etc. which have been reported only from the west coast (Kathiresan and Qasim 2005; Kathiresan

Table 7.2 Mangrove cover assessment 2021 (in km^2)

State/UT	Very dense mangrove	Moderately dense mangrove	Open mangrove	Total
Andhra Pradesh	0	213	192	405
Goa	0	21	6	27
Gujarat	0	169	1006	1175
Karnataka	0	2	11	13
Kerala	0	5	4	9
Maharashtra	0	90	234	324
Odisha	81	94	84	259
Tamil Nadu	1	27	17	45
West Bengal	994	692	428	2114
Andaman and Nicobar Islands	399	168	49	616
Daman and Diu	0	0	3	3
Puducherry	0	0	2	2
Total	**1475**	**1481**	**2036**	**4992**

Source: India State of Forest Report, Forest Survey of India (2021)

2010). The associated mangrove flora is quite common to both the coasts, with minor variations in distribution. The floral diversity of mangroves of India is comprised of 38 core mangrove species (Kathiresan 2003). Mangrove ecosystems along India's east coast and the Andaman and Nicobar Islands showcase notable density and floral variety. They flourish in diverse conditions: the Sundarbans benefit from high humidity and ample rainfall, supporting a wealth of biodiversity, whereas Gujarat's mangroves face arid climates with limited rainfall, leading to reduced biodiversity. Bhitarkanika, located in Odisha, is renowned as a global haven for mangrove genetics. It also hosts the largest populations of birds and crocodiles, including albino crocodiles.

There are different types of faunal communities in mangrove waters which are dependent on the water component in one way or the other (Table 7.3). The planktonic and benthic animal communities play a very important role in the mangrove ecosystem just like the terrestrial animals. There are a total of 105 species of fish which are typical mangrove dwellers in India (Kathiresan 2010). The fishes are represented by several species like the mud skippers, carangids, clupeids, serranids, mullets, hilsa, sea bass, milkfish, and others. There are different species of crustaceans like *Penaeus indicus*, *P. merguiensis*, and *P. monodon*, while the crabs are represented by *Uca* sp., *Scylla serrata*, *Thalassina*, etc. The mangrove swamps of India are favoured by a variety of birds, both migratory and resident.

7.6.2 West Bengal

The Sundarbans, located on the east coast of India, constitute a significant portion of the country's mangrove forests, spanning approximately 9600 km^2 of mangrove

Table 7.3 Number of floral and faunal species in mangrove ecosystems of India

	Group	Number of species
Faunal groups	Prawns and lobsters	55
	Crab	145
	Insects	661
	Molluscs	337
	Other invertebrates	745
	Fish parasites	7
	Fin fish	554
	Amphibians	13
	Reptiles	84
	Birds	513
	Mammals	68
	Total faunal species	**3182**
Floral groups	Mangroves	44
	Mangrove-associated plants	86
	Seagrass vegetation	11
	Marine algae	557
	Bacteria	69
	Fungi	103
	Actinomycetes	23
	Lichens	32
	Total floral species	**925**
Total number of species		**4107**

Source: Kathiresan (2018)

forest and water of which 65% is in Bangladesh (FAO 2003). This area is characterized by numerous islands formed by sediments deposited by three major rivers: Ganga, Brahmaputra, and Meghna along with a dense network of smaller rivers, channels, and creeks. Indian Sundarbans has an area of 1781 km^2 under water and 4225 km^2 in reserved forest (Mathew et al. 2010). Sundarbans is internationally recognized as the World Heritage Site of UNESCO. The Gangetic Delta in which Sundarbans is situated is the largest wetland with the tightest sedimentation in the world. Sundarbans in India and Bangladesh put together is the largest mangrove forest, and it is also the only mangrove colonized by the royal Bengal tigers, in the world. This is the dense mangrove forest, unique in extending 100 km from the seafront, and the flora and fauna residing here have amazingly adjusted to the rigorous ever-fluctuating tidal environment. Sundarbans supports globally threatened species such as tigers, fishing cat, Gangetic dolphin, estuarine crocodile, horseshoe-crabs, monitor lizard, and river terrapins. Mangroves dominate the Sundarbans, with around 36 true mangrove species belonging to 21 genera and 14 families which are identified in the area according to Debnath and Naskar (1999). One hundred seventy-seven species of fishes are reported from the area along with estuarine crocodiles (*Crocodylus porosus*), marsh crocodiles (*Crocodylus palustris*), fishing cat, deer, wild pig, and marine turtles (*Lepidochelys olivacea*, *Eretmochelys imbricata*, and *Chelonia mydas*). Salinization is one of the major causes of mangrove degradation in Sundarbans resulting in alteration of species composition of mangroves. The population of salt-sensitive species, like *Heritiera fomes*, *Nypa fruticans*, *Xylocarpus*, and *Phoenix paludosa*, are replaced by salt-tolerant species like *Avicennia*, *Excoecaria agallocha*, and members of Rhizophoraceae in the area (Chowdhury et al. 2019; Kathiresan 2024).

7.6.3 Gujarat

Gujarat has historically been home to a vast and diverse mangrove ecosystem. It has the second largest mangrove cover along the Indian coast, following the Sundarbans. Approximately 23% of India's estimated mangrove cover, which amounts to 1175 km^2, is found in Gujarat. The coastal district of Kachchh alone accounts for nearly 90 percent of the total mangrove cover in the state. The Gujarat coast is reported to harbour approximately 16 mangrove species (Bhatt et al. 2009).

7.6.4 Andaman and Nicobar Islands

The mangrove occurs mostly along fringing the creeks, muddy shores, and backwaters, constituting 9.4% of land area (Mathew et al. 2010). The mangroves present here are the most diverse with 38 species belonging to 13 families (Ragavan et al. 2016). Ragavan et al. (2015) documented 35 species from Andaman Islands and 21 true mangrove species from Nicobar Island. The December 2004 tsunami has adversely affected the area.

7.6.5 Lakshadweep

The Lakshadweep Islands are devoid of mangroves except at Minicoy. Mangroves occur in patches in area. Unlike the mangroves of the mainland, the mangroves of Minicoy are in the formative stage and free from serious human pressure. The species reported are *Bruguiera cylindrica*, *Ceriops tagal*, *Pemphis acidula*, and *Avicennia marina* (Nasser et al. 1999; Ragavan 2016).

7.6.6 Maharashtra

Mangroves in Maharashtra extend across the tidal river creeks and backwaters of various regions including Achara, Devgad, Vijaydurg, Ratnagiri, Kundalika, and Mumbra Diva in the Mumbai area (Fig. 7.10). Additional locations include Veldur, Vikroli, Shreevardhan, Vaitarna, Vasai-Manori, and Malvan. Maharashtra is home to a total of 22 species identified as authentic mangrove species, representing 15 genera and 11 families (Kathiresan and Qasim 2005; Kulkarni and Bhosale 2021).

Fig. 7.10 Mangrove patch in Ratnagiri, Maharashtra

7.6.7 Karnataka

In Karnataka, mangroves are found primarily in Uttar Kannada and Dakshina Kannada districts. In Dakshina Kannada, key estuarine areas supporting mangroves include Netravathi-Gurupur, Mulki-Pavanje, Udyavara-Pangala, Swarna-Sita-Kodi, Chakra-Haladi-Kollur, Baindur hole, and Shiroor hole. In Uttar Kannada, mangroves are located in estuarine complexes such as Honovar, Venkatapur, Sharavathi, Aghanashini, Gangavali, and Kali rivers. A total of 16 species, spanning 11 genera and 7 families, are identified as authentic mangrove species in Karnataka (Ragavan et al. 2016). Notably, *Bruguiera cylindrica*, *Lumnitzera racemosa*, and *Acrostichum aureu*m were recently documented in Karnataka by Nayak and Andrade (2008).

7.6.8 Odisha

In Orissa, the mangrove area spans approximately 200 km, with recent estimates indicating a degradation of about 20 km over a decade. The mangroves of the Odisha coast are distributed in three zones, i.e. Mahanadi Delta, Brahmani, and Baitarani Delta, i.e. the Bhitarkanika mangrove zone and Balasore coast. Totally 35 true mangrove species belonging to 20 genera and 14 families have been recognized in mangroves of the Odisha (Ragavan et al. 2016). These mangrove forests receive water from three rivers, are known for their rich species diversity, and are characterized by dense, tall trees akin to those found in the Sundarbans (Selvam et al. 2003). Bhitarkanika mangrove is the most important mangrove zone in Odisha with an area of 145 km^2 having 64 plant species, 28 true mangroves, 4 mangrove associates, and 32 others (Singh et al. 2012). Saltwater crocodile and olive ridley reside in this area. The total sanctuary area in the state is 672 km^2, of which mangrove forests make up 130 km^2.

7.6.9 Puducherry

In Puducherry, 7 true mangrove floral species from 3 families have been identified, while approximately 16 mangrove-associated floral species from 12 families are reported in the inundated and adjacent regions (Ragavan et al. 2016). Regarding macrofauna, the mangroves in Pondicherry harbour a diverse fish population, with 39 species belonging to 24 families under 7 orders. (Kathiresan and Qasim 2005).

7.6.10 Kerala

Mangrove patches are mainly distributed in intertidal areas of Kochi, Vembanad, Kollam, Thiruvananthapuram, Kannur, Kozhikode, and Kottayam. A total of 19 species belonging to 12 genera and 8 families have been recognized as true mangroves in Kerala (Singh et al. 2012; Ragavan et al. 2016). The mangroves on the southernmost west coast of India are dominated with species tolerant of low salinity such as *Acanthus ilicifolius*, *Avicennia officinalis*, *Excoecaria agallocha*, *Bruguiera cylindrica*, and *B. gymnorhiza* (Selvam et al. 2003). In the Ayiramthengu mangroves of Kollam district, Kerala, located on the west coast, researchers have identified 27 fish species from 17 families (Ragavan et al. 2016). Among these families, Mugilidae has the highest representation with four species, followed by Carangidae and Cichlidae, each with three species. Ambassidae, Gerreidae, and Siganidae are represented by two species each, while the remaining 11 families are each represented by a single species (Jisha et al. 2004).

7.6.11 Andhra Pradesh

Andhra Pradesh has approximately 582 km of mangrove area, with the Coringa Wildlife Sanctuary serving as a refuge during calamities, as noted by MSSRF (2005) and WWF (2005). The mangrove forests in Andhra Pradesh are located in the estuaries of the Godavari and the Krishna rivers, in East Godavari, Krishna, and Guntur districts. Apart from these, mangroves are also found in small patches along the coast of Visakhapatnam, West Godavari, and Prakasam districts. The mangroves in Andhra Pradesh exhibit moderate diversity, with 35 mangrove species identified in the Godavari and Krishna estuaries (Singh et al. 2012; Ragavan et al. 2016). Among these, 22 are true mangroves requiring both sea and river water, while 13 associated species are capable of surviving in both terrestrial and estuarine conditions (Ragavan et al. 2016). These wetlands are abundant in crustaceans, molluscs, and finfish resources.

7.6.12 Goa

Of the 130 km coastal wetlands in Goa, approximately 67.30 km comprises mudflats and mangroves. These mangroves grow along the narrow intertidal mudflats along estuary banks and have a fringing nature, attributed to the rising topography of the coast. Unfortunately, these habitats are facing threats from reclaiming for urbanization and agricultural purposes. In terms of mangrove flora, Goa is home to 15 species belonging to 11 genera and 7 families, with dominance by species like *Rhizophora mucronata*, *Avicennia alba*, *Sonneratia alba*, *S. caseolaris*, *Excoecaria agallocha*, and *Acanthus ilicifolius*. *Kandelia candel*, found in the oligohaline region, has better formation in Goa compared to other states (Jagtap and Singh 2004).

7.6.13 Tamil Nadu

Tamil Nadu, one of India's nine maritime states, has the second longest coastline in the country, stretching over 1076 km. The primary mangrove wetlands in Tamil Nadu include Pichavaram (Fig. 7.11) and Muthupet, with the river Cauvery serving as the main source of fresh water and the Gulf of Mannar. The total area covered by mangrove ecosystems in Tamil Nadu is approximately 225 km^2. A total of 17 true mangrove species belonging to 12 genera and 8 families have been recognized from Tamil Nadu (Ragavan et al. 2016). The common dominant mangroves found in Tamil Nadu are *Rhizophora apiculata*, *Rhizophora mucronata*, *Sonneratia apetala*, *Avicennia marina*, *Avicennia officinalis*, *Bruguiera cylindrica*, *Ceriops decandra*, *Aegiceras corniculatum*, and *Lumnitzera racemose* (Arunprasath and Gomathinayagam 2014). Among these, one of the largest and most pristine mangrove forests is located at Pichavaram in Cuddalore district, spanning an area of 1100 km^2 (Venkataraman 2006, 2008). The Vellar-Coleroon estuarine complex

Fig. 7.11 Mangrove patch in Pichavaram, Tamil Nadu

forms the Killai backwater and Pichavaram mangrove. Thirteen species of mangroves, with *Avicennia marina* and *Rhizophora* as predominant species, are reported along with 177 fish species, 40 species of meiobenthos, 52 species of macrobenthos, and 200 species of birds from Pichavaram (Kathiresan 2000; Kathiresan and Qasim 2005). Eight species of mangroves are reported from Muthupet region and nine species from Gulf of Mannar region (Arunprasath and Gomathinayagam 2014). The floral community structure of the Muthupet mangroves and other small patches to its south is overwhelmingly dominated by a single species, *Avicennia marina*, as noted by Azariah et al. (1992). This dominance is primarily due to the hypersaline conditions prevalent along the southernmost east coast of India (Selvam et al. 2003).

7.7 Threats to Mangrove Ecosystem

The pressure on wetlands has intensified due to rapid population growth, technological advancements, urbanization, and economic expansion. Natural factors like droughts, hurricanes, occasional tsunamis, intense grazing, infestation by pests, and erosion, coupled with human activities such as over-exploitation, construction of jetties, ports, encroachment, pollution, and reclaiming large wetland areas for agriculture, aquaculture, commercial, and residential purposes, have significantly altered wetland functioning in recent years (Fig. 7.12). Oil pollution poses a

Fig. 7.12 Major threats to mangrove ecosystem

significant threat to mangroves as it can cover pneumatophores, leading to their suffocation and ultimately causing mortality (Jagtap and Nagle 2007). Despite their critical ecological importance, mangroves were historically viewed as undesirable vegetation primarily exploited for timber and charcoal, resulting in substantial depletion over the past two decades. Recent surveys suggest that deforestation has led to the loss of approximately 44% and 26% of mangroves along the west and east coasts, respectively, highlighting the urgent need for a comprehensive understanding of mangrove ecosystem biodiversity (Mathew et al. 2012; Ragavan et al. 2016). Changes in the distribution and abundance of mangroves are primarily driven by habitat loss due to sea-level rise (Woodroffe 1990). Mangrove destruction disrupts the sediment budget, leading to coastal erosion. In Kerala, the construction of embankments to prevent the entry of seawater and to raise coconut trees has resulted in mangrove destruction. Similarly, shrimp farming activities have caused severe mangrove destruction in Andhra Pradesh, Orissa, and Tamil Nadu. The Sundarbans mangroves are degrading rapidly due to silt deposition, increased salinity, low water tables, issues associated with acid sulphate soils, coastal erosion, discharge of industrial wastes with arsenic pollutants from the Hooghly River, and the dumping of plastic waste, resulting in a decline in fishery resources and loss of biodiversity (Kathiresan 2024). Olive ridley turtles, once abundant in the area, have decreased recently due to island submergence (Kathiresan 2024). The population of invasive species like *Prosopis juliflora* and *Derris trifoliata* is increasing in the area leading to loss of the mangrove biodiversity (Kathiresan 2024). Another factor that threatens the mangroves of South India and Sri Lanka is invasive species. In South India, *Prosopis juliflora*, an exotic species of shrub (family Fabaceae) that can tolerate a wide range of environmental conditions, including high temperature, low fresh water availability, and high salinity, is making entry into mangroves gradually, though currently they are present on a large scale in the periphery of mangroves.

Mangrove ecosystems play a crucial role as nurseries for economically important fish and prawn species, producing a significant amount of detritus food that sustains a diverse range of young fish and shrimps. However, mangroves in private areas or khazan lands face severe threats from the conversion of these lands for aquaculture, agriculture, and other development activities. Reclamation efforts further harm saplings and small mangrove plants in general.

7.8 Conservation and Management of Mangroves

India is at the forefront of bioprospecting mangroves, which are a valuable source of salt-resistant genes, novel chemicals, and high-value products such as (1) mosquito repellents, (2) black tea beverages, (3) polysaccharides for preventing human immunodeficiency virus (HIV) infection, (4) lignins for combating oral and cervical cancers, (5) anti-diabetic extracts, (6) rapid nanoparticle synthesis, and (7) hair growth stimulants (Kathiresan and Ravikumar 2010). Continued research in this area is expected to lead to the development of patents, processes, and valuable products, including medicines.

Initiatives to conserve and reforest mangroves along the central west coast of India were launched to raise public awareness about their importance, control intertidal mud banks, promote forestry and social forestry activities, increase biomass along estuaries to enhance biological productivity, and improve bird and animal habitats (Untawale 1996). National Mangroves Committee was established in 1976 to advise the government on matters related to the conservation and development of mangroves. The committee recommended surveying existing mangrove areas, and subsequently, a scheme was formulated for mangrove conservation and protection.

The Goa government has prohibited the felling of 15 species of mangroves under the Goa, Daman and Diu Preservation of Trees Act, 1984. These areas were declared ecologically sensitive under the Environment (Protection) Act, 1986, and development activities and waste disposal were banned following the Coastal Regulation Zone (CRZ) Notification 1991. In 1986, the Ministry of Environment, Forest and Climate Change formulated a plan-scheme for the conservation and management of mangroves and coral reefs and established a national committee to advise the government on relevant policies and programmes. Based on their recommendations, 15 mangrove areas in the country were identified for intensive conservation (Anon 1997; Jagtap et al. 2002). A key consideration in developing a management strategy for mangroves is creating buffer zones to limit anthropogenic activities around the wetland corridor, thus restoring its natural functioning (Castelle et al. 1994). The government has taken steps to protect, sustain, conserve, and enhance forests in the country through both promotional and regulatory measures. Promotional measures are implemented via a central sector scheme under the National Coastal Mission Programme on 'Conservation and Management of Mangroves and Coral Reefs'. This programme includes the formulation and implementation of annual management action plans (MAP) for the conservation and management of mangroves in all coastal states and union territories. Regulatory measures are enforced through the Coastal Regulation Zone (CRZ) Notification (2019) under the Environment (Protection) Act, 1986; the Wild Life (Protection) Act, 1972; the Indian Forest Act, 1927; the Biological Diversity Act, 2002; and the rules under these acts, as amended periodically. The World Wide Fund for Nature (WWF) India has engaged citizens in nine states—Maharashtra, Goa, Gujarat, Andhra Pradesh, Tamil Nadu, Kerala, Odisha, West Bengal, and Karnataka—in mangrove conservation through the Magical Mangroves campaign. The government, under the Centrally Sponsored Scheme for Conservation & Management of Mangroves, provides assistance to coastal states and union territories for implementing action plans, including surveys and demarcation, alternative and supplementary livelihoods, protection measures, and education and awareness activities. The ministry piloted an integrated coastal zone management project in coastal stretches of three states, namely, Gujarat, Odisha, and West Bengal, with objective of conservation and protection of coastal resources which included the plantation of mangroves as one of the major activities.

Until the 1960s, Gujarat's mangroves were considered economically unproductive and faced destruction due to economic and industrial development (Hirway and Goswami 2007). However, in recent years, significant efforts have been made by the state government and international agencies to restore and regenerate mangrove stocks in Gujarat. There are reports of an increase in mangrove cover in Gujarat primarily due to plantation and protection measures implemented by the state. Gujarat has undertaken a major mangrove plantation drive along the coastal regions of Kutch and Jamnagar under the Gujarat Forestry Development Project (Upadhyay et al. 2015).

It is essential to conduct thorough monitoring to prevent illegal activities like poaching of mangrove fruits, fishing, and movement of barges, which can damage young plants and deplete fish germplasm. Fisheries should be managed with proper vigilance and legislation to avoid harm to existing mangroves, and the speed of barges should be regulated. Aquaculture should be practiced in thinly populated mangrove areas, and proper planning for activities such as agriculture, construction, mining, and industrialization should include surveys, staff for monitoring, and restoration efforts.

In Sundarbans the government of West Bengal has undertaken an innovative initiative of developing a 'green-grey hybrid' by combining a hard engineering structure with soft mangroves as embankments to prevent seawater intrusion under the aegis of the World Bank (Kathiresan 2024). This hybrid model is a nature-based solution for the natural regeneration of mangroves, besides supporting fish production, carbon sequestration and coastal protection, as well as reducing the maintenance cost of embankments. This model has been proven successful in Indonesia, Thailand, Guyana, Suriname, and Vietnam (SAIME 2023).

The participation of people is vital for the conservation of mangrove ecosystem. The management plans of the National Mangroves Committee under the Ministry of Environment and Forestry includes afforestation, regeneration of mangrove areas, protective measures, and eco-development. Natural and artificial regeneration have to be initiated in degraded areas. Afforestation involves planting trees in new areas, while restoration focuses on replanting in damaged, degraded, or destroyed areas. Among the types of restoration, natural restoration and hydrological restoration are preferred over artificial restoration through planting (Kathiresan 2018). In most cases fast-growing species such as *Avicennia* and *Sonneratia* are preferred (Kathiresan 2018). Mangrove planting efforts have largely been unsuccessful due to the high mortality rate of planted seedlings. Consequently, 'ecological restoration' is also recommended, which involves creating the right conditions for mangroves to regenerate naturally (Kathiresan 2018). This approach requires a thorough understanding of local ecological conditions and hydrological patterns to restore hydrology and eliminate barriers to natural regeneration.

In Tamil Nadu and Andhra Pradesh, high salinity of dry soil due to lack of regular tidal flushing is another cause of mangrove degradation. The canal bank planting technique was used to overcome this. Formation of canals helps in regular tidal inundation. Canal bank planting with 'fish bone' design for mangrove restoration was carried out by the M. S. Swaminathan Research Foundation (Chennai) and the forest department in Muthupet and Pichavaram in Tamil Nadu as well as East Godavari and Krishna districts of Andhra Pradesh (Ramasubramanian and Ravishankar 2004; Baruah 2004; Kathiresan 2018). In this technique, canals are formed so that the high-saline soil gets regular tidal inundation.

The Community-Based Ecological Mangrove Restoration (CBEMR) method has achieved global success by offering an effective and sustainable approach to mangrove restoration (Howard et al. 2023). Unlike many traditional planting projects, CBEMR emulates natural processes to rehabilitate degraded mangrove areas. This method focuses on building skills and empowering local stakeholders and communities, teaching them how to restore mangrove forests. Involving local communities is crucial to the success of these projects, ensuring that those living in the area actively participate in the restoration efforts. By engaging coastal communities, CBEMR fosters a sense of stewardship, encouraging locals to take ownership of the restoration work and sustain the long-term benefits of the ecosystem (Howard et al. 2023). In India, the states of Tamil Nadu, Odisha, Andhra Pradesh, West Bengal, and Gujarat prominently manage mangroves through community-based co-management. With financial backing from the India-Canada Environmental Facility (ICEF) in New Delhi, the M. S. Swaminathan Research Foundation implemented Joint Mangrove Management (JMM) for the restoration and conservation of mangroves, engaging local communities alongside forest departments (Kathiresan 2018). This project involved 5240 families across 28 villages along India's east coast and restored approximately 1475 hectares of mangroves by planting 6.8 million saplings. To empower local communities, 194 self-help groups were established to carry out poverty alleviation programmes, including supplementary income-generating activities for firewood, fodder, fencing, and house construction (Kathiresan 2018). A similar initiative was undertaken by the Gujarat Ecology Commission, also with ICEF's financial support, promoting community-based regeneration and management of about 5000 hectares of mangroves in the Gulf of Kachchh and the Gulf of Khambhat over 5 years, from 2001 to 2006 (Kathiresan 2018). Similar kind of mangrove restoration programmes were conducted in the Vellar Estuary involving students and local stakeholders. In this, mature propagules of local mangroves (*Rhizophora* sp. and *Avicennia* sp.) were planted along intertidal areas of the Vellar Estuary. There was a large increase in fish resources because of mangrove restoration, especially shrimps and crabs which attracted more community participation in the project. Such efforts involving community participation is needed for conservation and management of mangroves. Many mangrove restoration projects have failed because of a lack of community involvement, inappropriate governance structures, and a failure to align objectives and goals of external agents with those of local stakeholders. Restoration is a social enterprise and local leadership is the key. Projects often fail without sufficient community and political support to sustain management in the long term. Community-level stakeholder participation and co-management are the key to success (Howard et al. 2023).

The government of Maharashtra constituted 'Mangrove Cell' and then the 'Mumbai Mangroves Conservation Unit' in 2013 to protect mangroves in Mumbai and the adjoining areas and elevated the status of mangrove forests on government land from 'protected forests' to 'reserved forests'. This led to the notification of 15,088 hectares of mangroves on government land as 'reserved forests' in 7 districts of Maharashtra. The Mangrove Cell also achieved plantation in more than 200 ha of mangrove-degraded areas in Greater Mumbai. As a result of all these efforts, the mangrove cover increased phenomenally by 20% between 2013 and 2015 (Vasudevan 2017; Kathiresan 2018). Further, the Mangrove Cell undertook the UNDP-GEF project on 'Mainstreaming of Coastal and Marine Biodiversity' into production sectors in Sindhudurg district during the mid-2012. Under this project, the cell implemented several programmes such as the deployment of artificial reefs, transplantation of corals. Identification and protection of sea turtle nesting sites, release of the turtle hatchlings into the sea, and survey of corals, coastal birds, sea snakes, otters, monitor lizards, and marine mammals (dolphins and finless porpoises) (Vasudevan 2017; Kathiresan 2018). The Mangrove Cell has also promoted crab, oyster, and sea bass fish farming practices in the mangrove waters, as supplementary livelihood for the benefit of local people. In addition, the Mangrove Cell implemented GIZ project 'Sustainable Management of Coastal and Marine Protected Areas' in 2014 under 'International Climate Initiative Agreement' which led to the notification of the 'Thane Flamingo Sanctuary', which covers an area of 1690 ha and supporting 10 mangroves and 200 bird species. The Mangrove Cell has successfully demonstrated a cross-sectoral approach towards marine and coastal conservation, in partnership with leading national institutions, agencies, NGOs, and with a wide range of government departments. ICAR-NBFGR in association with the Mangrove Cell has a project on 'Setting up of marine ornamental fish village at Maharashtra: Way forward to promote livelihood to mangrove dwellers and biodiversity conservation'. As a part of the project, ICAR-National Bureau of Fish Genetic Resources established a master breeding facility at the premises of the Coastal and Marine Biodiversity Centre, Mangrove Foundation, government of Maharashtra at Airoli, for mass-scale seed production of indigenous clownfishes for the livelihood development of the fishers of coastal Maharashtra; ten cluster mode rearing units of clownfishes are under operation in five coastal districts. One hundred two families have adopted clownfish-rearing protocols successfully and are earning regular income. About 1.5–2-month-old clownfish seeds produced from the facility are being supplied to the beneficiary rearing units free of cost. The supplied seeds are reared for 2 months by the beneficiaries and marketed for enhancing their income (ICAR-NBFGR 2022). The government of Maharashtra has recently set up a registered body of the 'Mangrove and Marine Biodiversity Conservation Foundation of Maharashtra', to promote conservation and management (Vasudevan 2017). In Tamil Nadu, the integration of sea bass culture and crab fattening in Integrated Mangrove Fishery Farming System (IMFFS) is a major livelihood intervention introduced among tribal fisherfolk in the backwaters of the Vellar-Coleroon estuary. Fish for All Centre of the M. S. Swaminathan Research Foundation (MSSRF) in collaboration with ICAR-NBFGR has undertaken an initiative for the livelihood development of the tribal community which involves fattening of the mud crab, *Scylla serrata,* near Pichavaram backwaters providing additional income to the local community (ICAR-NBFGR 2022).

In addition to restoration work, awareness campaigns and educational materials should be utilized to improve understanding of mangrove habitats, resources, relevant legislation, policies, and conservation strategies through media such as magazines, films, posters, pamphlets, documentaries, exhibitions, bird watching tours, study tours, and competitions on mangrove knowledge. Research into mangrove taxonomy, diversity, distribution, current health status, conservation measures, restoration efforts, and assessments of human and natural impacts should be conducted to facilitate learning and conservation efforts. Information centres should be established to disseminate resources effectively.

Addressing poverty, which drives people to engage in illegal collection of fuel material, wood, and timber, by providing these resources at nominal prices could help reduce illegal activities. Mangroves often become dumping grounds for industrial and municipal waste without consideration for the consequences such as declining fisheries and vulnerability to natural calamities like tsunamis and climate change-induced flooding.

In efforts to conserve mangroves, the foremost priority should be the preservation of remaining areas of natural mangrove forests, especially those supporting mature trees capable of producing seedlings. Wetland habitats deemed particularly valuable from ecological and biodiversity perspectives can be best conserved by granting them special status, recognized either nationally or internationally (Umali et al. 1987). Such designations may include national parks, nature reserves, or gazetted forests at the national level and biosphere reserves, Ramsar sites, or World Heritage Sites at the international level. Some researchers have concluded that Ramsar sites have effectively maintained balanced ecosystems by promoting the restoration of wetlands and raising public awareness. There is significant recognition of Ramsar's role in mangrove protection and rehabilitation efforts across various levels. The effectiveness of Ramsar is not solely measured by raising public awareness but also by tangible outcomes such as the health and status of mangroves and the livelihoods of people dependent on these resources. Establishment of 'mangrove germplasm preservation centres' is another approach to protect the mangrove species which are at high risk of extinction (Kathiresan 2018). Mangrove restoration can serve as an innovative countermeasure against global warming by significantly reducing carbon emissions. The future of mangroves in India hinges on restoring their ecosystem services with robust community involvement to mitigate climate change impacts.

7.9 Conclusion

In conclusion, habitat degradation in coral reef ecosystems and mangroves in India presents a significant environmental challenge that requires immediate and comprehensive management measures. The current status of these habitats shows considerable deterioration due to various anthropogenic pressures, including climate change, pollution, overfishing, and coastal development. Despite the critical condition of these ecosystems, efforts in restoration and conservation have been relatively limited and are in their early stages. Successful initiatives, such as those in the Gulf of Kachchh and Tuticorin islands, highlight the potential for recovery if appropriate actions are taken.

To combat this degradation, India must implement a multifaceted approach that includes expanding restoration projects, enforcing stringent protection regulations, fostering community involvement, and enhancing scientific research. Restoration efforts should be scaled up and integrated with comprehensive monitoring and adaptive management practices. Additionally, sustainable practices and strong collaborations among government agencies, non-governmental organizations, local communities, and researchers are essential for the long-term health and resilience of coral reefs and mangrove ecosystems in India. By addressing both the immediate threats and underlying causes of habitat degradation, it is possible to preserve these vital ecosystems for future generations and maintain the biodiversity and ecological services they provide.

References

Adhavan D (2015) Impact of Algal Bloom on Mangrove and Coral Reef Ecosystem in the Marine National Park. Gulf of Kachchh, Gujarat, India. J Mar Biol Aquacult 2:1–2

Adhavan D, Kamboj RD, Chavdaand DV et al (2014) Status of intertidal biodiversity of Narara Reef Marine National Park, Gulf of Kachchh, Gujarat. J Mar Biol Ocean 3(3):3

Anon (1997) Estuaries of India: State of the Art Report. ENVIS Publication Series 2/97

Arora M, Chaudhury NR, Gujrati A et al (2019) Coral bleaching due to increased sea surface temperature in Gulf of Kachchh Region, India, during June 2016. Curr Sci 117(2):242–246

Arunprasath A, Gomathinayagam M (2014) Distribution and composition of true mangroves species in three major coastal regions of Tamilnadu, India. Int J Adv Res 2:241–247

Azariah J, Selvam V, Gunasekaran S (1992) Impact of the past management practices on the present status of the Muthupet mangrove ecosystem. Hydrobiol 247:253–259

Bakus GJ (1994) (ed.) Coral reef Ecosystem Oxford and IBH Publishing Co. India.

Baruah AD (2004) Muthupet mangroves and canal bank planting technique. Tamil Nadu Forest Department Publication, Chennai

Bhatt S, Shah DG, Desai N (2009) The mangrove diversity of Purna Estuary, South Gujarat. India Tropical ecology 50(2):287–293

Brown B (1997) Proceedings of the VIII International Coral Reef Symposium. In: Lessios HA, Macintyre IG (eds) Smithsonian Tropical Research Institute. Balboa, Panama, pp 65–74

Caeiro S (1999) Coral Fauna of Lakshadweep with special reference to Agatti Atoll. PhD Thesis, submitted to Goa University

Castelle AJ, Johnson AW, Conolly C (1994) Wetland and stream buffer size requirements—a review. J Environ Qual 23(5):878–882

Chanda A (2022) An overview of the Indian Ocean coral ecosystems. In: Blue carbon dynamics of the Indian Ocean: the Present State of the Art. Springer, Switzerland, pp 203–225

Chowdhury R, Sutradhar T, Begam M et al (2019) Effects of nutrient limitation, salinity increase, and associated stressors on mangrove forest cover, structure, and zonation across Indian Sundarbans. Hydrobiol 842:191–217

De K, Venkataraman K, Ingole B (2020) The hard corals (Scleractinia) of India: a revised checklist. Indian J Geo-Mar Sci 49(10):1651–1660

Dhaneesh KV, Ghosh S, Kumar TTA et al (2009) Lakshadweep Islands: Coral paradise of Arabian Sea—A Glimpse on its awaiting catastrophe. J Indian Oce Stud 17(3):409–418

Edward JKP, Mathews G, Raj KD et al (2012) Coral reefs of Gulf of Mannar, India: signs of resilience. In: Proceedings of the 12th International Coral Reef Symposium, Cairns, Australia, 18F Does monitoring lead to improved coral reef management

Edward JK, Mathews G, Raj KD et al (2015) Study on the status of corals in Palk Bay, Southeast coast of India during 2007-2013. Indian J Geo Mar Sci 44(9):1449–1462

FSI (2021) India State of Forest Report. Forest Survey of India, Ministry of Environment and Forests, Dehradun

Ghosh AK (1991) Fauna of Lakshadweep: an overview. In: Fauna of Lakshadweep. Zoological Survey of India, Kolkata, pp 1–4

Goreau TJ (2010) Coral reef and fisheries habitat restoration in the Coral Triangle: The key to sustainable reef management. In: Proceedings of the coral reef management symposium on the Coral Triangle Area, Manado, Sulawesi, Indonesia, pp 11–16

Hirway I, Goswami S (2007) Valuation of coastland resources: the case of mangroves in Gujarat. Academic Foundation, New Delhi

Hoon V (1997) Coral reef of India: review of their extent, condition research and management status. In: Proc. Regional Workshop on the Conser. Sustain. Manag. Coral Reefs (Vineeta Hoon, ed.), organised by M.S. Swaminathan Research Foundation and BOBPof FAO/UN, 1–27

Howard J, Lovelock C, Beeston M, Cameron C, Sippo J, Hagger V, van Bijsterveldt C, van Eijk P, Tonneijck FH (2023) Best practice guidelines for mangrove restoration. Global Mangrove Alliance, p 278

ICAR-NBFGR (2022) Annual Report 2022. ICAR-National Bureau of Fish Genetic Resources, Lucknow, p 180

Jagtap TG, Komarpant DS (2003) Evaluation of mangrove ecosystem of India for assessing its vulnerability to projected climatic changes. In: Dash SK, Prakash Rao (eds) Assessment of climate change in India and mitigation policies. pp 39-51

Jagtap TG, Singh C (2004) Mangroves-Nursery for fishes. WWF for Nature-India, Bangalore

Jagtap TG, Murthy PS, Komarpant DS (2002) Mangrove ecosystem of India: conservation and management. Pointer Publication, Jaipur

Jagtap TG, Nagle V (2007) Response and adaptability of mangrove habitats from the Indian subcontinent to changing climate, Ambio 36(4):328–334

Jasmine S, George RM, Manisseri MK, Kingsly HJ (2009) Hard coral diversity along southwest coast of India. J Mar Biol Assoc India 51(2):189–193

Jisha S, Aravindan CM, Ritakumari SD (2004) Checklist of fish fauna of Ayiramthengu mangroves, Kollam district, Kerala, India. Seshaiyana 12(2):4–5

Jose S, Kumar TTA, Lal KK (2023) Marine ornamental shrimps in Lakshadweep, India: obscure to the limelight. World Aquac 54:59–61

Joshi D, Munjpara S, Banerji U, Parasharya D (2014) Coral bleaching observations in the Gulf of Kachchh, India–a climate induced stress on the Scleractinians. J Aquat Biol Fish 2:106–113

Joshi D, Banerji U, Mankodi PC (2015) Delayed recovery in Porites spp. following mass coral bleaching: a case study from the Gulf of Kachchh, Gujarat. India J Glob Biosci 4:2326–2331

Kathiresan K (2000) A review of studies on Pichavaram mangrove, southeast India. Hydrobiologia 430:185–205

Kathiresan K (2003) How do mangrove forests induce sedimentation? Revista de Biologia Tropical 51:355–360

Kathiresan K (2018) Mangrove forests of India. Curr Sci:976–981

Kathiresan K (2010) Importance of mangrove forests of India. J Coast Environ 1(1):11–26

Kathiresan K (2024) Save the Sundarbans. Curr Sci 126:536–538

Kathiresan K, Qasim SZ (2005) Biodiversity of mangrove ecosystems. Hindustan Publishing Corporation

Kathiresan K, Ravikumar S (2010) Marine pharmacology: an overview. Mar Pharmacol 1:1–37

Kulkarni NA, Bhosale LJ (2021) Mangroves of Maharashtra State (India): diversity and sustainability. Plant Sci 4(3):178–207

Kumar TTA, Ghosh S, Murugan et al (2011a) Marine ornamental fishes of the Gulf of Mannar. Ministry of Environment, Forests and Climate Change, Govt. of India, New Delhi

Kumar TTA, Vinoth R, Prakash S et al (2011b) Reef fishes of Lakshadweep archipelago. Ministry of Earth Sciences, Govt. of India, New Delhi

Kumar TTA, Prakash S, Balasubramanian T (2012) Coral bleaching at Lakshadweep. Geography You 12(75):21–23

Kumar TTA, Prakash S, Jafer Hisham T (2013) Status of coral reefs of Lakshadweep: A measure towards conservation. Coastal Ecosystems of India—Special publication. In: Kathiresan K, Ajmal Khan S (eds) ENVIS, CAS in Marine Biology, Faculty of Marine Sciences, Annamalai University, vol 1. ENVIS Publication Series, pp 129–136

Kumar JY, Marimuthu N, Geetha S, Satyanarayana C, Venkataraman K, Kamboj RD (2014) Longitudinal variations of coral reef features in the Marine National Park, Gulf of Kachchh. J Coast Conserv 18:167–175

Kumar JSY, Satyanarayana C, Venkataraman K et al (2017) Coral reefs transplantation and restoration experience in Pirotan Island, Marine National Park, Gulf of Kachchh, India. Ind J Mar Sci 46(02):299–303

Kumara PTP, Kumara WAAU, Sandaruwan HAA et al (2008) Impacts of reef related resource exploitation on coral reefs: some cases from Southern Sri Lanka, vol 23. Coastal Oceans Research and Development in the Indian Ocean, Mombasa

Obura DO, Tamelander J, Lindén O (2008) Coastal oceans research and development in the Indian Ocean: status report 2008. In: CORDIO (Coastal Oceans Research and Development in the Indian Ocean)/Sida-SAREC. CORDIO, Mombasa

Majumdar SD, Hazra S, Giri S et al (2018) Threats to coral reef diversity of Andaman Islands, India: a review. Reg Stud Mar Sci 24:237–250

Manikandan B, Ravindran J, Mohan H, Periasamy R, Manimurali R et al (2016) Community structure and coral health status across the depth gradient of Grande Island, central west coast of India. Reg Stud Mar Sci 7:150–158

Mallik TK (2017) Coral atolls of Lakshadweep. Arabian Sea, Indian Ocean. MOJ Ecol Environ Sci 2(2):68–83

Mathew G, Jeyabaskaran R, Prema D (2010) Mangrove ecosystems in India and their conservation. In: Coastal Fishery Resources of India-Conservation and Sustainable Utilization. Society of Fisheries Technologists, Kerala, pp 186–196

Mondal T, Raghunathan C, Venkataraman K (2016) Diversity of Scleractinian corals in Great Nicobar Island. Andaman and Nicobar Islands, India, Proc Zool Soc 69(2):205–216

Mondal T, Raghunathan C, Chandra K (2017) Checklist of Scleractinian corals of India with their IUCN status: A special reference to Andaman and Nicobar Islands. (Lambert Academic Publishing), pp 1–96

Mondal T, Raghunathan C (2022) Differential thermal stress with associated bleaching on stony corals in the Andaman Sea and Bay of Bengal: yet another impact of the El Niño. Cah Biol Mar 63(2):103–117

Muley EV, Venkataraman K, Alfred JRB et al (2000) Status of coral reefs of India. In: Proceedings of the Ninth International Coral Reef Symposium, Bali, 23-27 October 2000, vol 2, pp 847–853

Nair RR, Qasim SZ (1978) Occurrence of a bank with living corals off the south-west coast of India. Indian J Mar Sci 7(1):55–58

Nanajkar M, De K, Ingole B (2019) Coral reef restoration-a way forward to offset the coastal development impacts on Indian coral reefs. Mar Pollut Bull 149:110504

Nasser AKV, Kunhikoya VA, Aboobaker PM (1999) Mangrove ecosystems of Minicoy Island, Lakshadweep. Mar Fish Infor Serv T E Ser 159:8–10

Nayak VN, Andrade LV (2008). Floral diversity and distribution of mangroves in the Kali Estuary, Karwar, West Coast of India. National Workshop on 'Mangroves in India: Biodiversity, Protection and Environmental Services'. Institute of Wood Science and Technology, Bangalore. Abstracts, p. 8.

Patterson Edward JK, Kulkarni S, Jeyabaskaran R et al (2005) The effects of the 2004 Tsunami on mainland India and the Andaman and Nicobar Islands. In: Wilkinson C, Souter D, Goldberg J (eds) Status of Coral reefs in Tsunami affected countries: 2005. Australian Institute of Marine Science, Cleveland

Pillai CG (1996) Coral reefs of India their conservation and management. In: Pillai CG (ed) Menon, N.G. Marine Biodiversity Conservation and Management, CMRI, pp 16–31

Prakash S, Kumar TTA, Raghavan R et al (2017) Marine aquarium trade in India: Challenges and opportunities for conservation and policy. Mar Policy 77:120–129

Qasim SZ, Wafar MVM (1979) Occurrence of living corals at several places along the west coast of India. Mahasagar 2(1):53–58

Ragavan P, Saxena A, Mohan PM et al (2015) Diversity, distribution and vegetative structure of mangroves of the Andaman and Nicobar Islands. India J Coast Conserv 19(4):417–443

Ragavan P, Saxena A, Jayaraj RSC et al (2016) A review of the mangrove floristics of India. Taiwania 61(3):224–242

Raghukumar C, Raghukumar S (1991) Fungal invasion of massive corals PSZNI. Mar Ecol 12:251–260

Raghunathan R, Rajkumar R (2012) Current status of coral reefs in the Andaman and Nicobar Islands. In: Bhatt JR, Patterson Edward JK, Macintosh DJ, Nilaratna BP (eds) Coral reefs in India status, threats and conservation measures. IUCN India, New Delhi

Raghuraman R, Sreeraj CR, Raghunathan C et al (2012) Scleractinian coral diversity in Andaman Nicobar Island in comparison with other Indian reefs. In: Proceedings of international day for biological diversity, pp 75–92

Raj KD, Mathews G, Malleshappa H et al (2015) Reproductive success of restored coral colonies in Vaan Island, Gulf of Mannar, Southeastern India. Indian J Geomar Sci 44(4):589–598

Rajan R, Rajan PT, Mishra SS et al (2021) Fishes of Lakshadweep archipelago: new records, review and a revised checklist. Mar Biodivers Rec 14(1):14

Rajan R, Satyanarayan C, Raghunathan C et al (2015) Status and review of health of Indian coral reefs. J Aquat Biol Fish 3:1–14

Rajasuriya ARJAN, Perera N, Karunarathna C et al (2005) Status of coral reefs in Sri Lanka after the tsunami. Status of coral reefs in Tsunami affected countries, 99–110

Ramaiyan V, Ramesh DA, Subramaniam (1995) Coral Reef Formations of India. Seshaniyana 3(1):4–8

Ramasubramanian R, Ravishankar T (2004) Mangrove Forest Restoration in Andhra Pradesh. MS Swaminathan Research Foundation, India, Chennai

Ramesh CH, Koushik S, Shunmugaraj T et al (2020) Seasonal studies on in situ coral transplantation in the Gulf of Mannar Marine Biosphere Reserve, Southeast coast of Tamil Nadu, India. Ecol Eng 152:105884

Ravindran J, Kannapiran E, Manikandan B et al (2012) Bleaching and secondary threats on the corals of Palk Bay: a survey and proactive conservation needs. Indian J Geo Mar Sci 41(1):19–26

Sadhukhan K, Chatragadda E, Shunmugaraj T et al (2022) Health monitoring status of corals in Gulf of Mannar and Palk Bay (India) with special reference to coral bleaching event. Indo Pac J Ocean Life 6(2):72–84

SAIME (2023) Proceedings of National Symposium on building synergies for Healthy Mangroves and Robust Value Chain Around the Mangrove Ecosystem in Sundarbans, Kolkata, 26–27 July 2023, pp 9–11

Saranya JS, Roxy MK, Dasgupta P et al (2022) Genesis and trends in marine heatwaves over the tropical Indian Ocean and their interaction with the Indian summer monsoon. J Geophys Res Oceans 127(2):e2021JC017427

Saroj J, Gautam RK, Joshi A, Tehseen P (2016) Review of coral reefs of India: distribution, status, research and management. Int J Sci Environ Technol 5(5):3088–3098

Selvam V, Ravichandran KK, Gnanappazham L et al (2003) Assessment of community-based restoration of Pichavaram mangrove wetland using remote sensing data. Curr Sci 85:794–798

Singarayan L, Rethnaraj C (2016) Occurrence of azooxanthellate Scleractinian corals off Goa, mid-west coast of India. Mar Biodiv Rec 9:78

Singh AK, Ansari A, Kumar D (2012) Status, biodiversity and distribution of mangroves in India: an overview. In: Uttar Pradesh State Biodiversity Board. Marine biodiversity: one ocean, many worlds of life. ICAR, Lucknow, pp 59–67

Sivakumar K, Mathur VB, Pande A et al (2013) Coastal and marine protected areas in India: challenges and way forward. ENVIS Bull 15:292–298

Srivastava PK (1965) A note on the Quaternary geology of Saurashtra Peninsula. Quater J Geol Minerol Metallurg Soc India 1:55–63

Turner JR, Vousden D, Klaus R et al (2009) Coral reef ecosystem of Andaman Islands—remote sensing and rapid site assessment survey. Rec Zool Surv India Occ Paper 301:1–132
Umali RM, Zamora PM, Gatera RR et al (1987) Mangroves of Asia and the Pacific: status and management. UNDP-UNESCO Research and Training Pilot Program on Mangrove Ecosystem in Asia and the Pacific, Metro Manila, pp 471–511
Untawale AG (1996) Restoration of mangroves along the central west coast of India. International Tropical Timber Org, Okinawa
Upadhyay R, Joshi N, Sampat AC et al (2015) Mangrove restoration and regeneration monitoring in Gulf of Kachchh, Gujarat State, India, using remote sensing and geo-informatics. Int J Geosci 6(4):299–310
Vasudevan N (2017) Safeguarding the sentinels: the story of mangrove conservation in Maharashtra. Mangal-Van, Mangrove Society of India, Goa, pp 55–65
Venkataraman K, Satyanarayana, Alfred JRB, Wolstenholme J (2003) Handbook on hard corals of India. Zoological Survey of India, Kolkata
Venkataraman K (2006) Coral reefs in India. National Biodiversity Authority, Chennai
Venkataraman K (2008) Coastal and marine wetlands in India. In: Proceedings of the Taal 2007: the 12th World Lake Conference, pp 392–400
Venkataraman K (2011) In: Hopley D (ed) Coral reefs of India in Encyclopedia of modern coral reefs. Springer), (Dordrecht
Venkataraman K, Rajan R (2013) Status of coral reefs in Palk Bay. Rec Zool Surv India 113(1):1–11
Vinoth R, Gopi M, Kumar TTA et al (2012) Coral reef bleaching at Agatti Island of Lakshadweep atolls, India. J Ocean Univ China 11(1):105–110
Viswambharan D, Sreenath KR, Anto A, Raju AK, Mohan S, Jasmine S, Joshi KK, Rohit P (2021) New distributional record of Acroporids along the eastern Arabian Sea. Reg Stud Mar Sci 41:101550
Wafar MVM (1999) Status report India. In: Linden O, Sporrong (eds) Coral reef degradation in the Indian Ocean CORDIO, Status report 1999:25-26
Wafar MVM, Muley EV, Yennawar P (2000) Status of the coral reefs of the Gulf of Kachchh, India. In: 9th International Coral Reef Symposium, Bali, Indonesia
Wallace CC, Muir PR (2005) Biodiversity of the Indian Ocean from the perspective of staghorn corals (Acropora spp). Indian J Geo Mar Sci 34:42–49
WWF (2005) Tsunami update. http://www.wwf.org.uk/News/n_0000001426.asp
Zacharia PU, Krishnakumar PK, Dineshbabu AP et al (2008) Species assemblage in the coral reef ecosystem of Netrani Island off Karnataka along the southwest coast of India. JMarBiolAssIndia 50(1):87–97

Climate Change and Fish Genetic Resources: Impact, Status and Perspectives

8

Uttam Kumar Sarkar, A. Kathirvelpandian, and Ganesan Kantharajan

Abstract

India's dynamic fishing sector significantly contributes to the agricultural economy and foreign exchange earnings, with freshwater and marine environments playing pivotal roles. Nonetheless, climate change poses numerous challenges, including escalating temperatures, shifts in rainfall patterns, and intensified extreme weather events, all of which profoundly affect fish populations, biodiversity, and the livelihoods of fishing communities. The reproductive sensitivity of inland fishes was studied to implement conservation management and create strategies for adaptation. Climate resilience aquaculture systems were developed for prioritized fishes from an aquaculture perspective. Insights from genomic analyses and research on climate-resilient species shed light on the genetic mechanisms underpinning adaptation to environmental stressors. Furthermore, adaptation and restoration strategies for riverine fisheries are examined, underscoring the importance of community engagement, policy interventions, and ecosystem-based approaches. Looking forward, further studies are needed to evaluate ecological conditions, prioritize conservation efforts, and formulate climate-resilient adaptation strategies. It underscores the necessity for granular analysis across various spatial scales to guide targeted interventions and bolster the adaptive capacity of India's fisheries and aquaculture sector.

Keywords

Fisheries dynamics · Climate impacts · Biodiversity · Mitigation · Aquaculture resilience

U. K. Sarkar (✉) · A. Kathirvelpandian · G. Kantharajan
ICAR-National Bureau of Fish Genetic Resources, Lucknow, Uttar Pradesh, India
e-mail: Uttam.Sarkar@icar.gov.in; A.Pandian@icar.gov.in; kantharajan.g@icar.gov.in

U. K. Sarkar et al. (eds.), *Sustainable Management of Fish Genetic Resources*,
https://doi.org/10.1007/978-981-97-5250-8_8

8.1 Introduction

The total fish production of the country is 16.24 million metric tonnes; the freshwater sector contributes 12.12 million metric tonnes, whereas 4.12 million metric tonnes is derived from the marine sector (2021–2022). The export value of the sector is Rs 639.69 billion, which contributes >6.5% to the agricultural economy and >1% to the Indian economy (2022–2023). The Aquatic Genetic Resources Information System (AqGRISI) of India, a database developed by ICAR-National Bureau of Fish Genetic Resources, documented 3193 native fish species belonging to 254 families, 53 orders, and 1036 genera (ICAR-NBFGR 2022). As per IUCN categorization, 21 species are critically endangered, 81 endangered, 116 vulnerable, and 1204 least concern, and 219 data-deficient species were documented.

Climate change is mostly evident by global warming and more frequent weather events like heat waves and cyclones (Field et al. 2014; Lin 2019). While "climate change" only refers to worldwide changes in weather patterns that last for decades or longer, climate variability encompasses all changes in the climate that last longer than individual weather events. An extended period of extreme heat and humidity in inland water bodies, typically during the summer, is referred to as an "inland heat wave (IHW)" (Niedrist et al. 2018) which will create an impact on primary producers and water quality (Sarkar et al. 2022a, b). As a result of climate change, sea surface temperature is expected to rise by as much as 3 °C by 2100 (Collins et al. 2013), and it will have serious implications on all types of ecosystems. There is growing concern about the impacts of climate change on marine organisms and ecosystems, yet the potential for acclimation and adaptation to this threat is poorly understood. It threatens health, food production, fisheries, aquaculture, and the fundamental necessities of life on Earth (Stern 2007). Strategies for climate change adaptation and resilience will improve with the help of a greater understanding of climatology's role in fisheries and aquaculture at the meso and micro levels. This will require a more detailed investigation to identify long-term patterns and changes. Potential indicators of perceived vulnerability were divided into seven categories, including resource usage, health, livelihood, social, food, ecological, and ecological aspects. For instance, wetlands' vulnerability indices will make it easier to plan climate-resilient fisheries along with developing location-specific adaptation plans (Velumani et al. 2021). At the gene level, transcriptional modulation of genes in key biological pathways is one way that organisms can respond to new or changing environmental conditions and may be at the heart of capacity for phenotypic plasticity and underpin responses to environmental stressors and adaptive resilience.

Inland aquatic environments play a major role in the dynamics of fish populations (Shuter et al. 1980; Schlesinger et al. 1982) and are considered potential indicators of climate change (Adrian et al. 2009). In addition to its negative effects, climate change is anticipated to have positive effects on inland fisheries by increasing fish production as a result of altered rainfall patterns and an increase in the number of rainy days, which will likely lead to an increase in the number of breeding days (Panikkar and Sarkar 2016). According to Barange and Perry (2009), the evolution of completely new species will probably benefit from climate change.

Impacts can be seen in India in the form of increased frequency of heavy rainfall events and decreased frequency of low- and medium-level rainfall events (Goswami et al. 2006). Additionally, there will be a significant effect on traditional fisheries (Borah 2021) as well as changes in the geographic distribution, extinction of species, breeding season, life cycle, and physiological behaviour of inland fishes (Sharma et al. 2015; Sarkar and Das 2021).

Microalgae or phytoplankton form a valuable biological parameter in climate models (Ghosal et al. 2000). The concentration of chlorophyll a serves as a stand-in for phytoplankton biomass, which is directly related to indicating the production functions of aquatic ecosystems. To improve water governance and design strategy protocols for sustainable water management that will result in better ecosystem services, it is crucial to understand their spatiotemporal dynamics in lotic environments (Sarkar et al. 2021a, b, c).

8.2 Impacts on Fish Biodiversity

India possesses rich aquatic biodiversity spread across different ecosystems with 4 of the 34 global biodiversity hotspots, i.e. the Western Ghats, the North-East Himalayas Region, and the Nicobar Islands. The fisheries and aquaculture sector plays an important role in providing food and nutritional security for the country's huge population and also contributes significantly to the economic upliftment of fishers and farmers. In addition, the sector provides a considerable contribution to the country's foreign exchange earnings.

8.2.1 Inland Ecosystem

Climate change has impacted wetland fish production, altering processes at the individual, population, community, and ecosystem levels (Ficke et al. 2007; Heino et al. 2009; Sarkar et al. 2020b). It is necessary to develop creative tools and methodologies that integrate several ways to distinguish between changes in inland waters caused by anthropogenic or other stressors and those caused by the climate.

Environmental parameters (13 parameters) were considered potential cues for breeding: air temperature, monthly rainfall, transparency, dissolved oxygen, free carbon dioxide, pH, total dissolved solids, conductivity, total alkalinity, hardness, salinity, nitrate, and phosphate (Sarkar et al. 2021a, b, c).

It is necessary to evaluate the reproductive sensitivity of inland fish before implementing conservation management and creating strategies for adaptation. The pre-monsoon rainfall and rising temperatures have prolonged the spawning season of Indian major carps (*Labeo catla*, *Labeo rohita*, and *Cirrhinus mrigala*) in hatcheries (Sharma et al. 2015). Snow trout (*Schizothorax richardsonii*) demonstrated that breeding phenology has stabilized and predicted that the species is constantly adjusting to the shifting climate in the lower Himalayas (Joshi et al. 2018). More research on this migratory, regionally significant fish species is necessary since the water temperature of the Ganges affects the gonadal maturation and spawning of hilsa (*Tenualosa ilisha*) (Das et al. 2016; CIFRI-NICRA 2016).

Extreme weather events like flash floods and prolonged dry spells brought on by climate change provide many obstacles and difficulties for fish farmers. Programmes for managing watersheds have several advantages, including resource conservation, groundwater recharge, drought mitigation, and sustainable production (Brooks et al. 1991). The flood followed by the cyclone had a higher economic risk rating, and its magnitude was greater in coastal regions than in inland ones. A comprehensive list of adaptation measurements does not exist; each aquaculture system must be assessed individually. Farmers, researchers, and policymakers should execute planned and autonomous adaptation strategies to enable climate-resilient aquaculture. To improve farmers' ability to adapt, regulations must be taken to develop planned adaptive measures, which are either lacking or at the primary level (Muralidhar et al. 2021).

It is possible to carefully manage the eutrophic wetlands to preserve fish productivity and ecosystem benefits. In the context of the changing climate, it is advised that sewage should be disposed of and treated as planned to preserve the ecological health of floodplain wetlands for sustainable fisheries management (Puthiyottil et al. 2021). Recent research on the ecological vulnerability of wetlands has shown that, in comparison to rainfall, the effects of water temperature have a more significant influence on ecological parameters (Sarkar et al. 2020a). Three universal prokaryote primers are used to evaluate the variation in community composition of prokaryotes in response to ecosystem variables which served as the most sensitive bioindicator of wetland ecosystem assessment and management (Kumari et al. 2021).

The climate change-related factors (e.g. abnormal rains, floods, or drought-like situations) and other factors—both natural (e.g. hyper-nitrification) and anthropogenic (e.g. construction of roads, houses, and other structures)—in the adversely affected lower Brahmaputra Valley, Assam, were assessed. Mitigation strategies were recommended to help fishers better manage these issues and boost their capacity for adaptation: the practice of enclosure aquaculture in beels; the construction of floating macrophyte refuges (katal); deep pool refuges; and stock enhancement programmes with climate-resilient indigenous fishes (Debnath et al. 2021).

Planning strategically and considering the effects at both the temporal and spatial scales determine how effective mitigation and adaptation measures work. The adaptation strategies identified, promoted, and implemented by ICAR-CIFRI under the NICRA project to increase the adaptive capacity of fishers in changing climate scenarios are (a) climate-resilient pen systems (CRPS), (b) climate smart strategy for cage culture, (c) climate adjustable pens or pre-summer enclosures, (d) deep water pools as climate adaptation, (e) periphyton-based refuge system, (f) macrophyte-based shelters, (g) fences/screens on the embankments, (h) autumn stocking, (i) responsible fishing methods, and (j) shifting from capture-based fisheries to culture-based ones (Sarkar et al. 2021a, b, c). For improving fish production in shallow, multi-stake floodplain wetlands choked by macrophytes, pen culture can be a useful climate-resilient technique, especially in the eastern and northeastern states of India (Karnatak et al. 2021).

8.2.2 Marine Ecosystem

Marine fish production in India has increased sixfold in the last six decades. However, sustainability concerns include production approaching the potential yield, overcapacity in the fishing sector, open access to the fishery, degradation of habitats, and trade-related issues. Climate change exacerbates the situation.

Sea surface temperature has increased by 0.2–0.3 °C along the Indian coast in the last 45 years and is projected to increase by 2.0–3.5 °C by 2099. The projected sea level rise is 30 cm in 50 years. During the southwest monsoon, the wind speed, coastal upwelling strength, and chlorophyll concentration also change the Indian seas. There is growing evidence that these changes will impact the composition and operation of marine ecosystems. Different species react differently to high temperatures, exhibiting alterations in the abundance and composition of the base of the food chain (Vivekanandan 2011).

The small pelagic fish such as the oil sardine and the Indian mackerel have extended their distributional boundary to northern and eastern latitudes contributing to fisheries in the last two decades. They have extended their distribution to midwaters as well. The threadfin breams are found to shift their spawning towards cooler months off Chennai. These distributional and phenological changes may have an impact on the nature and value of fisheries. Large-sized, high-value fish species already exhibiting signs of decline from fishing and other non-climatic reasons may be replaced by small-sized, low-value fish species with rapid generation turnover if they can adapt to the changing environment.

In mariculture, elevated water temperature and changes in salinity and pH will likely to affect the spawning season, spawning strength, and larval growth of candidate species. Certain changes could be advantageous, such as accelerating the growth of fish and shellfish and expanding their range into previously unproductive areas. Climate change is predicted to have two main effects on fisheries: those that affect the biological health and viability of fish stocks and those that impact the safety or the social, cultural, and economic sustainability of fishers and fishing communities (Zacharia et al. 2015).

Numerous coral bleaching events have been reported in the major reef ecosystems in India due to elevated sea surface temperature. The branching corals were more vulnerable to elevated SST than the massive coral colonies (Krishnan et al. 2011). Climate change factors including sea level rise, increased frequency of cyclones, altered rainfall, and increasing temperature impact the structure and functions of the mangrove ecosystem. The decrease in the density of mangrove vegetation in the Sunderban region is associated with an increase in temperature and a decrease in rainfall (Mukherjee and Paul 2021). The impact of climate change on various systems is mentioned in Table 8.1.

Table 8.1 The impact of climate change on various systems

Systems/factors	Impacts	References
Inland open-water fisheries	• Production and ecology • Farming and fishing operations • Community and livelihoods • Wider society and economy	Sarkar and Das (2021)
Shrimp aquaculture	• Environmental, biological, and economic impacts	Muralidhar et al. (2021), Paul et al. (2024)
River	• Direct cascading effects on key environmental parameters include water temperature, total dissolved solid, salinity, total alkalinity, and pH	Sarkar et al. (2021a, b, c)
Floodplain wetlands	• Hydrodynamics • Ecological resilience of the eutrophic wetlands • Catch and diversity of wild fish stocks reduced • Occurrence of some common fish diseases • An extended or shifted spawning season	Puthiyottil et al. (2021), Debnath et al. (2021), Sarkar et al. (2022a, b), Kumari et al. (2021)
Marine	• Shifts in the seasonal timing (phenology) of crucial events such as spawning by fishes • The spring blooms of phytoplankton and zooplankton • Ontogenetic change • High rate of local extinction and reduction in maximum catch potential	Zacharia et al. (2015)
Aquatic ecosystems	• Hydrological and biological variables • Variation with the abundance of diatoms and a diverse pattern of variation with blue-green and green algae	Sarkar et al. (2022a, b), Panikkar and Sarkar (2016)
Biotic factors	• Primary productivity; phytoplankton/zooplankton biomass and species composition	Sarkar et al. (2022a, b)
Fish and fisheries	• Fish catch structure • Change in species distribution • Size of fish at first maturity • Phenology of spawning and reproductive vulnerability • Migration pattern • Catch per unit effort • Diet and trophic level • Variation in fish abundance • Traditional fisheries	Yengkokpam et al. (2022), Sarkar et al. (2021a, b, c)

8.3 Climate-Resilient Species

Climate-resilient fish species are anticipated to become increasingly necessary as climatic stressors rise daily. Certain native fish species, such as *Clarias magur*, *Heteropneustes fossilis*, *Anabas testudineus*, *Channa* spp., *Puntius* spp., and *Notopterus notopterus*, demonstrated greater tolerance to climate change in India's inland waters (Sarkar et al. 2020a, b, c). *Amblypharyngodon mola* (mola carplets), a species that inhabits wetlands in the the lower Indo-Gangetic floodplains, was studied to validate the reproductive resilience of minnows to climate fluctuation (Paul et al. 2024).

In marine ecosystems, sardines, *Sardinella gibbosa*, *S. fimbriata*, and *S. longiceps*, ranked high in vulnerability assessment. Similarly, threadfin breams, *Nemipterus japonicus* and *N. randalli*, possess high resilience under changing climatic conditions (Zacharia et al. 2015).

8.4 Genomic Analyses in Fishes Against Biotic and Abiotic Factors

ICAR-National Bureau of Fish Genetic Resources (NBFGR) has developed a database of "Abiotic Stress Responsive Genes in Fishes", which presently covers information on 250 abiotic stress-responsive genes of 38 fish species. Identification and characterization of genes associated with abiotic stress from whole genome sequence data of *Clarias magur* was carried out.

The freshwater species *Labeo catla*, *Channa striata*, *Labeo rohita*, and *Clarias magur* were studied to estimate the critical limits of abiotic stressors in fishes and investigate the biological changes taking place in the organisms during exposure to these abiotic factors. The study identified proteins that could be used in future stress management and adaptation studies.

Identification of genes involved in adaptability to varying salinity was studied in hilsa, *Tenualosa ilisha*, through differential expression analysis. Understanding the genomic mechanisms of thermal tolerance using the golden mahseer *Tor putitora* as a model was carried out to facilitate gene prospecting and mining alleles responsible for adaptation to thermal stress in the species from two diverse habitats. In *C. magur*, two nuclear genes (Nupr1 and Park2) were identified as responsive to water temperature (Kumar et al. 2022).

Sebastian et al. (2021) utilized ddRAD sequencing to investigate the adaptive genetic variation of the Indian oil sardine in response to the diverse habitats of the Indian Ocean. The thermal adaptability of Indian pompano, *Trachinotus mookalee*, was investigated through experimentation at six distinct acclimation temperatures. This study pioneered the derivation of the acclimation response ratio (ARR) for the species, highlighting its resilient response and demonstrating the species' capacity to adapt to diverse thermal environments (Divu et al. 2024).

8.5 Adaption and Restoration

Adaptation in riverine fisheries refers to adjusting gear and vessels to keep fishing pressure at sustainable levels. Based on changes in recruitment, growth, survival, and spawning success, setting standards on catch limits is to be done through adaptive management and monitoring. A further strategy to reduce the potential effects of climate change on fisheries is to maintain market development, control fisheries, and economic incentives for the fishing community (Macfadyen 2010).

Since the consequences of climate change on the riverine system are observable and quantifiable, information sharing between managers, researchers, policymakers, and other stakeholders is necessary for effective management (Seavy et al. 2009). The most common view for transforming the riverine ecosystem into other ecosystems is to lessen stresses like pollution and habitat degradation and increase resilience (Palmer et al. 2008) and vegetation like riparian aids in moderating the water temperature and buffering the warming effects of climate change (Bond et al. 2015). In natural environments, pre-spawning girth should be considered when assessing the vulnerability of declining fish species (Sarkar et al. 2021b, c).

Implementing better risk management techniques, such as fish brood stock insurance and policy intervention in watershed management programmes (Bhattacharjya et al. 2021), can mitigate the impact of drought-like conditions.

There is a need for community mobilization so that fishermen have more knowledge about climate change and can participate in the planning and preparation process for disaster management. Therefore, a bottom-up strategy incorporating the community and the primary stakeholders—fishers—will appropriately position them for climate change adaptation and mitigation. Policymakers and wetland managers can use the findings based on the overall vulnerability score to improve fishermen's ability to adapt and establish climate-smart fisheries management. Various vulnerability score forms will be useful in creating an action plan for the restoration of fisheries and habitat. Enhancing the planning and implementation of climate-smart adaptation at the local level can reduce the risks that climate change poses to future and sustainable fisheries (Paul et al. 2024). Climate change impacts and adaptation strategies are detailed in Fig. 8.1.

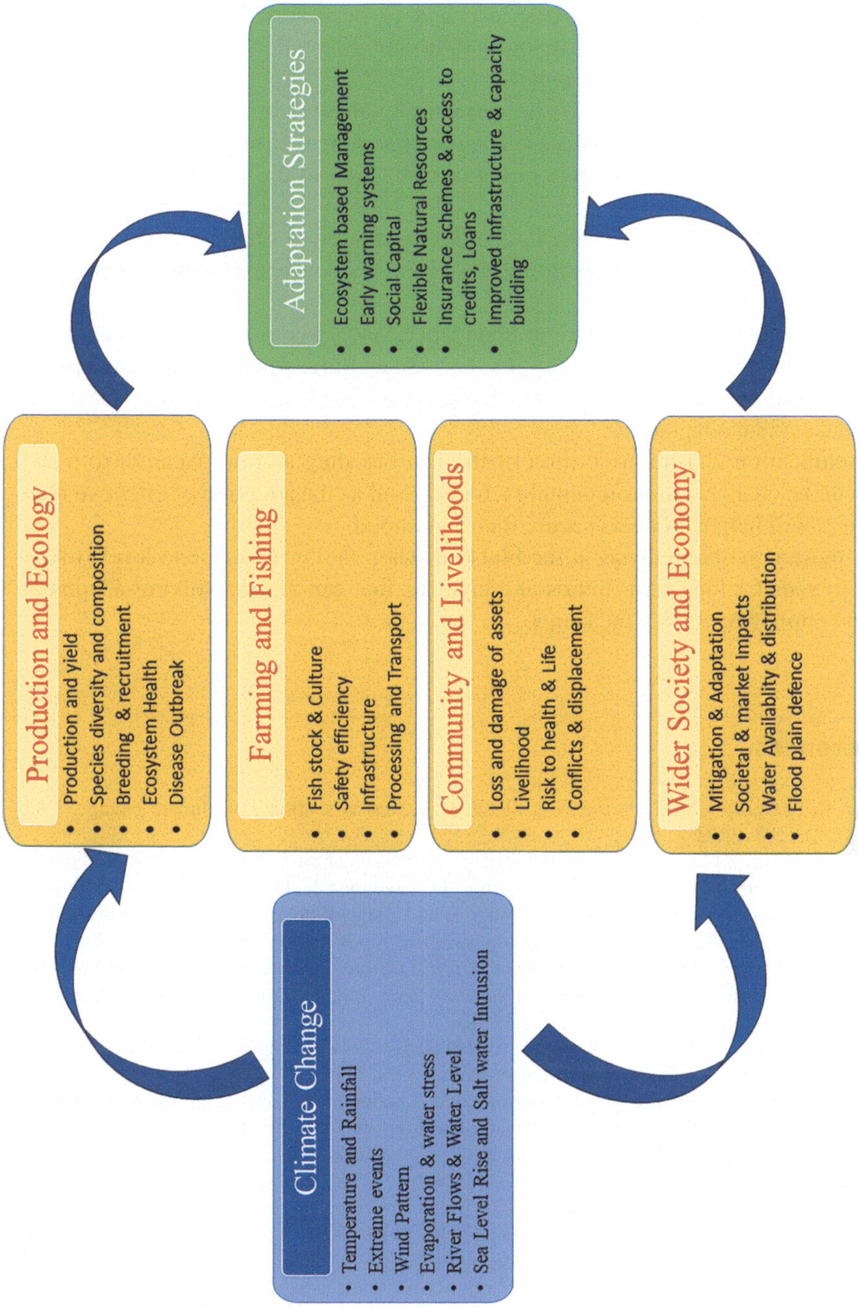

Fig. 8.1 Climate change impacts and adaptation strategies

8.6 Way Forward/Future Studies

- Ecological conditions and fisheries need to be assessed concerning climate change to formulate effective mitigation and adaptation strategies to combat losses caused by climate change.
- It is also essential to identify the optimal climatic conditions for the successful recruitment of natural fish stocks. The identification of climate-sensitive/climate-resilient species will also help in the prioritization of the conservation needs of species.
- The identification, development, and demonstration of climate-resilient adaptation strategies will help in building ecosystem resilience and adaptive capacity of fishers.
- Projections on fish distribution, abundance, and catches need to be developed under changing climatic conditions. For aquaculture, suitable candidate species that would benefit from elevated temperature need to be identified for better management adaptations.
- Modification and improvisation of the fish breeding and propagation to match with the changed environmental factors as well as the adoption of effective measures will help the fishers secure their livelihood.
- A more granular analysis at the block, district, regional, and state levels to discern specific long-term trends and changes that can aid in evolving adaptation and climate resilience measures.

References

Adrian R, O'Reilly CM, Zagarese H et al (2009) Lakes as sentinels of climate change. L&O 54(6/2):2283–2297

Barange M, Perry RI (2009) Physical and ecological impacts of climate change relevant to marine and inland capture fisheries and aquaculture. In: Cochrane K, De Young C, Soto D, Bahri T (eds) Climate change implications for fisheries and aquaculture: overview of current scientific knowledge. FAO fisheries and aquaculture technical paper, 530. FAO, Rome, pp 7–106

Bhattacharjya BK, Yadav AK, Debnath D et al (2021) Effect of extreme climatic events on fish seed production in lower Brahmaputra Valley, Assam, India: constraint analysis and adaptive strategies. Aquat Ecosyst Health 24(3):39–46

Bond RM, Stubblefield AP, Van Kirk RW (2015) Sensitivity of summer stream temperatures to climate variability and riparian reforestation strategies. J Hydrol 4:267–279

Borah BC (2021) Impact of climate change-induced challenges on fisheries in the North Eastern Region of India and the way ahead. Aquat Ecosyst Health 24(3):94–102

Brooks KN, Ffolliott PF, Gregersen HM et al (1991) Hydrology and the management of watersheds. Iowa State University Press, Ames, Iowa

CIFRI-NICRA (2016) Consolidated report of project National Innovations on climate resilient agriculture (NICRA) 2012–2017 from ICAR-CIFRI, Barrackpore. ICAR-Central Inland Fisheries Research Institute, Barrackpore. India

Collins M, Knutti R, Arblaster J et al (2013) Long-term climate change: projections, commitments, and irreversibility. The Physical Science Basis. Contribution of Working Group I to the Fifth Assessment Report of the Intergovernmental Panel on Climate Change. Cambridge University Press, Cambridge, UK/New York, NY, USA, pp 1029–1136

Das MK, Srivastava PK, Rej A et al (2016) A framework for assessing vulnerability of inland fisheries to impacts of climate variability in India. Mitig Adapt Strateg Glob Chang 21:279–296

Debnath D, Bhattacharjya BK, Yengkokpam S et al (2021) Ecosystem vulnerability of floodplain wetlands of the lower Brahmaputra Valley to climatic and anthropogenic factors. Aquat Ecosyst Health 24(3):57–70

Divu DN, Mojjada SK, Anil MK et al (2024) Thermal tolerance and environment adaptability of Indian pompano: discovery of a resilient candidate species for sustainable mariculture production in a climate change scenario. Aquaculture 584:740665

Ficke AD, Myrick CA, Hansen LJ (2007) Potential impacts of global climate change on freshwater fisheries. Rev Fish Biol Fish 17(4):581–613

Field CBVR, Barros DJ, Dokken KJ et al (2014) Technical summary. In: Climate change 2014: impacts, adaptation, and vulnerability. Part a: global and sectoral aspects. Contribution of Working Group II to the fifth assessment report of the intergovernmental panel on climate. Change Cambridge University Press, Cambridge, United Kingdom and New York, NY, USA, pp 35–94

Ghosal S, Rogers M, Wray A (2000) The turbulent life of phytoplankton. Center for Turbulence Research, pp 31–45

Goswami BN, Venugopal V, Sengupta D et al (2006) Increasing trend of extreme rain events over India in a warming environment. Science 314(5804):1442–1445

Heino J, Virkkala R, Toivonen H (2009) Climate change and freshwater biodiversity: detected patterns, future trends and adaptations in northern regions. Biol Rev 84(1):39–54

ICAR-NBFGR (2022) Annual Report (2022) ICAR-National Bureau of Fish Genetic Resources, Lucknow

Joshi KD, Das SCS, Pathak RK et al (2018) Pattern of reproductive biology of the endangered golden mahseer *Tor putitora* (Hamilton, 1822) with special reference to regional climate change implications on breeding phenology from lesser Himalayan region, India. Appl Anim Res 46(1):1289–1295

Karnatak G, Sarkar UK, Puthiyottil M et al (2021) Enhancing adaptive capacity of wetland fishers through pen culture in the face of changing climate: a study from a tropical wetland, India. Aquat Ecosyst Health 24(3):83–93

Krishnan P, Roy SD, George G et al (2011) Elevated Sea surface temperature during May 2010 induces mass bleaching of corals in the Andaman. Curr Sci 100(1):111–117

Kumari K, Naskar M, Aftabuddin M et al (2021) Evaluation of three prokaryote primers for identification of prokaryote community structure and their abode preference in three distinct wetland ecosystems. Front Microbiol 12:643945

Kumar MS, Baisvar VS, Kushwaha B et al (2022) Thermal stress induces expression of Nuclear protein and Parkin genes in endangered catfish, *Clarias magur*. Gene 825:146388

Lin N (2019) Tropical cyclones and heatwaves. Nat Clim Chang 9:579–580

Macfadyen G (2010) Addressing potential impacts of climate change on fisheries trade. Bioresources 4(4):28–35

Mukherjee D, Paul AK (2021) Mangrove sensitivities to climate change and its impacts in the Sundarbans: a case study in the Patibania Island of south western Sundarbans, India. In: Modern cartography series, vol 10. Academic Press, pp 353–385

Muralidhar M, Kumaran M, Jayanthi M et al (2021) Impacts of climate change and adaptations in shrimp aquaculture: a study in coastal Andhra Pradesh, India. Aquat Ecosyst Health 24(3):28–38

Niedrist GH, Psenner R, Sommaruga R (2018) Climate warming increases vertical and seasonal water temperature differences and inter-annual variability in a mountain lake. Clim Chang 151:473–490

Palmer MA, Reidy Liermann CA, Nilsson C et al (2008) Climate change and the world's river basins: anticipating management options. Front Ecol Environ 6:81–89

Panikkar P, Sarkar UK (2016) Impact of climatic variations on inland fisheries in Peninsular India. In: Proceedings of national workshop on mainstreaming climate change and adaptation in agriculture and allied sector, p 6–10

Paul TT, Sarkar UK, Salim SS et al (2024) Assessing multi-scale vulnerability of fisheries of Vembanad lake, peninsular India, due to climate change: a stakeholders-based approach. Environ Dev Sustain 26(3):6719–6749

Puthiyottil M, Sarkar UK, Lianthuamluaia L et al (2021) Floodplain wetlands of eastern India in a changing climate: trophic characterization, ecological status, and implications for fisheries. Aquat Ecosyst Health 24(3):47–56

Sarkar UK, Das BK (2021) Research advances in climate and environmental change impacts on inland fisheries of India: status, vulnerability and mitigation strategies. Aquat Ecosyst Health 24(3):7–17

Sarkar UK, Bakshi S, Lianthuamluaia L et al (2020a) Understanding the enviro-climatological impact on fish biodiversity of the tropical floodplain wetlands for their sustainable management. Sustain Water Resour Manag 6(5):1–12

Sarkar UK, Mishal P, Borah S et al (2020b) Status, potential, prospects, and issues of floodplain wetland fisheries in India: synthesis and review for sustainable management. Rev Fish Sci Aquac 29(1):1–32

Sarkar UK, Saha S, Mishal P et al (2020c) Fisheries management of a sewage-fed Gangetic floodplain wetland in the context of changing climate: a case study using multiple approaches. J Inland Fish Soc India 52(1):28–41

Sarkar SD, Sarkar UK, Naskar M et al (2021a) Effect of climato-environmental parameters on chlorophyll a concentration in the lower Ganga basin, India. Rev Biol Trop 69(1):60–76

Sarkar UK, Roy K, Karnatak G et al (2021b) Reproductive environment of the decreasing Indian river shad in Asian inland waters: disentangling the climate change and indiscriminative fishing threats. Environ Sci Pollut Res 28:30207–30218

Sarkar UK, Saha SSS, Puthiyottil M et al (2021c) Documenting adaptation strategies for building up climate resilient inland fisheries practices in developing countries: status, synthesis and overview. J Inland Fish Soc India 53(1–2):08–21

Sarkar UK, Karnatak G, Lianthuamluaia L et al (2022a) Combining stakeholder perception and ecological approaches for assessing vulnerability of floodplain wetlands in changing climate: a regional study. Int J Biometeorol 66(7):1415–1427

Sarkar UK, Paul TT, Puthiyottil M et al (2022b) Inland heat waves (IHWs) and associated impacts on hydro-biology of aquatic ecosystems in lower ganga basin, India. Int J Biometeorol 66:235–245

Schlesinger DA, Henry A, Regier HA (1982) Climatic and morphoedaphic indices of fish yields from Natural Lakes. Trans Am Fish Soc 111(2):141–150

Seavy NE, Gardali T, Golet GH et al (2009) Why climate change makes riparian restoration more important than ever: recommendations for practice and research. Restor Ecol 27:330–338

Sebastian W, Sukumaran S, Abdul Azeez S et al (2021) Genomic investigations provide insights into the mechanisms of resilience to heterogeneous habitats of the Indian Ocean in a pelagic fish. Sci Rep 11(1):20690

Sharma AP, Joshi KD, Naskar M et al (2015) Impact of climate variation on breeding of major fish species in inland waters. Central Inland Fisheries Research Institute, Barrackpore, India

Shuter BJ, Regier HA, Fry FEJ et al (1980) Simulation of temperature effects on first-year survival of smallmouth bass. Trans Am Fish Soc 109(1):1–10

Stern N (2007) The economics of climate change: the stern review. Cabinet Office, HM Treasury, UK

Velumani T, Ananthan PS, Sarkar UK et al (2021) Climate variability and fisheries vulnerability: a case study of Thanjavur district. Tamil Nadu J Indian Fish Ass 48(1):81–93

Vivekanandan E (2011) Climate change and Indian Marine fisheries. CMFRI special publication No. 105, Marine fisheries policy brief - 3, ICAR-CMFRI Publication

Yengkokpam S, Debnath D, Bhattacharjya BK et al (2022) Exploring polyculture of small indigenous fishes with major carps in pens as a climate-resilient adaptation strategy for northeastern wetlands of India. Mitig Adapt Strateg Glob Chang 27:1–19

Zacharia PU, Dineshbabu AP, Thomas S et al (2015) Relative vulnerability assessment of Indian marine fishes to climate change using impact and adaptation attributes. CMFRI Special Publication No. 125, Central Marine Fisheries Research Institute, Kochi, India

Harmonizing Conservation with Livelihood: Aquaculture of Marine Ornamentals, A Model

9

T. T. Ajith Kumar

Abstract

The marine ornamental aquaculture and aquarium trade is a daybreak industry in the aquaculture sector, and the captive culture of them is a growing enterprise worldwide. As a result, the trade is expanding and has grown into a multi-dollar initiative. The increasing demand for marine organisms in the market appears to be a very lucrative venture. Most of the trading animals are being collected from the coral reef regions. This is one of the major threats to the marine environment, as coral reef regions provide various ecosystem services and also serve as a breeding and rearing ground for many treasured organisms. This warrants an alternative option such as captive propagation for meeting the ever-increasing demand. The action will reduce the pressure on the wild stock and also provide an alternate source of income for the coastal and island communities across the country. In this context, the article focuses on the importance and status of marine ornamentals and on the research initiatives on captive breeding, culture, technology dissemination, and adaptation. Moreover, special attention has been given to the Indian context.

Keywords

Marine ornamentals · Captive propagation · Community aquaculture · Livelihood upliftment · Conservation

T. T. A. Kumar (✉)
ICAR-National Bureau of Fish Genetic Resources, Lucknow, Uttar Pradesh, India
e-mail: tt.ajith@icar.gov.in

U. K. Sarkar et al. (eds.), *Sustainable Management of Fish Genetic Resources*,
https://doi.org/10.1007/978-981-97-5250-8_9

9.1 Introduction

Marine ornamentals, the colourful organisms associated with coral reefs, play a vital role in sustaining the ecological balance of the reef ecosystem, besides increasing the value of the aquarium trade and providing livelihood to the coastal and island communities. The marine ornamentals including fishes and invertebrates are widely collected from the coral reef habitats throughout the Indo-Pacific, as well as the Red Sea and Caribbean regions. Fishes and corals including invertebrates and live rocks contribute to the bulk of the trade in terms of quantity and value. The demand for marine invertebrates such as soft corals, sea anemones, shrimps, crabs, and sea lilies are increasing, as a result of the growing interest in keeping home aquaria and technological developments.

Ornamental aquaculture possesses huge export potential and increasing local demand, which makes it a lucrative industry for income generation in India. The country possesses diversified fresh, brackish, and marine ornamental organisms, mainly from natural waterbodies such as rivers, estuaries, lagoons, and reef ecosystems (Pabhu 2021).

The beautiful shapes, attractive colours, and marvellous patterns of marine ornamentals make them appropriately called living jewels (Ajith Kumar et al. 2007). These organisms fetch a higher price than their freshwater counterpart, due to their attractive colour and high attention-grabbing behaviour (Raja et al. 2019). Unlike freshwater ornamental fishes, marine fishes available in the trade are mostly collected from the wild, and less than 10% of the species in the trade are estimated to be captive-bred (Wabnitz et al. 2003).

The collection of marine ornamentals leads to the destruction of coral reef ecosystem and reduction of biodiversity (Dammannagoda 2018). Hence, development of captive propagation technology for the fragile reef ecosystem is essential for its safeguarding and also for guaranteeing the sustainable supply of marine ornamentals to the trade, thereby minimizing the negative impacts on the natural environment (Lin et al. 2002).

Researchers, breeders, traders, and fish collectors are involved in a global effort to minimize the growing pressure on the natural population of marine ornamentals and also to promote the sustainable use of these high-valued resources. The captive breeding program of marine ornamental fish species is the only way to support the marine aquarium trade (Olivotto et al. 2017). Research initiatives to develop breeding and rearing methods for marine ornamentals are essential for the development of sustainable ornamental aquaculture (Ajith Kumar and Balasubramanian 2009; Ajith Kumar et al. 2010; Dhaneesh et al. 2011). The major steps involved in the captive breeding of ornamental fishes are broodstock development, breeding, larviculture, grow-out culture, live feed culture, disease management, packing, transportation, etc. Increasing the research efforts on marine ornamental fish culture is one of the ways to overcome the current hurdles to increase the availability of captive-bred species for trade.

In India, marine ornamental fishes are distributed in the reef and lagoon regions of the Gulf of Mannar, Palk Bay, the Gulf of Kutch, the Andaman and Nicobar

Islands, and the Lakshadweep Islands. A total of 400+ species of marine ornamental finfishes belonging to Indian waters have been recorded with the domination of families such as Acanthuridae, Chaetodontidae, Balistidae, Pomacanthidae, Labridae, Haemulidae, and Pomacentridae (Pandey and Mandal 2017; Raja et al. 2019).

9.2 Global Scenario

More than two million people in the world are involved in the ornamental trade as collectors and hobbyists, including government agencies, airlines, other associates, etc. In general, the trade is dominated by freshwater species; however, the increasing popularity of reef aquaria has become a leading trend in recent days. Current global trade of the marine ornamental organisms from wild collections reveals ecologically unsustainable practices that require immediate policy interventions. It is estimated that around 90% of the freshwater fishes are captive-raised and that the remaining 10% are wild-caught. However, in marine ornamentals, about 95% are harvested from natural waters, while only 5% are hatchery-produced. The mortality of tropical organisms prior to reaching the aquarium market (25–80%) is associated with a range of factors, including poor or even destructive collection and husbandry practices, stress, and deprived shipping, adding to the loss of marine ecosystem, particularly corals. It is difficult to estimate the long-term effects of this wild exploitation from the vulnerable and fragile reef ecosystem, which already faces the serious challenge of climate change, ocean acidification, and bleaching.

Marine ornamental aquaculture provides employment opportunities to the rural poor and is also an earner of foreign exchange in many developing countries. The global ornamental fish trade is estimated to be US$ 18–20 billion, which is supported by about 100 million hobbyists around the world, and the enterprise has been expanding at a rapid pace in recent years (Shinoj et al. 2021). Accurate data for the ornamental fish industry is difficult to obtain, as statistics vary in terms of data collected, format, and reality in different nations. It was estimated that around 2278 fishes and 724 invertebrates were involved in the global trade. Further, about 500 species of invertebrates are popular, including molluscs, echinoderms, crustaceans, polychaetes, and poriferans (Rhyne et al. 2012).

The magazine *CORAL* (https://www.coralmagazine.com) is the source for an updated list of marine fishes, which are captive-bred since January 2013. The yearly update is being carried out by *CORAL* magazine to document the tropical marine fish breeding success and accessibility of captive-bred marine fishes among hobbyists and industries (CORAL 2018).

Currently, a total of 338 marine ornamental fish species belonging to 37 families are captive-bred. Among them, only 134 species are commercially available, and only 18% of them are very common in the market. The major fishes are clowns, few damsels (Pomacentridae), blenny (Blennidae), dottyback (Pseudochromidae), and cardinalfish (Apogonidae) (Pouil et al. 2020).

9.3 Indian Scenario

In India, the aquarium trade is gaining popularity due to the strong demand from domestic as well as export markets. The diversity of marine ornamentals of the country comprises around 400 indigenous species, besides more than 100 exotic species (Lal and Ajith Kumar 2021). Prakash et al. (2017) reported that 87 fish species belonging to 55 genera and 22 families from the Gulf of Mannar Marine Biosphere Reserve are being collected for aquarium trade. Besides, invertebrates belonging to 21 marine taxa including anemones, shrimps, starfishes, hermit crabs, lobsters, and nudibranchs are available in the marine aquarium trade, which are also the residents of the Gulf of Mannar. According to Anil et al. (2018), more than 200 species of marine ornamental fishes including angels (Pomacanthidae), triggers (Balistidae), butterfly fishes (Chaetodontidae), damsels and clowns (Pomacentridae), cardinals (Apogonidae), puffers (Diodontidae), squirrels (Holocentridae), wrasses (Labridae), parrots (Scaridae), and batfishes (Ephippidae) are being traded in India. However, despite the huge potential, India contributes only 2.5% to the total Asian ornamental fish export (Ajith Kumar et al. 2015). The industry possesses a vast probability to scale up the trade in a bigger way. The exploitation of wild-caught ornamentals may support the local economy of the region; however, it is a major conservation challenge in biodiversity.

9.4 Captive Breeding of Marine Ornamentals: Indian Perspective

9.4.1 Fishes

It was widely accepted that the collection of marine ornamental fishes in India is mostly from coral reef regions, which often leads to habitat destruction. In this context, the only way for sustainable aquarium trade of marine ornamental fishes is through hatchery production of particular fishes to meet the demand and conserve the wild germplasm resources. Successful larviculture mainly depends on the size and nutritional quality of live feeds. Among the marine ornamental fishes, the breeding and seed production of clown fishes were first achieved successfully, and the technology was also standardized (Hoff 1996). In India, the techniques on the captive culture of clownfish and a few damselfishes were achieved (Gopakumar et al. 2001, 2002; Ignatius et al. 2001; Madhu and Madhu 2002; Ajith Kumar et al. 2007, 2008, 2010, 2015).

The establishment of marine ornamental fish hatcheries was accomplished to develop propagation protocols for marine ornamental fishes and transfer of technology to the coastal and island communities for their sustainable livelihood development and conservation of marine biodiversity. The hatchery technology for marine ornamental fishes was developed for *Amphiprion chrysogaster* (Mauritian anemone fish); *A. percula* (true percula); *A. sandaracinos* (yellow skunk clown); *A. ocellaris* (common clown/false clown); *A. sebae* (sebae clown); *A. clarkii* (Clark's anemone fish); *A. nigripes* (Lakshadweep clown); *A. frenatus* (tomato clown); *A. akallopisos* (nosestripe clownfish); *A. perideraion* (pink anemone fish); *A. polymnus* (saddleback clown); *A. ephippium* (red saddleback anemone fish); *Premnas biaculeatus* (maroon clown/spine cheek anemone fish); damsels, viz. *Dascyllus trimaculatus* (three-spot damsel), *D. aruanus* (striped damsel), *D. carneus* (bluetop dascyllus), *D. reticulatus* (two-stripe damselfish), *Pomacentrus caeruleus* (blue damsel), *P. pavo* (sapphire or peacock damsel), *Neopomacentrus nemurus* (yellow-tail damsel), *N. filamentosus* (filamentous tail damsel), *N. cyanomos* (regal demoiselle), *Neoglyphidodon oxyodon* (bluestreak damselfish), *Chrysiptera unimaculata* (one-spot damsel), and *C. cyanea* (sapphire devil); dottyback *Pseudochromis dilectus* (redhead dottyback); and *Nemateleotris decora* (purple fire goby) (Madhu and Madhu 2006; Madhu et al. 2006a, b, c; Gopakumar et al. 2007; Ajith Kumar et al. 2007; Madhu et al. 2007a, b; Ajith Kumar et al. 2008; Madhu et al. 2008; Ajith Kumar et al. 2010, 2015; Madhu et al. 2016). The final solution for the long-term sustainable trade of marine ornamentals depends on the trade developed from hatchery-produced animals. Due to the high unit value of marine ornamentals, the economic viability is also more lucrative than other mariculture species.

9.4.1.1 Captive Propagation of Clown and Damsel: A Glance

Sub-adult fishes of clown and damsel normally collected from the wild and also procured from traders were accommodated in the quarantine facility of the hatchery. Healthy clown fishes were transferred to broodstock development tanks after a period of 1.5–2 weeks (two nos./tank) along with an anemone. The tanks are of FRP with 600–750 lit capacity. A biological and canister filtration system needs to be provided for the continuous spawning. Damsels were stocked in FRP tanks with 750–1000 l capacity, with an individual canister filter. Water exchange of 10–25% was carried out twice a month. The water quality parameters such as water temperature (27 ± 01 °C), salinity (29–30 PPT), pH (7.9–8.1), and ammonia (0.001–0.003 PPM) were maintained in the rearing systems. The fishes were fed at 5% of the body weight thrice a day, mainly with boiled prawn, clam, *Acetes*, and frozen *Artemia*. A photoperiod of 14 h L:10 h D was maintained throughout the culture period.

Spawning was noticed in clowns, after 2.5–3 months of rearing, and the male exhibited parental care for the eggs, by fanning with fins. Fecundity ranged between 400 and 2000 nos., and it varied depending on the species and size of the female. Spawning frequency was recorded mostly twice in a month and on rare occasions thrice per month, and that too varied with species. Hatching took place, at 6.00–7.00 p.m., after 7–8 days of incubation, which was influenced by the surrounding temperature. Larvae were transferred to the larval rearing tanks, filled with micro algae *Chlorella* sp. Rotifer (*Brachionus rotundiformis*) was given from the second day (6–8 nos./ml) of hatching and continued till the tenth day. Following, newly hatched *Artemia* nauplii were fed at the rate of 3–5 nos. ml^{-1}, and gradually the concentration was increased till the larvae reached metamorphosis, which took 25–30 days. Further, the babies were moved to grow-out tanks along with sea anemones, where squashed mussels, prawns, and frozen *Artemia* were given as feed.

The damsels laid eggs after 5 months on any hardy substratum available in the tank, with a preference for dead corals. They exhibited the spawning frequency with an average of 2–3 per month, with intervals of 1–2 weeks. The fecundity was around 2000–3000 nos. per spawning. Hatching took place in the late evening hours after 3–4 days with a hatching rate of around 90%. The larval body was fully formed after the third day. When compared with the clown larvae, the mouth size of damsels is small; hence, they are fed with copepod nauplii in the initial days followed with rotifer, *B. plicatilis*, and *Artemia* nauplii. The larvae took around 45 days to reach a miniature adult.

9.5 Marine Ornamental Invertebrates

9.5.1 Shrimps

Marine ornamental invertebrates are the other group preferred by aquarists. Similar to the fishes, these resources are also mostly captured from the reef environment for the aquarium trade. Technology for marine ornamental crustaceans has been limited to a few numbers, and among them, the species belonging to *Lysmata*, *Periclimenes*, *Hymenocera*, and *Stenopus* received special attention, due to the growing demand and aquarium trade (Lin et al. 2001).

Live ornamental shrimps were collected from different islands, and a broodstock was developed for six marine ornamental shrimps, namely, *Stenopus hispidus*, *Gnathophyllum americanum*, *Saron marmoratus*, *S. neglectus*, *Ancylocaris brevicarpalis*, and *Thor hainanensis*. Three years of experimental trials have been yielded with captive propagation technology for *T. hainanensis* and *A. brevicarpalis*, and the technology has been commercialized.

9.6 Successful Model in India

The ICAR-National Bureau of Fish Genetic Resources (NBFGR) in collaboration with the Mangrove Foundation and Mangrove Cell, Department of Forest, Government of Maharashtra, established a demonstration ornamental fish hatchery at the premises of the Coastal and Marine Biodiversity Centre, Thane, Maharashtra, India. This facility was established for the supply of hatchery-bred clown fishes to the beneficiaries for further rearing in cluster mode rearing units, established at five coastal districts of Maharashtra, with the aim of establishing a model marine ornamental fish village in the country. In similar lines, a programme is being operated by the institute at Pichavaram mangrove region, the recently declared Ramsar site in Tamil Nadu, India, for the livelihood development of the local community.

The other initiative of the NBFGR is at Agatti Island of Lakshadweep, India, on the breeding and rearing of marine ornamental invertebrates with a people-participatory approach through community aquaculture units. This is a unique and successful venture, where science and societal development are converging together by utilizing native organisms and resources for the livelihood development of the local community. Exploratory surveys conducted in different pristine reef islands of Lakshadweep resulted in the discovery of five new species/new to science and a few new distributional records from Indian waters, which may be promising candidate species for marine ornamental aquaculture in the near future. The captive breeding protocol was standardized and disseminated for two marine ornamental shrimps, *T. hainanensis* and *A. brevicarpalis*, for the first time globally (ICAR-NBFGR 2019). The rearing technology of the said two species was disseminated to the women beneficiaries of Agatti Island, Lakshadweep, for livelihood development through community aquaculture. Four community aquaculture units, an eco-friendly model, are being operated successfully by the local women and generating regular income. This is rather a unique venture, wherein science and societal development are taken up hand in hand with indigenous resources.

9.7 Scope and Way Forward

Marine ornamental fish culture provides livelihood to millions of people in the world, and culturing ornamentals is alternatively reducing the pressure on coral reefs. Integration of technology with societal benefit needs to be prioritized for the effective utilization of biodiversity for the benefit of the local people. Further, scientific evidence-based plans and policies are the need of the hour for ushering the growth of the evergreen and profitable marine ornamental sector.

Culturing aquatic organisms in a confined system is an alternate way to conserve wild resources and maintain the ecological balance. Further, hatchery production, adaptation, and supply of organisms for aesthetic purposes will open new avenues in the rural regions, and it will elevate the hope of the people. In this context, the establishment of community aquaculture units of marine ornamentals at various locations of the country will pave the way for the utilization of native resources for income generation by the local community.

The functional target of the model is to conserve wild collected parents in captivity and utilize F2 generation, to help generate supplementary income for the families. Mostly, such communities, residing close to the fragile ecosystems, have limited livelihood options and skills and are more prone to the adverse impacts of changing climates. The diversified livelihood options and supplementary income can help such people to remain resilient and at the same time sensitive to protecting the ecosystems.

This model involves exploration and captive propagation with community participation as two major pillars for native species grown by local people in an environment-friendly manner, besides abiding by the laws of biodiversity. The whole process is environment-friendly (utilizing local indigenous resources) and does not release any obnoxious emissions, and more importantly the power energy required to run the facilities is primarily met through solar units.

Moving forward, this model has the potential to develop Lakshadweep as a marine ornamental hub with native species for the benefit of local islanders and build a case history of successful access-benefit sharing. The paybacks are capacity building, improved income of locals, and gender mainstreaming. The program aims to comply with national and international goals such as the Aichi Biodiversity Targets, Nagoya Protocol, and Sustainable Development Goals (SDGs).

Acknowledgements The authors thank the Director of the ICAR-National Bureau of Fish Genetic Resources (NBFGR) for the support and encouragement. The Centre for Marine Living Resources and Ecology (CMLRE, MoES), the Department of Science and Technology (SEED, DST), and the Mangrove Foundation, Govt. of Maharashtra, are acknowledged for funding support to implement the programme at Lakshadweep.

References

Ajith Kumar TT, Balasubramanian T (2009) Broodstock development, spawning and larval rearing of the false clown fish, *Amphiprion ocellaris* in captivity using estuarine water. Curr Sci 97(10):1483–1486

Ajith Kumar TT, Jeniffer PN, Murugesan P et al (2007) Marine ornamentals in India: challenges and opportunities for sustainability. Fish Chimes 27:44–51

Ajith Kumar TT, Dhaneesh KV, Arumugam M et al (2008) Stability of marine ornamental fishes in captivity: a case study in marine research aquarium of Annamalai University. Glob J Mol Sci 3(1):35–41

Ajith Kumar TT, Setu SK, Murugesan P et al (2010) Studies on captive breeding and larval rearing of clownfish *Amphiprion sebae* (Bleeker, 1853) using estuarine water. Indian J Mar Sci 39(1):114–119

Ajith Kumar T, Gunasundari V, Prakash S (2015) Breeding and rearing of marine ornamentals. Advances in Marine Brackishwater Aquaculture:101–107

Anil MK, Gomathi P, Raheem PK et al (2018) Captive broodstock development, breeding and seed production of Anthid fish (family: Serranidae) Marcia's anthias, *Pseudanthias marcia* in recirculation aquaculture system (RAS). Aquaculture 492:265–272

CORAL (2018) CORAL magazine's captive-bred marine aquarium fish & invertebrate list—project homepage. https://www.reef2rainforest.com/coral-magazines-captive-bred-marine-aquarium-fishlist-project-homepage/

Dammannagoda ST (2018) Sustainable fishing methods in Asia pacific region. Sustainable Aquaculture 95–122

Dhaneesh KV, Ajith Kumar TT, Vinoth R et al (2011) Influence of brooder diet and seasonal temperature on reproductive efficiency of clownfish *Amphiprion sebae* in captivity. Recent Res Sci Technol 3(2):95–99

Gopakumar G, George RM, Jasmine S (2001) Hatchery production of the clownfish *Amphiprion chrysogaster*. In: Menon NG, Pillai PP (eds) Perspectives in mariculture., J Mar Biol Ass India, pp 305–310

Gopakumar G, Sreeraj T, Ajith Kumar TT et al (2002) Breeding and larval rearing of three species of damselfishes (Family: Pomacentridae). Mar Fish Infor Serv T & E Ser 171:3–5

Gopakumar G, Madhu K, Madhu R (2007) Seed production of marine ornamental fishes for trade. CMFRI Newsletter No. 114 April–June 2007, 114:1–3

Hoff FH (1996) Conditioning, spawning and rearing of fish with emphasis on marine clownfish. Aquaculture Consultants, Incorporated

ICAR-NBFGR (2019) Annual Report 2022. ICAR-National Bureau of Fish Genetic Resources, Lucknow, India

Ignatius B, Rathore G, Jagadis I et al (2001) Spawning and larval rearing technique for tropical clown fish *Amphiprion sebae* under captive condition. J Agric Trop 16(3):241–249

Lal KK, Ajith Kumar TT (2021) Marine ornamental aquaculture: measure towards biodiversity conservation and livelihood promotion. J Indian Soc Coast Agric Res 39(2):49–54

Lin J, Zhang D, Creswell R (2001) Aquaculture of marine ornamental shrimps: an overview. Aquaculture 2001 book of abstracts. World Aquaculture Society, Orlando, p 378

Lin J, Zhang D, Rhyne AL (2002) Broodstock and larval nutrition of marine ornamental shrimp. Avances en Nutrici6n Acuicola 589

Madhu K, Madhu R (2002) Successful breeding of common clownfish *Amphiprion percula* under captive conditions in Andaman and Nicobar Island. Fish Chimes 22(9):16–17

Madhu K, Madhu R (2006) Hatchery seed production of marine ornamental clown fishes: a remunerative option for livelihood in India. In: National Official language Seminar Livelihood issues in Fisheries and Aquaculture Special Publication-No. 90, p 5155

Madhu K, Madhu R, Krishnan L, Sasidharan C, Venugopal K (2006a) Spawning and larval rearing of *Amphiprion ocellaris* under captive condition. Mar Fish Inform Serv Tech Ext. 188

Madhu K, Madhu R, Gopakumar G, Sasidharan CS (2006b) Breeding, larval rearing and seed production of maroon clown *Premnas biaculeatus* under captive conditions. Mar Fish Infor Serv T & E Ser 190:1–5

Madhu K, Madhu R, Krishnan L, Sasidharan CS, Venugopal KM (2006c) Spawning and larval rearing of *Amphiprion ocellaris* under captive condition. Mar Fish Infor Serv T & E Ser 188:1–5

Madhu K, Madhu R, Sasidharan CS et al (2007a) Success in captive breeding of maroon clown *Premnas biaculeatus* (Bloch, 1790). Fish Chimes 27(2):30–57

Madhu R, Madhu K, Gopakumar G (2007b) Success in captive breeding of redhead dottyback, *Pseudochromis dilectus*, CMFRI Newsletter (October to December), 24

Madhu K, Madhu R, Gopakumar G (2008) Hatchery production of the Spine Cheek Anemone fish *Premnas biaculeatus* (Bloch, 1790) Tropical Fish Hobbyist, USA 5:104–108

Madhu K, Madhu R, Retheesh T (2016) Spawning, embryonic development and larval culture of redhead dottyback, *Pseudochromis dilectus* Lubbock, 1976 under captivity. Aquaculture 459:73–83

Olivotto I, Chemello G, Vargas A et al (2017) Marine ornamental species culture: from the past to "finding dory". Gen Comp Endocrinol 245:116–121

Pabhu C (2021) Ornamental fisheries of India. Adv Fish Aquac Hydrobiol 9(2):3–4

Pandey PK, Mandal SC (2017) Present status, challenges, and scope of ornamental fish trade in India. In: Conference on Aqua Aquaria India, Mangalore, pp 1–10

Pouil S, Tlusty MF, Rhyne AL et al (2020) Aquaculture of marine ornamental fish: overview of the production trends and the role of academia in research progress. Rev Aquac 12(2):1217–1230

Prakash S, Ajith Kumar TT, Raghavan R et al (2017) Marine aquarium trade in India: challenges and opportunities for conservation and policy. Mar Policy 77:120–129

Raja K, Aanand P, Padmavathy S et al (2019) Present and future market trends of Indian ornamental fish sector. Int J Fish Aquat Stud 7(2):6–15

Rhyne AL, Tlusty MF, Schofield PJ et al (2012) Revealing the appetite of the marine aquarium fish trade: the volume and biodiversity of fish imported into the United States. PLoS One 7(5):e35808

Shinoj P, Baiju KK, Vijayagopal P (2021) Status and prospects of ornamental fish and fish feed industry in southern India. Mar Fish Infor Serv T & E Ser 248:7–11

Wabnitz C, Taylor M, Green E et al (2003) From ocean to aquarium. Cambridge: United Nations Environment Programme. World Conservation Monitoring Center. https://www.unep-wcmc.org/resources/publications/UNEP_WCMC_bio_series/17.html

Genotoxicity Biomarkers for Environmental Health Assessment Towards Sustainable Management of Fish Genetic Resources

10

Basdeo Kushwaha, Rashmi Srivastava, and Ravindra Kumar

Abstract

High industrialization, urbanization, and agricultural practices generate several types of chemicals, pesticides, microplastics, dusts, and heavy metal particles that are severely accumulating in the environment, including in water bodies, causing environmental pollution and genotoxicity in living organisms. Genotoxicity biomarkers play a dynamic role in assessing the impact of environmental stressors and are also considered instrumental in formulating strategies for sustainable management. Micronuclei, the small additional nuclei formed during cell division, resulting from chromosomal damage are widely used to assess the genotoxicity in fish populations exposed to pollutants. The comet assay is a very sensitive gel electrophoresis-based method that measures the DNA strand breaks and allows the detection of damaged DNA at the individual cell level. The sister chromatid exchange (SCE), indicative of the exchange of genetic material between the sister chromatids, is another method that provides information on chromosomal instability and genotoxic effects. The changes in the activity of antioxidant enzymes, such as SOD and CAT, indicate oxidative stress and potential genotoxic effects. Elevated levels of malondialdehyde (MDA) suggest damage to cell membranes due to oxidative stress. Regular monitoring of these biomarkers contributes to the conservation of fish genetic

Basdeo Kushwaha and Rashmi Srivastava contributed equally with all other contributors.

B. Kushwaha (✉) · R. Kumar
ICAR-National Bureau of Fish Genetic Resources, Lucknow, Uttar Pradesh, India
e-mail: basdeo.kushwaha@icar.gov.in; Ravindra.Kumar1@icar.gov.in

R. Srivastava
School of Life Sciences, Department of Zoology, Babasaheb Bhimrao Ambedkar University, Lucknow, Uttar Pradesh, India

U. K. Sarkar et al. (eds.), *Sustainable Management of Fish Genetic Resources*,
https://doi.org/10.1007/978-981-97-5250-8_10

resources and the overall health of aquatic ecosystems. The future of genotoxicity biomarkers has great promise for evaluating and improving the health of both fish and humans. In the present chapter, we have discussed here the genotoxicological endpoints of various environmental pollutants in fish, highlighting the biomarkers for environmental health assessment towards the sustainable management of fish genetic resources.

Keywords

Fish · Genotoxicity · Mutagenicity · Oxidative stress · Gene expression · Pollutants

10.1 Introduction

Genotoxicological endpoints in fish play a crucial role in environmental health assessments and also provide insights into the impact of environmental contaminants on the genetic material of aquatic organisms (Lakra and Nagpure 2009). Genotoxicity refers to the ability of substances to cause damage to an organism's genetic material, which includes DNA, RNA, etc. Fish are sensitive indicators of environmental stress due to their direct exposure to waterborne contaminants (Pandey et al. 2005, 2006). Some of the genotoxicological endpoints and biomarkers used in fish for environmental health assessment are DNA damage and repair, comet assay (single-cell gel electrophoresis), micronucleus test, oxidative stress biomarkers including antioxidant enzymes, gene expression, gill and liver histopathology, adduct formation and transcriptomics (Nagpure et al. 2008; Tiwari et al. 2010; Nwani et al. 2013; Rashid et al. 2019a, b, 2020). Among these parameters/assays, the comet assay technique assesses the DNA damage at the individual cell level by measuring the migration of damaged DNA fragments in an electric field. The presence of micronuclei in fish erythrocytes indicates chromosomal damage and can serve as a biomarker for genotoxicity (Chaudhary et al. 2006). Increased levels of antioxidant enzymes, like catalase, superoxide dismutase, etc., can indicate oxidative stress resulting from exposure to genotoxic agents. The measurement of lipid peroxidation products, such as malondialdehyde (MDA), can reflect oxidative damage to cell membranes (Dabas et al. 2011, 2014a, b).

Genotoxicity biomarkers at the molecular level involve the assessment of gene expression that can indicate DNA damage and repair processes such as the tumor suppressor gene ($_p$53 gene). Sister chromatid exchange (SCE) is a cytogenetic technique that reveals the exchange of genetic material between sister chromatids and is used as an indicator of genotoxicity. At the physiological levels, the examination of fish gill and liver tissues can reveal abnormalities such as hyperplasia, necrosis, and tumors, which may be indicative of genotoxic effects (Sharma et al. 2007; Srivastava et al. 2016). In addition, biomarkers of exposure to specific contaminants are the adduct formation formed by the binding of contaminants to DNA and can be specific to certain genotoxic agents. Several genomic instability markers cause changes in microsatellite instability and chromosomal aberrations and may indicate genomic instability, caused by exposure to genotoxicants.

Gene expression profiling involving transcriptomics analyzes changes in gene expression patterns and can provide information on the molecular pathways, via fish genomics, affected by genotoxic agents (Tiwari et al. 2010, 2011a, b; Pathak et al. 2019). Some reproductive biomarkers, which cause genotoxic effects on gametes, help in the assessment of DNA damage in sperm and eggs and can provide insights into the potential impact on reproductive success and offspring health. Monitoring these genotoxicological endpoints in fish populations contributes to a comprehensive understanding of the environmental health of aquatic ecosystems and aids in the identification of potential sources of pollution. These biomarkers play a vital role in regulatory frameworks for assessing the safety of aquatic environments and developing strategies for environmental protection and conservation.

10.2 Fish and Environment

The relationship between fish and the environment is intricate and dynamic. Fish are highly sensitive to environmental changes, and their health and survival are directly influenced by various ecological factors (Pathak et al. 2019). The key aspects of the interaction between fish and their environment include water quality, pH level, temperature, dissolved oxygen (DO), aquatic habitats, anthropogenic activities, biotic interactions, pollution and contaminants, microplastics, fish migration, climate change, ocean acidification, overfishing and fisheries management, sustainable harvesting, etc.

Fish exhibit specific pH requirements, and variations outside their preferred range can be stressful or even fatal. Acidification or alkalinity of water bodies can affect fish physiology. Different fish species thrive in specific temperature ranges. Temperature influences metabolic rates, reproduction, and overall behavior. Rapid temperature changes can be detrimental to fish health. Adequate oxygen levels are crucial for fish respiration. Pollution, eutrophication, or other factors leading to low oxygen concentrations can harm fish populations. Fish depend on diverse habitats such as rivers, lakes, oceans, and wetlands, whereas human activities, habitat destruction, and alterations to water flow can disrupt these ecosystems, affecting fish populations. Fish interact with other species in their ecosystems, including prey, predators, and competitors. Changes in the abundance of these species can have cascading effects on fish populations. Industrial discharges, agricultural runoffs, and other human activities introduce pollutants into water bodies or riverine ecosystems (Srivastava et al. 2005). Fish are particularly vulnerable to contaminants like heavy metals, pesticides, and industrial chemicals, which can lead to bioaccumulation. The presence of microplastics in aquatic environments poses a threat to fish. Ingestion of microplastics can have adverse effects on fish health, affecting their digestive systems and potentially leading to the transfer of pollutants up the food chain. Many fish species undertake migrations for spawning or feeding purposes. Dams, barriers, and other obstructions can disrupt these migration patterns, impacting fish reproduction and population dynamics. Connectivity between different water bodies is essential for maintaining healthy fish populations, and fragmentation of water bodies can isolate populations and reduce genetic diversity (Lakra et al. 2010).

Global climate change affects water temperatures, impacting fish distribution and their behavior. Some species may need to migrate to cooler waters to survive, leading to shifts in their geographic ranges. Increased carbon dioxide levels in the atmosphere contribute to ocean acidification, which can affect the development and survival of fish larvae and other marine organisms. Overfishing can deplete fish populations, disrupting ecosystems and jeopardizing the livelihoods of communities dependent on fisheries. Sustainable fisheries management is crucial to maintaining healthy fish stocks. Understanding the complex interactions between fish and their environment is essential for effective conservation and management strategies. Conservation efforts often involve habitat restoration, pollution control, sustainable fishing practices, and the establishment of protected areas to safeguard fish populations and preserve aquatic ecosystems. Additionally, research on the impacts of climate change and other stressors helps inform adaptive management strategies to ensure the long-term health of fish and their environments.

10.3 Environmental Pollutants as "Genotoxic Agents"

Environmental pollutants can act as genotoxic agents, causing damage to the genetic material (DNA) of living organisms. Genotoxicity refers to the ability of substances to induce genetic damage, which can lead to mutations, chromosomal aberrations, and other alterations in the DNA structure. Various environmental pollutants have been identified as genotoxic agents, posing risks to both human and ecological health. Some common classes of environmental pollutants with genotoxic properties include polycyclic aromatic hydrocarbons (PAHs) that are produced during incomplete combustion of organic matter, such as fossil fuels, wood, and tobacco. PAHs are known to form DNA adducts and induce mutations. They are associated with an increased risk of cancer and can affect both aquatic and terrestrial organisms. Heavy metals such as lead, cadmium, mercury, and arsenic are dumped into the environment after industrial activities, mining, and burning of fossil fuels (Dabas et al. 2011, 2014a, b). Heavy metals can induce DNA damage, interfere with DNA repair mechanisms, and lead to chromosomal abnormalities, and long-term exposure is associated with cancer and other health problems (Tiwari et al. 2010, 2011a, b).

On the other hand, agricultural and residential use of pesticides introduces chemical compounds such as organochlorines, organophosphates, and pyrethroids into the environment (Ali et al. 2008, 2009; Kumar et al. 2010). Some pesticides are genotoxic and can cause DNA damage, leading to mutations and chromosomal aberrations (Pandey et al. 2011a, b, c). They may also contribute to the development of cancer in exposed organisms. Endocrine-disrupting chemicals (EDCs) include substances like bisphenol A (BPA), phthalates, and certain pesticides, which interfere with hormonal systems in organisms and may induce genotoxic effects by disrupting the endocrine system, leading to alterations in reproductive tissues, hormonal imbalances, and potential impacts on the genetic material. Combustion of fossil fuels, industrial processes, and other human activities release particulate matter into the air. Inhalation of airborne particulate matter has been linked to DNA damage, oxidative stress, and inflammation in respiratory cells. Some particles may also reach other organs and tissues, contributing to systemic genotoxic effects.

Persistent organic pollutants (POPs), including chemicals like polychlorinated biphenyls (PCBs), dioxins, and certain pesticides, are resistant to environmental degradation and can bioaccumulate in the food chain. POPs can induce genotoxic effects, leading to DNA damage and disruption of cellular processes. They are associated with various health issues, including cancer and reproductive abnormalities. Certain metabolic byproducts of industrial processes, such as reactive oxygen species (ROS) and free radicals, can be generated during the breakdown of pollutants. ROS and free radicals can cause oxidative stress, leading to DNA damage and mutations (Nwani et al. 2013). Chronic exposure may contribute to the development of cancer and other health problems. Understanding the genotoxic effects of environmental pollutants is crucial for assessing the risks associated with exposure and implementing strategies to mitigate environmental contamination (Fig. 10.1). Monitoring and regulation of pollutants, along with the development of sustainable practices, are essential for minimizing the impact of genotoxic agents on ecosystems and human health.

10.3.1 Pesticides

Pesticides are substances or mixtures of substances that are used to control, repel, or kill pests. Pests can include insects, weeds, fungi, rodents, and other organisms that negatively impact crops, livestock, or human health. Pesticides play a crucial role in modern agriculture by helping to increase crop yields and protect crops from pests that can reduce productivity. There are various types of pesticides: insecticides are designed to control insects that damage crops; herbicides are used to control or eliminate unwanted plants, commonly known as weeds; fungicides target fungi and prevent or control fungal diseases in plants; rodenticides are designed to control rodents such as mice and rats; bactericides and virucides are used to control bacterial and viral diseases, respectively; and nematicides aim to kill nematodes, which are microscopic worms and can harm plants. On one side, pesticides have contributed towards increased agricultural productivity, but on the other side there are concerns and challenges associated with their use. Pesticides can have unintended effects on nontarget organisms, leading to environmental contamination and harming beneficial insects, birds, and aquatic life (Nwani et al. 2010a, b, 2011, 2014; Kumar et al. 2010; Pandey et al. 2014). The presence of pesticide residues in food raises concerns about potential health risks for consumers. Pests can develop resistance to pesticides over time, rendering them less effective, which in turn necessitates the development of new and more effective pesticides. Pesticide exposure can pose health risks to farmers, farmworkers, and consumers, if not used properly, wherein chronic exposure has been associated

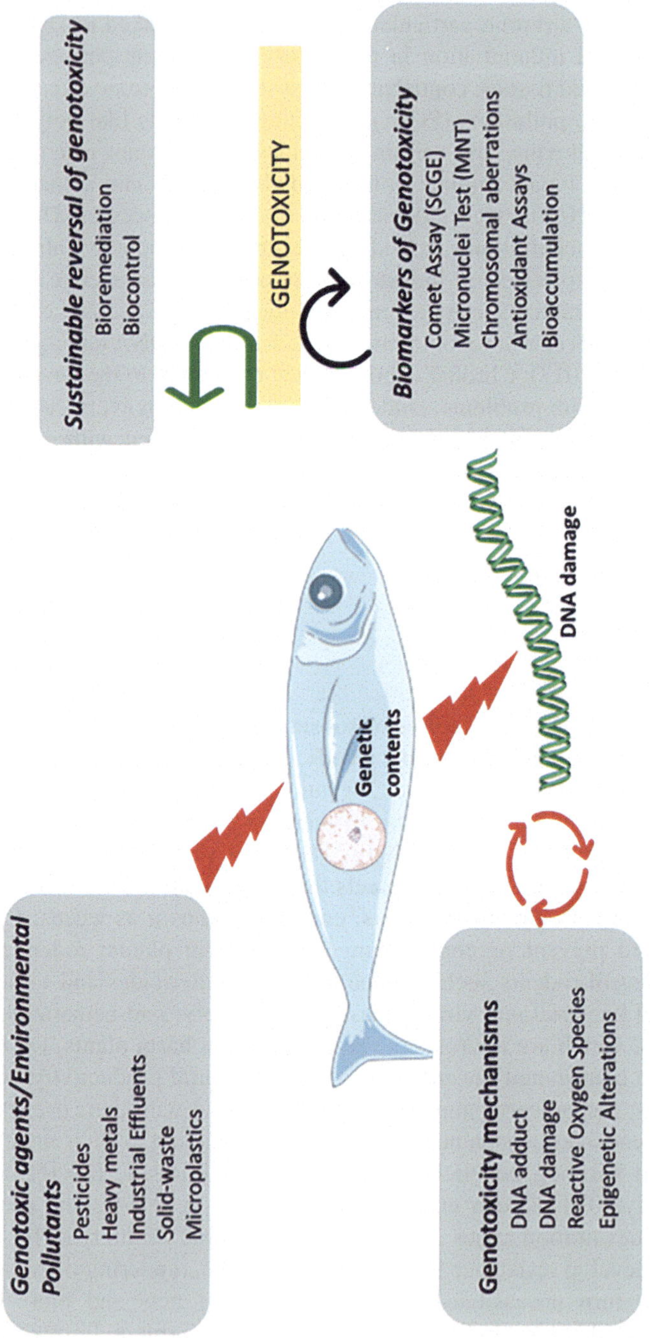

Fig. 10.1 A general overview of genotoxic effects induced by environmental pollutants and its sustainable reversal in fish

with various health issues. Efforts are ongoing to develop and promote sustainable and environmentally friendly alternatives to traditional pesticides, such as integrated pest management (IPM) strategies, that combine biological control, cultural practices, and judicious pesticide use. Additionally, the development of genetically modified crops with built-in resistance to pests aims to reduce the need for chemical pesticides. Regulations and guidelines are also in place to ensure safe and responsible use of pesticides.

10.3.2 Heavy Metals

Heavy metals are metallic elements with high atomic weights and densities. They can be natural components of the Earth's crust or introduced into the environment through human activities such as industrial processes, mining, and agricultural practices. Some heavy metals are essential for biological processes in trace amounts, but excessive exposure can lead to environmental pollution and pose risks to human health. Some of the common heavy metals that cause deteriorative effects on the environment when their exposure crosses the permissible limits are lead (Pb), mercury (Hg), cadmium (Cd), arsenic (As), chromium (Cr), nickel (Ni), etc. To mention, Pb is found in lead-based paints, batteries, and some industrial processes. Lead exposure is associated with neurological and developmental issues, especially in children. Hg is released into the environment through industrial processes, coal combustion, and certain natural sources and can accumulate in fish, posing a risk to human health when consumed in large quantities. It can have adverse effects on the nervous system. Cd is present in some fertilizers, mining activities, and certain industrial processes. Long-term exposure to cadmium can lead to kidney damage and other health issues. Arsenic is found in certain minerals and used in some industrial processes and can contaminate water sources, leading to serious health problems, including skin lesions and cancers. Cr is used in various industrial applications, including metal plating and leather tanning, whereas hexavalent chromium, a more toxic form, can pose health risks, including lung cancer (Kumar et al. 2012, 2013; Nagpure et al. 2015b). Ni is prevalent in some industrial processes and alloys, and exposure to it can cause skin allergies and respiratory issues.

The presence of heavy metals in the environment raises concerns due to their persistence, bioaccumulation in the food chain, and potential toxicity. Exposure to heavy metals can occur through inhalation, ingestion, or skin contact. The health effects of heavy metal exposure depend on various factors, such as the type of metal, its chemical form, concentration, and duration of exposure. Efforts to address heavy metal pollution include implementing environmental regulations, adopting cleaner industrial processes, and developing technologies for the remediation of contaminated sites. Monitoring and testing for heavy metals in water, soil, and air are crucial for assessing environmental quality and protecting human health. Additionally, public awareness and education play a key role in reducing exposure to heavy metals, especially in vulnerable populations.

10.3.3 Industrial Effluents

Industrial effluents refer to the liquid waste streams generated as a result of industrial processes. These effluents can contain a variety of pollutants, including chemicals, heavy metals, nutrients, and other substances that may be harmful to the environment if not properly treated before discharge (Kushwaha et al. 2012). Managing industrial effluents is essential to prevent environmental contamination and protect ecosystems as well as human health. The composition of industrial effluents varies widely depending on the industry and the specific processes involved. Common components of industrial effluents include suspended solids, organic compounds, heavy metals, acids, alkaline substances, and sometimes toxic chemicals (Nagpure et al. 2015a, b, 2017). Various industries contribute to industrial effluent generation, including manufacturing, mining, chemical production, food processing, and textile manufacturing. Each industry has its specific set of pollutants based on the materials and processes used. Improperly treated or untreated industrial effluents can have adverse effects on water quality, soil, and aquatic ecosystems. Pollutants in effluents can disrupt the balance of ecosystems, harm aquatic life, and pose risks to human health, if they contaminate drinking water sources or agricultural areas. In the direction of mitigating the environmental impact, industrial effluents often undergo treatment before discharge. Treatment processes may include physical, chemical, and biological methods to remove or reduce pollutants. Effluent treatment plants (ETPs) are commonly used in industries to treat wastewater. Most countries have regulations and standards in place to control the discharge of industrial effluents. These regulations specify permissible pollutant levels, require industries to obtain permits, and outline the necessary treatment processes. Noncompliance may result in penalties and legal consequences. Some industries implement strategies for recycling and reusing treated effluents within their processes. This not only reduces environmental impact but also conserves water resources.

Efforts to address industrial effluents involve a combination of regulatory measures, technological advancements, and industry best practices. Sustainable practices, such as cleaner production methods and the use of eco-friendly technologies, aim to minimize the generation of harmful effluents in the first place. Industries must adopt responsible waste management practices and invest in effective effluent treatment technologies to minimize their environmental footprint and contribute to sustainable development. Public awareness and engagement are also essential in promoting environmentally responsible industrial practices.

10.3.4 Solid Waste

Solid waste refers to any discarded or abandoned materials that are not liquid or gas. It includes a wide range of items, such as household waste, industrial waste, construction and demolition debris, and agricultural waste. Solid waste can be categorized into various types based on its source, composition, and characteristics. The management of solid waste is a crucial aspect of environmental sustainability and public health. Solid waste can be classified into several forms and modes of transmission. One is municipal solid waste (MSW) generated from households, commercial establishments, and institutions. Others include industrial solid waste produced by manufacturing and industrial processes; construction and demolition waste generated during construction, renovation, and demolition activities; and hazardous waste that contains materials harmful to human health and the environment. Contamination from solid waste can be eradicated by practicing solid waste management strategies that involve gathering and transporting waste from its source to a treatment or disposal facility, composting, incineration, and recycling to reduce the volume and harmful effects of waste. The final destination of waste can include landfills, incineration plants, or other specialized facilities. Recycling is a key component of solid waste management, aiming to recover valuable materials from waste and reduce the environmental impact such as paper, glass, metal and plastics. Some encouraging practices such as waste reduction, reuse, and sorting waste at its origin to facilitate recycling and proper disposal help minimize the amount of waste sent to landfills. Improper solid waste management can lead to environmental pollution, soil and water contamination, and adverse health effects. Landfills, if not properly designed and managed, can contribute to the release of greenhouse gases. Many countries have regulations and policies in place to govern the proper management and disposal of the solid waste. These regulations often aim to promote recycling, reduce landfill use, and minimize environmental impact. Effective solid waste management is essential for sustainable development, environmental conservation, and public health. Community awareness and participation play a crucial role in achieving successful waste management practices.

10.3.5 Microplastics

Microplastics are small plastic particles, generally defined as plastic fragments with a size ranging from 1 micrometer to 5 mm. They can be categorized into two primary types: primary microplastics and secondary microplastics. Primary microplastics are intentional microplastics that are deliberately produced for certain products, such as microbeads in personal care products (e.g., exfoliating scrubs, toothpaste) or pellets used in industrial processes, while nurdles are small, pre-production plastic pellets used as raw material in the manufacturing of plastic products. Secondary microplastics can break down from larger to smaller plastics over time, as in the case of larger plastic items like bottles and bags that break down into smaller particles due to weathering, UV radiation, and other environmental processes. Microscopic plastic fibers shed from synthetic textiles during washing can contribute to microplastic pollution. Microplastics can enter the environment through various pathways, like direct disposal involving improper disposal of plastic waste into the oceans, rivers, and other water bodies. Stormwater runoff that occurs due to rainwater has the potential to transport microplastics from urban areas into water systems. Microplastics can become airborne and settle into the environment. In addition, wastewater treatment plant effluents may not effectively capture all microplastics, allowing them to enter water bodies.

Microplastics can harm aquatic ecosystems, affecting marine life such as fish, shellfish, and other organisms. They can absorb and accumulate toxic substances, leading to potential bioaccumulation in the food chain. The impact of microplastics on human health is an area of ongoing research. Microplastics have been detected in various food items and drinking water sources, raising concerns about potential ingestion by humans. Microplastics can interfere with natural processes and contribute to the overall degradation of ecosystems. Microplastic pollution can be addressed by the implementation of rules and regulations and bans on avoidable plastic products and their constituents. Some countries and regions have implemented bans on certain types of microplastics, such as microbeads in personal care products. Spreading public awareness and educating about the sources and impacts of microplastics to encourage responsible consumption and waste management are good steps in protecting the environment against microplastics. In the present scenario, research works should be encouraged for a better understanding of the sources, transport, and impact of microplastics, and the development of innovative technologies to capture and remove microplastics from the environment. Reducing the use of single-use plastics, improving waste management practices, and promoting sustainable alternatives are key strategies to mitigate the issue of microplastic pollution.

10.4 Genotoxicity Mechanisms

10.4.1 DNA Adducts

Genotoxicity, the potential of substances to harm genetic material within cells, is intimately linked to the formation of DNA adducts (Nagpure et al. 2015b). These adducts arise when external agents form covalent bonds with DNA, creating abnormal structures that disrupt essential functions (Dutta et al. 2024). This disruption can lead to various forms of genetic damage, including mutations, chromosomal aberrations, and altered gene expression. Several mechanisms contribute to the genotoxicity of DNA adducts. One key mechanism is mutagenesis, where mismatches or deletions occur during DNA replication due to the adducts' interference (Ahmed et al. 2015). Additionally, DNA repair mechanisms can be hampered by the presence of adducts, hindering the recognition and removal of damaged segments (Kumar et al. 2012, 2023). Some adducts directly or indirectly induce DNA strand breaks, leading to chromosomal aberrations and increased genetic instability (Xiong et al. 2022). Furthermore, DNA adducts can form covalent cross-links between DNA strands or between DNA and proteins, disrupting crucial processes like replication and transcription (Carlsson et al. 2022). These cross-links pose a significant challenge for repair mechanisms, further contributing to genotoxicity. Another mechanism involves the activation of cell cycle checkpoints, where DNA damage prompts the cell to temporarily halt the cell cycle for repair or, in cases of irreparable damage, triggers apoptosis to prevent the propagation of harmful mutations (Kumar et al. 2010). Moreover, certain DNA adducts can trigger oxidative stress, leading to the generation of reactive oxygen species (ROS) that inflict additional damage on DNA and other cellular components, creating a cascade of genotoxic effects (Dabas et al. 2014a).

10.4.2 DNA Damage

Genotoxicity mechanisms related to DNA damage are crucial for comprehending how certain substances can compromise the integrity of genetic material within the cells. DNA damage can take various forms, with genotoxic agents inducing structural modifications in the DNA. These modifications can interfere with the normal processes of DNA replication and repair, paving the way for a range of genotoxic effects (Canedo and Rocha 2021). One primary mechanism involves mutagenesis, where structural disruptions in DNA during cell division can result in errors in nucleotide incorporation or the skipping of damaged sites, leading to mismatched bases or deletions. Such mutations can give rise to genetic variations and abnormalities (Xu et al. 2020). Another crucial aspect of genotoxicity stems from the interference with DNA repair mechanisms. DNA repair processes are designed to rectify damage and maintain genomic stability. However, certain structural modifications can impede these repair mechanisms by hindering the recognition of damaged sites or acting as obstacles for repair enzymes. This interference may contribute to the accumulation of unrepaired DNA lesions, amplifying the risk of genetic instability (Vijitkul et al. 2022). DNA strand breaks represent a direct consequence of some structural modifications. The distortions induced by these modifications make the DNA helix more susceptible to cleavage, resulting in breaks that can lead to chromosomal aberrations and genomic rearrangements (Gupta and Verma 2022).

Furthermore, certain structural modifications can induce the formation of covalent cross-links. These cross-links may occur between adjacent DNA strands or between DNA and proteins. The formation of such cross-links poses a significant challenge for the cell, hindering essential processes like DNA replication and transcription. The difficulty in resolving these cross-links adds another layer to the genotoxicity of DNA damage (Prathiksha et al. 2023). Activation of cell cycle checkpoints is a protective response triggered by DNA damage. This mechanism provides the cell with an opportunity to halt the cell cycle temporarily, allowing time for DNA repair (Xu et al. 2020). However, persistent or severe DNA damage may activate apoptotic pathways, leading to programmed cell death as a preventive measure against the propagation of damaged genetic material (Rabelo et al. 2021).

10.4.3 Reactive Oxygen Species

Reactive oxygen species (ROS) play a crucial role in the genotoxicity mechanism, influencing the genetic material within cells and contributing to various cellular processes. ROS encompasses a group of highly reactive molecules, including superoxide anion ($O_2^{\bullet -}$), hydrogen peroxide (H_2O_2), and hydroxyl radical ($^{\bullet}OH$), which are generated as byproducts of cellular metabolism. While ROS is essential for normal cellular functions, an imbalance in their production and elimination can lead to genotoxic effects (Gupta and Verma 2022). The crucial mechanism by which ROS induces genotoxicity is through oxidative damage to DNA. The highly reactive nature of ROS allows them to interact with the purine and pyrimidine bases of DNA, causing modifications such as base oxidation, DNA strand breaks, and the formation of DNA adducts. These alterations can lead to mutations, chromosomal aberrations, and even genomic instability, ultimately contributing to the initiation and progression of various diseases, including cancer (Esmaeilbeigi et al. 2021). Moreover, ROS can activate signaling pathways that trigger cell cycle progression and apoptosis. Persistent oxidative stress can activate checkpoint kinases, influencing cell cycle checkpoints and promoting aberrant cell division. Additionally, ROS-mediated damage to mitochondria can lead to the release of pro-apoptotic factors, initiating programmed cell death as a protective mechanism to eliminate genetically compromised cells (Hong et al. 2020).

The genotoxicity of ROS is not limited to direct DNA damage; they can also interfere with DNA repair mechanisms. ROS-induced lesions can overwhelm the cellular repair machinery, leading to errors or incomplete repair. This compromised DNA repair capacity further contributes to the accumulation of genetic mutations, creating a conducive environment for the development of genotoxic effects (Mirzaee et al. 2021). Furthermore, ROS-mediated genotoxicity extends beyond nuclear DNA to mitochondrial DNA. Mitochondria are particularly vulnerable to oxidative damage due to their proximity to the primary source of ROS production—the electron transport chain. Accumulated damage to mitochondrial DNA can compromise cellular energy production, disrupt cellular homeostasis, and contribute to the overall genotoxic stress within the cell (Kumar et al. 2023).

10.4.4 Epigenetic Alterations

The correlation between genotoxicity and epigenetic alterations shows a fascinating intersection of genetic and environmental influences on cellular function. Epigenetic modifications refer to heritable changes in gene expression that do not involve alterations to the underlying DNA sequence. While genotoxicity traditionally focuses on direct damage to the DNA sequence, emerging research suggests that certain genotoxic agents can also induce epigenetic alterations, amplifying their impact on cellular processes. Genotoxic agents may induce epigenetic change by disrupting the enzymes responsible for maintaining the epigenetic code (Yin et al. 2020). DNA methyltransferases and histone-modifying enzymes, crucial players in regulating gene expression, are vulnerable targets. Exposure to genotoxic substances, such as certain chemicals or radiation, can perturb the delicate balance of these enzymes, leading to aberrant DNA methylation patterns or altered histone modifications. These epigenetic shifts can result in the misregulation of genes, potentially contributing to the initiation or progression of diseases, including cancer (Wang et al. 2023). Furthermore, genotoxicity-induced DNA damage can activate cellular repair mechanisms that, in turn, influence epigenetic regulation. DNA repair processes involve intricate crosstalk with epigenetic machinery, as the cell strives to restore genomic integrity. However, this repair response may inadvertently introduce changes in DNA methylation or histone modifications, creating a ripple effect that extends beyond the immediate repair site (Akcha et al. 2021). The impact of genotoxicity on noncoding RNA molecules also adds another layer to the intricate interplay between genotoxic stress and epigenetic alterations. MicroRNAs and long noncoding RNAs, known for their regulatory roles in gene expression, can be modulated by genotoxic insults. This modulation can lead to downstream effects on the expression of target genes, influencing cellular pathways associated with genomic stability, cell cycle control, and apoptosis (Wang et al. 2022).

10.5 Biomarkers of Genotoxicity

Biomarkers of genotoxicity are indicators or measurable changes that can be observed in biological systems and are associated with genetic damage. These markers are essential tools in toxicology and environmental health studies, providing insights into the potential harmful effects of various substances on genetic material. Some of the common biomarkers of genotoxicity are comet assay, DNA adducts, micronucleus test (MNT), sister chromatid exchange (SCE), chromosomal aberrations, urine and blood biomarkers, and P53 gene mutations. Apart from these, variations of the comet assay, such as the alkaline and neutral versions, allow the assessment of DNA damage under different conditions. The alkaline version is sensitive to single- and double-strand breaks, while the neutral version is more specific for double-strand breaks. Chemicals and carcinogens can form covalent adducts with DNA, leading to structural alterations. Detection and quantification of DNA adducts, such as those formed by polycyclic aromatic hydrocarbons, can serve as biomarkers of exposure and genotoxicity. These biomarkers collectively provide a comprehensive view of genotoxic effects, aiding in the assessment of potential risks associated with exposure to various substances. Their application is widespread in environmental monitoring, occupational health, and the evaluation of genotoxicity in pharmaceutical and chemical industries.

10.5.1 Comet Assay (SCGE)

The comet assay is a widely used technique to assess genotoxicity or DNA damage at the level of individual cells (Nagpure et al. 2008). It serves as a biomarker for genotoxic events, providing valuable information about the extent and nature of DNA damage in cells (Ali et al. 2008, 2009). The assay involves embedding isolated cells in a thin layer of agarose on a microscope slide. The cells are then lysed to remove cellular proteins, leaving behind the DNA in the shape of a comet with a distinct head and tail. The slide is subjected to electrophoresis, causing fragmented DNA to migrate away from the nucleus towards the anode, creating a comet-like appearance. The extent of DNA migration reflects the level of DNA damage, with longer tails indicating more damage. The comet assay is sensitive to various types of DNA damage, including single-strand breaks, double-strand breaks, and alkali-labile sites (Shukla et al. 2010). The assay can be adapted to detect specific types of DNA damage by using different alkaline or neutral conditions. The comet assay is used in various fields, including environmental toxicology, occupational health, and pharmaceutical research. It serves as a valuable tool for assessing genotoxicity induced by chemical agents, radiation, and environmental pollutants. The degree of DNA damage is quantified by measuring parameters such as tail length, tail intensity, and tail moment. These parameters provide a quantitative measure of DNA damage in individual cells. The comet assay is relatively simple and requires only a small number of cells. It allows for the detection of DNA damage at the level of individual cells, providing insights into cell-specific responses. The assay is influenced by various factors, including the type of cells used, experimental conditions, and the specific protocol employed. Interpretation of results requires careful consideration of factors influencing DNA migration. The comet assay is often used in combination with other biomarkers of genotoxicity to provide a comprehensive assessment of cellular responses to genotoxic agents. In summary, the comet assay is a valuable biomarker of genotoxicity that allows for the assessment of DNA damage at the single-cell level. Its versatility and sensitivity make it a widely used tool in genotoxicity testing and research (Nagpure et al. 2008).

10.5.2 Micronuclei Test

The micronucleus test (MNT) is another widely used technique in genotoxicity testing, serving as a biomarker for assessing chromosomal damage and instability in cells (Chaudhary et al. 2006). This test provides information about potential genetic damage induced by various factors, including exposure to genotoxic agents such as chemicals, radiation, or environmental pollutants. The MNT is usually performed on dividing cells, such as peripheral blood lymphocytes. After exposure to a genotoxic agent, cells are allowed to complete mitosis. Micronuclei are small, extranuclear bodies containing chromosomal fragments or whole chromosomes that were not incorporated into the main nucleus during cell division. The micronuclei may contain acentric fragments (chromosomal fragments lacking a centromere) or whole chromosomes that fail to segregate properly during cell division. The MNT is often conducted using peripheral blood samples, making it relatively noninvasive compared to other genotoxicity assays, where cells are stained to visualize the main nucleus and micronuclei. Microscopic analysis is then performed to count the number of micronuclei in a specified number of cells. The MNT is widely used in genotoxicity assessment for a variety of purposes, including environmental monitoring, occupational health studies, and drug safety testing (Nagpure et al. 2015a, b, 2017). It can be applied to different cell types, including human lymphocytes, rodent bone marrow cells, and other suitable cell lines (Goswami et al. 2014). An increase in the frequency of micronuclei is considered an indicator of genotoxicity. The test provides information about both clastogenic (causing chromosome breakage) and aneugenic (causing chromosomal missegregation) effects. MNT is relatively simple, cost-effective, and widely accepted for regulatory genotoxicity testing. It can be applied to a variety of cell types, allowing for flexibility in study design. The test is sensitive to the effects of certain cytotoxic compounds, and it may not distinguish between clastogenic and aneugenic effects. Interpretation of results requires consideration of factors that may affect the incidence of micronuclei. Like the comet assay, the MNT is often used in combination with other genotoxicity assays to obtain a comprehensive assessment of genetic damage. In summary, the MNT is a valuable tool for assessing genotoxicity by detecting chromosomal damage and instability. Its simplicity and versatility make it a commonly used assay in various fields of toxicology and environmental health (Kushwaha et al. 2012).

10.5.3 Chromosomal Aberrations

Chromosomal aberrations are structural or numerical abnormalities in the chromosomes of an organism's cells. These aberrations can have significant consequences for the health and development of an individual. Chromosomal aberrations can be categorized into two main types: numerical aberrations and structural aberrations. The numerical chromosomal aberrations include aneuploidy and polyploidy. Aneuploidy is a condition in which there is an abnormal number of chromosomes in a cell, for example, trisomy (three copies) or monosomy (one copy) of a particular chromosome. Down syndrome (trisomy 21) is a well-known example. Polyploidy is the presence of additional sets of chromosomes in a cell. Triploidy (three sets) and tetraploidy (four sets) are examples of polyploidy. Polyploidy is often lethal in humans but common in certain plants.

Structural chromosomal aberrations include deletion, duplication, inversion, translocation, ring chromosomes, etc. Deletion is the loss of a portion of a chromosome resulting in the loss of specific genes and leading to various genetic disorders. Duplication is the presence of an extra copy of a segment of a chromosome leading to an overexpression of certain genes and causing genetic abnormalities. Inversion is the reversal of the orientation of a segment within a chromosome and may not always have noticeable effects, but it can disrupt the normal functioning of genes if they occur within or near them. Translocation is the movement of a segment from one chromosome to another leading to the exchange of genetic material between nonhomologous chromosomes. Reciprocal translocations involve the exchange of segments between two chromosomes, while Robertsonian translocations involve fusion at the centromere. Ring chromosomes are depicted with the formation of a circular structure when the ends of a broken chromosome fuse leading to genetic disorders if critical genes are affected.

The potential reasons of chromosomal aberrations are spontaneous errors during DNA replication or repair. Exposure to certain chemicals, radiation, or drugs can increase the risk of chromosomal abnormalities. Inherited genetic mutations or chromosomal rearrangements can contribute to chromosomal aberrations. Chromosomal aberrations can lead to developmental abnormalities, intellectual disabilities, and various genetic disorders. Some chromosomal aberrations are incompatible with life and result in miscarriages or stillbirths. Chromosomal aberrations can be detected through techniques such as karyotyping, fluorescence in situ hybridization (FISH), and chromosomal microarray analysis (CMA). Prenatal screening and genetic testing are used to identify chromosomal abnormalities in developing fetuses. Understanding chromosomal aberrations is crucial for genetic counseling, diagnosis, and the study of genetic disorders. Advances in genetic testing and molecular techniques have improved our ability to detect and understand the implications of chromosomal abnormalities.

Sister chromatid exchange (SCE) is a cytogenetic technique that measures the exchange of genetic material between sister chromatids during DNA replication. An increased rate of SCEs can indicate chromosomal instability and genotoxicity.

10.5.4 Antioxidant Assays

Antioxidant assays are laboratory techniques used to measure the ability of substances to neutralize or inhibit the damaging effects of reactive oxygen species (ROS) or free radicals (Dabas et al. 2011, 2014a, b). Antioxidants play a crucial role in protecting cells and tissues from oxidative stress, which is associated with various diseases, aging, and other health conditions. There are several antioxidant assays designed to evaluate the antioxidant capacity of compounds.

DPPH (2,2-diphenyl-1-picrylhydrazyl) assay: DPPH is a stable free radical that accepts an electron or hydrogen atom from an antioxidant, causing a color change from purple to yellow. The antioxidant capacity is measured by monitoring the decrease in absorbance or discoloration of DPPH.

ABTS (2,2′-azino-bis(3-ethylbenzothiazoline-6-sulfonic acid)) assay: ABTS forms a blue-green radical when oxidized, and antioxidants reduce this radical, leading to a decrease in color intensity. The antioxidant activity is determined by measuring the reduction in absorbance at a specific wavelength.

Ferric reducing antioxidant power (FRAP) assay: FRAP measures the reduction of a ferric (Fe^{3+}) complex to its ferrous (Fe^{2+}) form by antioxidants. The reduction is monitored spectrophotometrically, and the antioxidant capacity is proportional to the amount of Fe^{2+} formed.

Total antioxidant capacity (TAC) assays: TAC assays measure the overall antioxidant capacity of a sample by assessing its ability to neutralize various free radicals. Trolox equivalent antioxidant capacity (TEAC) and oxygen radical absorbance capacity (ORAC) are examples of TAC assays.

Thiobarbituric acid reactive substances (TBARS) assay: TBARS assay measures lipid peroxidation by detecting malondialdehyde (MDA), a byproduct of lipid oxidation. Antioxidants inhibit lipid peroxidation, leading to a decrease in TBARS production, which can be quantified.

Superoxide dismutase (SOD) assay: SOD is an enzyme that catalyzes the dismutation of superoxide radicals into oxygen and hydrogen peroxide. The activity of SOD is measured by its ability to inhibit the reduction of nitroblue tetrazolium (NBT), and the inhibition is quantified spectrophotometrically.

Glutathione (GSH) assay: GSH is an essential cellular antioxidant, and its levels can be measured to assess the antioxidant status of cells or tissues. GSH can be quantified using enzymatic methods, including the glutathione reductase assay.

Catalase assay: The catalase assay is a laboratory technique used to measure the activity of the enzyme catalase, which plays a crucial role in protecting cells from oxidative damage by breaking down hydrogen peroxide (H2O2) into water (H2O) and oxygen (O2). Catalase is found in the peroxisomes of cells and is particularly abundant in tissues with high metabolic activity. The catalase assay is commonly employed in various fields, including biochemistry, cell biology, and clinical research. Catalase catalyzes the decomposition of hydrogen peroxide (H_2O_2) into water (H_2O) and oxygen (O_2). The catalase assay typically involves measuring the rate of oxygen evolution during the breakdown of hydrogen peroxide. The liberated oxygen is quantified using various methods, depending on the experimental setup. In this method, the rate of oxygen production is measured using a Clark-type oxygen electrode or other oxygen detection systems. The assay is initiated, and the rate of oxygen evolution or the decrease in hydrogen peroxide concentration is monitored over time. Catalase activity is expressed in units, where one unit is defined as the amount of enzyme that decomposes 1 micromole of hydrogen peroxide per minute under specified conditions. Studying catalase activity provides insights into cellular oxidative stress and antioxidant defense mechanisms. Changes in catalase activity may be associated with certain diseases or conditions, and the assay can be used in diagnostic tests. Catalase assays can be employed to assess the impact of drugs or compounds on cellular antioxidant capacity. The catalase assay is a valuable tool for researchers studying oxidative stress, cellular metabolism, and the role of antioxidants in various biological processes. The choice of assay method depends on the specific requirements of the experiment and the available equipment.

These assays are valuable tools for researchers studying the antioxidant properties of natural compounds, pharmaceuticals, and food products. They provide quantitative data on the ability of substances to scavenge free radicals or prevent oxidative damage, contributing to our understanding of their potential health benefits. The choice of assay depends on the specific properties of the antioxidant being studied and the experimental objectives.

10.5.5 Bioaccumulation

Bioaccumulation refers to the gradual accumulation of substances, such as pollutants or chemicals, in living organisms over time. This process occurs as organisms take in substances from their environment, primarily through ingestion or absorption, and the substances accumulate at a rate that exceeds their elimination or excretion. Bioaccumulation is a concern in environmental science and toxicology because it can lead to increased concentrations of harmful substances in organisms, potentially causing adverse effects at higher trophic levels in food chains. Bioaccumulation often begins when organisms take up substances from their surrounding environment. This can occur through various pathways such as the intake of contaminated food, water, or air, as well as dermal absorption. Bioaccumulation is particularly associated with persistent substances (do not readily break down) and lipophilic substances (soluble in fats and oils). These characteristics contribute to the substances' tendency to accumulate in fatty tissues. The process of bioaccumulation is magnified in food chains, where organisms at higher trophic levels consume other organisms that have already accumulated substances. This increases the concentration of substances as one moves up the food chain.

Biomagnification is a related concept where the concentration of certain substances increases at higher trophic levels in a food chain. This occurs as predators consume prey with higher concentrations of accumulated substances. Methylmercury, a highly toxic form of mercury, can bioaccumulate in fish and seafood, posing risks to organisms and humans that consume these products. Persistent organic pollutants (POPs) including certain pesticides, flame retardants, and industrial chemicals are examples of POPs that can bioaccumulate in organisms. Bioaccumulation can lead to adverse effects on the health of organisms and ecosystems. In some cases, the impacts may be particularly pronounced for top predators in food chains. For humans, bioaccumulation can result in exposure to elevated levels of contaminants through the consumption of contaminated food, posing health risks. Environmental monitoring programs assess the levels of bioaccumulative substances in various ecosystems and species. Regulatory measures aim to control the release of persistent and harmful substances into the environment. Understanding and mitigating bioaccumulation are essential for environmental protection and the conservation of biodiversity. Efforts to reduce the use of persistent and toxic substances, along with proper waste management practices, contribute to minimizing the impact of bioaccumulation on ecosystems and human health.

10.6 Impact of Trophic-Level Genotoxicity on Human Health

The impact of trophic-level genotoxicity on human health is a complex and multifaceted issue. Trophic levels represent the hierarchical levels in a food chain, where organisms are categorized based on their position in the energy transfer sequence. Genotoxicity at different trophic levels can have implications for human health in several ways. Genotoxic substances can accumulate in organisms at lower trophic levels through exposure to contaminated water, soil, or air. As these organisms are consumed by higher trophic-level organisms, the concentration of genotoxic substances can increase through biomagnification. Ultimately, humans at the top of the food chain may be exposed to higher levels of genotoxic agents. In aquatic ecosystems, fish and shellfish are commonly consumed by humans. If these organisms are exposed to genotoxic pollutants, there is a potential risk of human exposure to these substances through the consumption of contaminated seafood. This exposure may lead to adverse health effects, including genotoxicity, in humans. Genotoxic substances, especially those with carcinogenic and mutagenic properties, can induce DNA damage and increase the risk of cancer and genetic disorders in humans. Exposure to such substances through the consumption of contaminated food may contribute to these health risks. Some genotoxic agents, such as certain heavy metals and persistent organic pollutants, can induce oxidative stress in organisms. Oxidative stress is associated with various health issues, including inflammation, neurodegenerative diseases, and cardiovascular disorders. Humans consuming contaminated food may be exposed to these oxidative stress-inducing agents. Genotoxic substances can have adverse effects on reproductive health and fetal development, and exposure to such agents through contaminated food sources may pose risks to pregnant women, leading to developmental abnormalities in the fetus or reproductive issues.

Governments and regulatory bodies often set limits for the presence of genotoxic substances in food to protect public health. Monitoring trophic-level genotoxicity is crucial for assessing environmental quality and ensuring compliance with regulatory standards. Awareness of trophic-level genotoxicity can prompt the implementation of precautionary measures to reduce human exposure. This may include stricter regulations on industrial discharges, monitoring of water and soil quality, and promoting sustainable agricultural practices. Understanding the link between trophic-level genotoxicity and human health involves interdisciplinary research, including environmental science, toxicology, and public health. Continued efforts in monitoring and mitigating genotoxic substances in the environment, along with public awareness and policy interventions, are essential to minimizing the potential health impacts on human populations.

10.7 Sustainable Reversal of Genotoxicity

Addressing and reversing genotoxicity sustainably involve the following: implementing strategies that minimize or eliminate exposure to genotoxic agents, promote environmental and human health, and encourage sustainable practices; implementing and enforcing strict regulations on the use and release of genotoxic substances in industries and agriculture; promoting the development and use of safer alternatives to genotoxic agents; implementing remediation techniques to clean up contaminated sites and restore ecosystems affected by genotoxic pollutants; employing phytoremediation, where certain plants are used to absorb, accumulate, and detoxify genotoxic substances from soil and water; implementing proper waste disposal and recycling methods to reduce the release of genotoxic substances into the environment; encouraging the use of eco-friendly packaging materials to minimize the generation of genotoxic waste; encouraging organic farming practices that reduce reliance on genotoxic pesticides and fertilizers; promoting integrated pest management (IPM) strategies that minimize the use of chemical pesticides; educating the public about the sources and potential health risks associated with genotoxic agents; promoting awareness of sustainable practices and responsible consumer choices to reduce exposure; investing in research and development of green technologies that focus on reducing or eliminating the use of genotoxic substances in various industries; promoting the adoption of sustainable and environmentally friendly technologies in manufacturing processes; implementing robust environmental monitoring programs to assess genotoxicity levels in air, water, and soil; regularly monitoring the health of populations exposed to genotoxic agents and conducting biomonitoring studies to assess the effectiveness of interventions; strengthening and enforcing environmental regulations related to genotoxic substances and supporting the development and implementation of policies that prioritize sustainability, environmental protection, and human health; encouraging collaboration between countries, organizations, and research institutions to share knowledge and best practices for addressing genotoxicity on a global scale; supporting research aimed at understanding the mechanisms of genotoxicity and developing innovative solutions for its prevention and reversal; and investing in the development of advanced technologies for monitoring and mitigating genotoxic substances.

Sustainable reversal of genotoxicity requires a comprehensive and integrated approach that involves regulatory, technological, educational, and community-based interventions. By adopting sustainable practices and promoting environmental stewardship, it is possible to reduce the prevalence of genotoxic substances in the environment and protect the health of both ecosystems and human populations.

10.7.1 Bioremediation

Bioremediation is a process that uses living organisms, such as microorganisms, plants, or enzymes, to remove or neutralize pollutants from a contaminated environment. This environmentally friendly approach leverages the natural abilities of these organisms to degrade or transform contaminants, leading to the restoration of the contaminated site. Bioremediation is employed in various settings, including soil, water, and air, to address a wide range of pollutants. Microbial bioremediation involves the use of microorganisms like bacteria, fungi, and algae to degrade or transform contaminants. Phytoremediation utilizes plants to absorb, accumulate, or transform pollutants. Plants can enhance microbial activity in the rhizosphere, the soil zone influenced by root secretions. Enzymatic bioremediation involves the use of enzymes to break down pollutants. Enzymes can be produced by microorganisms or engineered for specific applications. Microorganisms can degrade oil and hydrocarbons through processes like aerobic and anaerobic biodegradation. Plants or microorganisms can accumulate heavy metals, and some microbes can transform them into less toxic forms (Singh et al. 2014a, b). Microbial communities can break down these chemicals into less harmful compounds. Bioremediation can address a variety of industrial pollutants, such as solvents, chlorinated compounds, and organic pollutants. Microorganisms play a central role in many bioremediation processes, including bacteria, fungi, and archaea that have specific enzymes that can break down complex organic molecules into simpler, less harmful substances. The success of microbial biodegradation depends on factors like temperature, pH, nutrient availability, and the specific microorganisms present. Plants absorb contaminants through their roots and accumulate them in their aboveground parts largely known as phytoremediation. Phytodegradation assists plants in releasing enzymes into the rhizosphere that break down contaminants in the soil. On the other side, rhizofiltration helps plants uptake and accumulate contaminants from water through their root systems. Biostimulation and bioaugmentation help enhance the growth and activity of indigenous microorganisms by providing nutrients, oxygen, or other growth-promoting factors and introducing specific microorganisms or enzymes to enhance the existing microbial community's ability to degrade contaminants.

Applications of bioremediation include land remediation that addresses contamination in soils, including brownfield sites, agricultural lands, and areas impacted by oil spills; water treatment that addresses contaminants in groundwater, surface water, and wastewater; and air quality improvement that involves the use of plants or microbes to degrade volatile organic compounds (VOCs) and other airborne pollutants. Bioremediation is an environmentally friendly and sustainable approach that is often less expensive compared to traditional remediation methods. It can be applied in situ, minimizing the need for excavation and transportation of contaminated materials. The effectiveness of bioremediation may depend on site-specific conditions and the type of contaminants. Bioremediation processes can be slow and may require long-term monitoring. The potential release of intermediates during degradation should be considered. Bioremediation offers a promising and eco-friendly approach to address environmental contamination by harnessing the power of natural processes and organisms. It is continually evolving through research and development, and its application is tailored to the specific characteristics of contaminated sites.

10.7.2 Biocontrol

Biocontrol, short for biological control, is a method of managing pests and diseases in agriculture, forestry, and other ecosystems using living organisms. The goal of biocontrol is to reduce the impact of pests and pathogens while minimizing the use of chemical pesticides. This approach involves introducing or promoting natural enemies or beneficial organisms that can regulate the population of harmful species. Predators are natural enemies that consume and control the population of pests including ladybugs, predatory beetles, and spiders. Parasitoids are small insects that lay eggs on or inside the host pests. The emerging larvae then feed on the host, eventually killing it, for example, parasitoid wasps. Pathogens are microorganisms, such as bacteria, fungi, and viruses, that infect and kill pests. Biopesticides based on these pathogens are used for biocontrol. Herbivores are certain herbivorous insects or animals that feed on invasive or unwanted plant species to control their population.

Biocontrol involves the introduction of natural enemies from the pest's native environment to control a specific pest in a new environment. The introduced organisms establish and reproduce, providing long-term pest control. Careful consideration is given to the potential impacts on nontarget species to avoid unintended consequences. Biocontrol is often a component of integrated pest management, which combines multiple strategies to manage pests sustainably. IPM integrates biological control with cultural practices, chemical control (when necessary), and monitoring to achieve effective and environmentally friendly pest management. Some examples of biocontrol agents include coccinellid beetles (ladybugs), parasitoid wasps, nematodes, *Bacillus thuringiensis* (Bt), etc.

Biocontrol relies on an environmentally friendly and sustainable approach, reduces reliance on chemical pesticides, minimizes potential environmental and health risks, and can be integrated into diverse agricultural and ecological systems. Effectiveness may depend on environmental conditions and the specific interactions between biocontrol agents and pests. There is a need to balance pest control with the preservation of nontarget organisms. Pests may evolve resistance to biocontrol agents, reducing their long-term effectiveness, therefore, channelizing their bio-availability and effectiveness is a primary requisite. Biocontrol is an integral part of sustainable agriculture and ecosystem management. By harnessing the natural interactions between organisms, biocontrol provides an environmentally friendly alternative to chemical pesticides, contributing to the conservation of biodiversity and the promotion of ecological balance.

10.8 Future Perspectives

In fish, observable biological markers that show the genetic harm brought on by exposure to different environmental stressors, such as pollutants or contaminants in aquatic ecosystems, are referred to as genotoxicological endpoints. These biomarkers are essential for determining how the environmental factors affect fish populations and giving useful data for the long-term sustainable management of fish genetic resources. Through the integration of these indicators and genotoxicological endpoints, scientists can obtain a full comprehension of how environmental stresses affect fish genetic resources. This information is essential for carrying out efficient environmental health evaluations and creating long-term management plans that protect fish populations and preserve the integrity of the ecosystem. Apoptosis biomarkers, such as the measurement of caspase enzymes involved in apoptosis, can indicate programmed cell death triggered by genotoxic events. Monitoring the expression of p53, a tumor suppressor gene involved in DNA repair and apoptosis, provides insights into genotoxic stress responses. The changes in gonadal tissue structure can indicate genotoxic effects on reproductive organs. Assessing changes in microsatellite DNA sequences can reveal genetic instability and potential genotoxic effects on population structure. Alterations in epigenetic marks or histone modification, such as DNA methylation, can influence gene expression and are increasingly recognized as important indicators of genotoxicity. Through the lens of these future perspectives, the researchers can further enhance the understanding of genotoxicity and help design more effective measures to safeguard human health and aquatic environments. The future of genotoxicity biomarker research will be greatly influenced by the incorporation of innovative technology, interdisciplinary collaborations, and ethical issues.

References

Ahmed MK, Habibullah-Al-Mamun M, Islam MM et al (2015) Toxicological assessment of arsenic-induced hematological alterations and chromosomal aberrations in tilapia *Oreochromis mossambicus*. Hum Ecol Risk Assess 21(1):146–156

Akcha F, Barranger A, Bachère E (2021) Genotoxic and epigenetic effects of diuron in the Pacific oyster: in vitro evidence of interaction between DNA damage and DNA methylation. Environ Sci Pollut Res 28:8266–8280

Ali D, Nagpure NS, Kumar S et al (2008) Genotoxicity assessment of acute exposure of chlorpyrifos to freshwater fish *Channa punctatus* (Bloch) using micronucleus assay and alkaline single-cell gel electrophoresis. Chemosphere 71(10):1823–1831

Ali D, Nagpure NS, Kumar S et al (2009) Assessment of genotoxic and mutagenic effects of chlorpyrifos in freshwater fish *Channa punctatus* (Bloch) using micronucleus assay and alkaline single-cell gel electrophoresis. Food Chem Toxicol 47:650–656

Canedo A, Rocha TL (2021) Zebrafish (*Danio rerio*) using as model for genotoxicity and DNA repair assessments: historical review, current status and trends. Sci Total Environ 762:144084

Carlsson MJ, Vollmer AS, Demuth P et al (2022) p53 triggers mitochondrial apoptosis following DNA damage-dependent replication stress by the hepatotoxin methyleugenol. Cell Death Dis 13(11):1009

Chaudhary R, Pandey S, Kushwaha B et al (2006) Fish micronucleus assay: a sensitive tool for ecogenotoxicity studies. J Ecophysiol Occup Health 6:143–147

Dabas A, Nagpure NS, Mishra RN et al (2011) Assessment of cadmium-induced alterations in superoxide dismutase and catalase activities of *Labeo rohita* (Hamilton 1822). J Ecophysiol Occup Health 11:205–211

Dabas A, Nagpure NS, Mishra RM et al (2014a) Investigation of cadmium-induced genotoxicity and oxidative stress response in Indian major carp, *Labeo rohita*. Hum Ecol Risk Assess 20:510–526

Dabas A, Nagpure NS, Mishra RM et al (2014b) Cadmium-induced lipid peroxidation and effects on glutathione dependent enzymes in tissues of *Labeo rohita*. Proc Natl Acad Sci India Sect B Biol Sci 84:981–988

Dutta D, Ray A, Ghosh B et al (2024) Assessment of imidacloprid induced genotoxicity in *Pethia conchonius* (Rosy barb), a common freshwater fish of India. Drug Chem Toxicol 47(1):101–114

Esmaeilbeigi M, Kalbassi MR, Seyedi J et al (2021) Intra and extracellular effects of benzo [α] pyrene on liver, gill and blood of Caspian White fish (*Rutilus frissi kutum*): cyto-genotoxicity and histopathology approach. Mar Pollut Bull 163:111942

Goswami M, Yadav K, Dubey A et al (2014) In vitro cytotoxicity assessment of two heavy metal salts in a fish cell line (RF). Drug Chem Toxicol 37(1):48–54

Gupta P, Verma SK (2022) Evaluation of genotoxicity induced by herbicide pendimethalin in fresh water fish *Clarias batrachus* (Linn.) and possible role of oxidative stress in induced DNA damage. Drug Chem Toxicol 45(2):750–759

Hong Y, Huang Y, Yang X et al (2020) Abamectin at environmentally-realistic concentrations cause oxidative stress and genotoxic damage in juvenile fish (*Schizothorax prenanti*). Aquat Toxicol 225:105528

Kumar R, Nagpure NS, Kushwaha B et al (2010) Investigation of the genotoxicity of Malathion to freshwater teleost fish *Channa punctatus* (Bloch) using the micronucleus test and comet assay. Arch Environ Conam Toxicol 58:123–130

Kumar P, Kumar R, Nagpure NS et al (2012) Genotoxic and mutagenic assessment of hexavalent chromium in fish following in-vivo chronic exposure. Hum Ecol Risk Assess 18:855–870

Kumar P, Kumar R, Nagpure NS et al (2013) Genotoxicity and antioxidant enzyme activity induced by hexavalent chromium in *Cyprinus carpio* after in vivo exposure. Drug Chem Toxicol 36(4):451–460

Kumar M, Singh S, Dwivedi S et al (2023) Copper-induced genotoxicity, oxidative stress, and alteration in transcriptional level of autophagy-associated genes in snakehead fish *Channa punctatus*. Biol Trace Elem Res 201(4):2022–2035

Kushwaha B, Pandey S, Sharma S et al (2012) In situ assessment of genotoxic and mutagenic potential of polluted river water in *Channa punctatus* and *Mystus vittatus*. Int Aquat Res 4:16

Lakra WS, Nagpure NS (2009) Genotoxicological studies in fishes: a review. Indian J Anim Sci 79(1):93–98

Lakra WS, Sarkar UK, Kumar RS et al (2010) Fish diversity, habitat ecology and their conservation and management issues of a tropical River in Ganga basin, India. Environmentalist 30:306–319

Mirzaee SA, Noorimotlagh Z, Ahmadi M et al (2021) The possible oxidative stress and DNA damage induced in Diclofenac-exposed non-target organisms in the aquatic environment: a systematic review. Ecol Indic 131:108172

Nagpure NS, Sharma S, Pandey S et al (2008) Use of comet assay as a sensitive tool for genotoxicity assessment in fishes from Gomti River. Indian J Fish 53(3):285–288

Nagpure NS, Srivastava R, Kumar R et al (2015a) Assessment of pollution of river Ganges by tannery effluents using genotoxicity biomarkers in murrel fish, *Channa punctatus* (Bloch). Indian J Exp Biol 53:476–483

Nagpure NS, Srivastava R, Kumar R et al (2015b) Assessment of the genotoxic and mutagenic potential of hexavalent chromium on the freshwater fish *Labeo rohita* (Hamilton, 1882). Drug Chem Toxicol 38(1):9–15

Nagpure NS, Rashmi S, Kumar R et al (2017) Mutagenic, genotoxic, and bioaccumulative potentials of tannery effluents in freshwater fishes of River Ganga. Hum Ecol Risk Assess 23(1):98–111

Nwani CD, Nagpure NS, Kumar R et al (2010a) Lethal concentration and toxicity stress of carbosulfan, glyphosate and atrazine to freshwater air breathing fish *Channa punctatus* (Bloch). Int Aquat Res 2:105–111

Nwani CD, Lakra WS, Nagpure NS et al (2010b) Mutagenic and genotoxic effects of carbosulfan in freshwater fish *Channa punctatus* (Bloch) using micronucleus assay and alkaline single-cell gel electrophoresis. Food Chem Toxicol 48:202–208

Nwani CD, Nagpure NS, Kumar R et al (2011) Mutagenic and genotoxic assessment of atrazine-based herbicide to freshwater fish *Channa punctatus* (Bloch) using micronucleus test and single cell gel electrophoresis. Environ Toxicol Pharmacol 31:314–322

Nwani CD, Nagpure NS, Kumar R et al (2013) DNA damage and oxidative stress modulatory effects of glyphosate-based herbicide in freshwater fish, *Channa punctatus*. Environ Toxicol Pharmacol 36(2):539–547

Nwani CD, Nagpure NS, Kumar R et al (2014) Induction of micronuclei and nuclear lesions in Channa punctatus following exposure to carbosulfan, glyphosate, and atrazine. Drug Chem Toxicol 37(4):370–377

Pandey S, Kumar R, Sharma S et al (2005) Acute toxicity bioassays of mercuric chloride and malathion on airbreathing fish *Channa punctatus* (Bloch). Ecotoxicol Environ Saf 61:114–120

Pandey S, Nagpure NS, Kumar R et al (2006) Genotoxicity evaluation of acute doses of endosulfan to freshwater teleost *Channa punctatus* (Bloch) by alkaline single-cell gel electrophoresis. Ecotoxicol Environ Saf 65:56–61

Pandey AK, Nagpure NS, Trivedi SP et al (2011a) Profenofos induced DNA damage in freshwater fish, *Channa punctatus* (Bhoch) using alkaline single-cell gel electrophoresis. Mutat Res Genet Toxicol Environ Mutagen 726(2):209–214

Pandey AK, Nagpure NS, Trivedi SP et al (2011b) Investigation on acute toxicity and behavioral changes in Channa punctatus (Bloch) due to organophosphate pesticide profenofos. Drug Chem Toxicol 34(4):424–428

Pandey AK, Nagpure NS, Trivedi SP (2011c) Haematological responses in a freshwater fish *Channa punctatus* (Bloch) exposed to profenofos. Nat Acad Sci Lett 34(9,10):359–363

Pandey AK, Nagpure NS, Trivedi SP (2014) Evaluation of genotoxicity of Profenofos freshwater fish *Channa punctatus* (Bloch) using the micronucleus assay. Afr J Biotechnol 13(39):3985–3988

Pathak AK, Rashid I, Nagpure NS et al (2019) FisOmics: a portal of fish genomic resources. Genomics 111:1923–1928

Prathiksha J, Narasimhamurthy RK, Dsouza HS et al (2023) Organophosphate pesticide-induced toxicity through DNA damage and DNA repair mechanisms. Mol Biol Rep:1–15

Rabelo JC, Hanusch AL, de Jesus LWO et al (2021) DNA damage induced by cylindrospermopsin on different tissues of the biomonitor fish *Poecilia reticulata*. Environ Toxicol 6(6):1125–1134

Rashid I, Pathak AK, Kumar R et al (2019a) Genome-wide comparative analysis of hif binding sites in *Cyprinus carpio* for in silico identification of functional hypoxia response elements. Front Genet 10:659

Rashid I, Baisvar VS, Singh M et al (2019b) Isolation and characterization of hypoxia-inducible gene connective tissue growth factor (CTGF) in *Labeo rohita*. Mol Biol Rep 46:1683–1691

Rashid I, Baisvar VS, Singh M et al (2020) Isolation and characterization of hypoxia-inducible heme oxygenase 1 (HMOX1) gene in *Labeo rohita*. Genomics 112(3):2327–2333

Sharma S, Nagpure NS, Kumar R et al (2007) Studies on the genotoxicity of endosulfan in different tissues of fresh water fish *Mystus vittatus* using the comet assay. Arch Environ Contam Toxicol 53:617–623

Shukla D, Nagpure NS, Kumar R et al (2010) Assessment of genotoxicity of Dichlorvos to Mystus vittatus (Bloch) by comet assay. Indian J Fish 57(2):39–44

Singh P, Dabas A, Srivastava R et al (2014a) Evaluation of genotoxicity induced by medicinal plant Jatropha gossypifolia in freshwater fish *Channa punctatus* (Bloch). Turk J Fish Aquat Sci 14:421–428

Singh P, Dabas A, Srivastava R et al (2014b) Mutagenic and genotoxic evaluation of medicinal plant Euphorbia royleana latex to freshwater fish *Channa punctatus* (Bloch). Int J Pharm Bio Sci 5(1):217–228

Srivastava SK, Kushwaha B, Nagpure NS et al (2005) Genotoxic studies in *Cyprinus carpio* (Linnaeus, 1758) of River Yamuna. Indian J Fish 52(2):235–239

Srivastava P, Singh A, Pandey AK (2016) Pesticides toxicity in fishes: biochemical, physiological and genotoxic aspects. Biochemic Cell Arch 16:199–218

Tiwari M, Nagpure NS, Saksena DN et al (2010) Time kinetic study of metallothionein mRNA expression due to cadmium exposure in freshwater murrel, *Channa punctata* (Bloch). J Ecophysiol Occup Health 10:85–96

Tiwari M, Nagpure NS, Saksena DN et al (2011a) Metallothionein mRNA expression in freshwater teleost, *Channa punctata* (Bloch) under the influence of heavy metal, cadmium—a dose kinetic study. Int Aquat Res 3:21–29

Tiwari M, Nagpure NS, Saksena DN et al (2011b) Evaluation of acute toxicity levels and ethological responses under heavy metal cadmium exposure in freshwater teleost, *Channa punctata* (Bloch). Int J Aquat Sci 2(1):36–47

Vijitkul P, Kongsema M, Toommakorn T et al (2022) Investigation of genotoxicity, mutagenicity, and cytotoxicity in erythrocytes of Nile tilapia (*Oreochromis niloticus*) after fluoxetine exposure. Toxicol Rep 9:588–596

Wang Y, Liao XL, Chen K et al (2022) Analysis of the miRNA expression profile of laboratory red crucian carp under low-dose caesium-137 irradiation. Ecotoxicology 31(8):1276–1286

Wang T, Meng Y, Tu Y et al (2023) Associations between DNA methylation and genotoxicity among lead-exposed workers in China. Environ Pollut 316:120528

Xiong Y, Kaw HY, Zhu L et al (2022) Genotoxicity of quinone: an insight on DNA adducts and its LC-MS-based detection. Crit Rev Environ Sci Technol 52(23):4217–4240

Xu C, Fan Y, Zhang X et al (2020) DNA damage in liver cells of the tilapia fish *Oreochromis mossambicus* larva induced by the insecticide cyantraniliprole at sublethal doses during chronic exposure. Chemosphere 238:124586

Yin X, Liu Y, Zeb R et al (2020) The intergenerational toxic effects on offspring of medaka fish *Oryzias melastigma* from parental benzo [a] pyrene exposure via interference of the circadian rhythm. Environ Pollut 267:115437

Aquatic Animal Disease Surveillance and Important Disease Affecting Finfish and Shellfish in India

11

Neeraj Sood, P. K. Pradhan, Anutosh Paria, Chandra Bhushan Kumar, Gaurav Rathore, Uttam Kumar Sarkar, and J. K. Jena

Abstract

Diseases continue to be a major constraint to sustainable growth of aquaculture, resulting in huge economic losses globally. Some of the diseases like white spot disease, acute hepatopancreatic necrosis disease and hepatopancreatic microsporidiosis have resulted in enormous economic losses in shrimp culture, whereas finfish diseases like epizootic ulcerative syndrome, tilapia lake virus disease and lice infestation have been responsible for significant losses in freshwater aquaculture. These losses underscore the need of aquatic animal disease surveillance for early detection and timely intervention. The disease surveillance can be broadly categorized into passive or farmer-based surveillance and active surveillance. Over the last few years, to improve the efficiency of disease surveillance, some of the approaches like syndromic surveillance, sentinel surveillance, indirect surveillance, risk-based surveillance etc. have been developed. Recognizing the importance of disease surveillance, the Government of India has been implementing the National Surveillance Programme for Aquatic Animal Diseases (NSPAAD) since 2013. The programme, currently funded under Pradhan Mantri Matsya Sampada Yojana (PMMSY), has been strengthened by the involvement of state fisheries departments and Marine Products Export Development Authority. Some of the important achievements of the NSPAAD include developing a strong network of aquatic animal health laboratories and diagnostic

N. Sood (✉) · P. K. Pradhan · A. Paria · C. B. Kumar · G. Rathore · U. K. Sarkar
ICAR-National Bureau of Fish Genetic Resources, Lucknow, Uttar Pradesh, India
e-mail: Neeraj.Sood@icar.gov.in; Pravata.Pradhan@icar.gov.in; Anutosh.Paria@icar.gov.in; Chandra.Kumar@icar.gov.in; Gaurav.Rathore@icar.gov.in; Uttam.Sarkar@icar.gov.in

J. K. Jena
Indian Council of Agricultural Research, Krishi Anusandhan Bhawan-II, Pusa, New Delhi, India
e-mail: jkjena@icar.gov.in

U. K. Sarkar et al. (eds.), *Sustainable Management of Fish Genetic Resources*, https://doi.org/10.1007/978-981-97-5250-8_11

capability for the detection of important aquatic animal pathogens, strengthening of passive disease surveillance, providing scientific advice to the farmers, detection of several new pathogens for the first time and establishing a transparent system for reporting aquatic animal diseases to international organizations. Furthermore, to strengthen the farmer-based reporting, the ReportFishDisease app has been developed which is available in Hindi, English and 11 regional languages. Using the app, the farmers can report disease cases and get scientific advice for quickly addressing the disease problem. The strengthening of the ongoing surveillance programme has been facilitating the early detection of diseases, accurate diagnosis and timely scientific advice to farmers and meeting country's disease reporting obligations.

Keywords

Surveillance · Aquatic diseases · Emerging pathogens · Economic loss · Disease reporting

11.1 Introduction

Fisheries and aquaculture production has shown tremendous growth globally over the last few decades, with production increasing from 19 million tonnes in 1950 to about 178 million tonnes in 2020 (FAO 2022). India ranks third in total fish production and second in aquaculture production in the world, with production reaching 17.5 million tonnes in 2022–2023. The fisheries and aquaculture sector has contributed over 6.7% to the agricultural gross domestic product (GDP) and about 1.09% to India's GDP in 2022–2023. The sector provides nutritional security and supports the livelihood of around 30 million people of India (PIB 2023). With the production from capture fisheries stagnating, aquaculture is expected to meet the growing demand of fish in the coming years. With the ongoing expansion, intensification and diversification of culture practices and the globalization of trade of live aquatic animals and their products, there is an associated risk of the emergence of new pathogens and the spread of existing pathogens to newer areas.

It is important to mention that diseases are the most significant threat to the growth of aquaculture. World over, there are a few reports of economic losses due to diseases. The estimated annual global economic impact of diseases on aquaculture was estimated to be about US$6 billion (World Bank 2014). Average losses due to various diseases in the Asian shrimp industry between 2009 and 2018 were estimated to be US$ 4 billion (Shinn et al. 2018). Among the shrimp diseases, infection with white spot syndrome virus (WSSV) continues to be a major threat since its first emergence from 1991. The losses due to white spot disease alone were estimated to be US$ 15 billion since its emergence, with an increasing trend of US$ 1 billion annually (Lightner et al. 2012; Stentiford et al. 2012). Losses due to WSSV in the Mekong Delta in 2015 were estimated at US$ 11.02 million, though other estimates by Hien et al. (2016) suggested losses could be as high as US$ 55.58 million (Shinn et al. 2018). In Thailand, the losses due to white spot disease and yellow head disease were approximately US$ 650 million (Bondad-Reantaso and Subasinghe 2010). Losses due to another devastating disease, i.e. acute hepatopancreatic necrosis disease (AHPND), have been estimated to be US$ 1 billion annually between 2010 and 2014 (Bayliss et al. 2017). It has been reported that the shrimp diseases in India resulted in an annual loss of 0.21 million tons of shrimp worth US$ 1.02 billion. The estimated production loss due to WSSV was lower, i.e. 0.033 million tons, with a revenue loss of US$ 238.33 million, whereas hepatopancreatic microsporidiosis accounted for a production loss of 0.077 million tons with a corresponding revenue loss of US$ 567.62 million (Patil et al. 2021).

Similar to shrimp diseases, fish diseases also have resulted in huge economic losses. The infectious salmon anaemia (ISA) outbreak in Chile in 2007 devastated the aquaculture industry, leading to bankruptcy with a debt of US$ 1.8 billion (Bayliss et al. 2017). In Australia, the pilchard herpesvirus outbreak was reported to result in economic losses exceeding A$ 27 million between 1995 and 1999 (Whittington et al. 2005). In Indonesia, another viral disease outbreak, i.e. infection with koi herpesvirus, in both koi and common carp has resulted in losses of over US$ 15 million in 2003 (Sunarto et al. 2005). In addition, parasitic infestations are estimated to cause annual global losses of US$ 1.05–9.58 billion (Shinn et al. 2015). Importantly, salmon lice infestation has had a significant impact on salmon aquaculture. The Norwegian salmon industry incurred expenses of around €500 million in 2019 for combating lice, with additional costs in Canada (€17 million), Chile (€304 million) and Scotland (€78 million) (see Guragain et al. 2021). In Indian carp culture, the infestation with a similar crustacean ectoparasite, *Argulus*, was reported to result in an annual loss of approximately US$ 37 million (Sahoo et al. 2013). Bacterial diseases also have been reported to cause huge economic losses in aquaculture globally. In China, bacterial septicaemia, caused by pathogens such as *Aeromonas hydrophila*, *Yersinia ruckeri* and *Vibrio fluvialis*, resulted in annual losses exceeding US$ 120 million in the early 1990s, with losses later reduced to about US$ 72.4 million due to improved control methods (Wei 2002). The catfish diseases in East Mississippi during the 2016 production season led to direct economic impacts totalling $16.9 million (Peterman and Posadas 2019). In Brazil, parasitic and bacterial diseases in freshwater farmed fish caused annual losses estimated at US$ 84 million (Tavares-Dias and Martins 2017). Similarly, infection with oomycete *Aphanomyces invadans* has resulted in widespread fish mortality, severely impacting the local aquaculture industries and contributing to substantial financial losses. The total losses due to EUS were more than US$ 15 million in only four countries (Thailand, Bangladesh, Sri Lanka and Nepal) for the period up to 1990 (see Lilley et al. 1992). A substantially higher estimate was given by Chinabut (1994), who suggested that losses due to EUS in Thailand could exceed US$ 100 million. Fungal diseases, such as those affecting rock oysters in New South Wales, Australia, have also caused significant economic impacts, with a single pathogen, *Marteilia sydneyi* (QX disease), resulting in production losses worth $30 million (Bondad-Reantaso and Subasinghe 2010).

The early detection of disease outbreaks through active vigilance and rapid identification of the aetiological agent is crucial for timely disease alerts concerning aquatic animals. This aids in the effective planning and monitoring of disease control initiatives, offering aquatic animal health guidance to farmers and fulfilling international disease reporting commitments. Establishing an aquatic animal disease surveillance programme is instrumental in achieving these objectives. Surveillance plays a pivotal role in understanding the health status of aquatic animals within a country, enabling the prompt identification of issues and the implementation of appropriate measures. Furthermore, disease surveillance provides cost-effective information for assessing and managing the risks associated with disease transmission linked to the trade of aquatic animals and their products. Timely and accurate disease reporting to international organizations helps to streamline aquatic animal trade while minimizing the potential for transboundary spread of aquatic animal pathogens, which can have significant serious socio-economic repercussions.

Disease is any deviation in structure or function exhibited by the living organism which is associated with a causative agent or a process. Diagnosis is the process of determining the nature of the disease through examination of clinical signs or diagnostic tests. Disease diagnosis can be presumptive, i.e. identifying the likely cause of disease, or definitive, where a cause of the disease is confirmed. Diagnostic tests for aquatic animal diseases include simple pond-side methods to more advanced laboratory-based techniques which require a high level of capital investment in infrastructure and also greater expertise. Diagnosis of diseases plays a crucial part in aquatic animal health management and disease control. The immediate and clearly defined role is to quickly and accurately determine the cause of unfavourable health condition or other abnormality in aquatic animals in order to implement appropriate management measures including treatment, emergency harvest, disinfection of farms etc. to minimize the negative impacts of the diseases. The other role of diagnostics is to screen healthy animals to ensure that they are not inapparent carriers of pathogens. The latter is important for disease prevention, as populations that have tested negative for specific pathogens are required for domestic or international movement of aquatic animals. Besides, this also assists in detecting infection at an early stage, thereby implementing timely intervention measures to prevent a scenario of disease outbreak and resulting economic losses. Therefore, diagnosis of diseases is an integral component of National Aquatic Animal Health Strategy.

11.2 Three Levels of Disease Diagnosis

There are three levels of diagnosis, i.e. levels I, II and III, have been recommended (FAO/NACA 2000). These three levels are interdependent and form a continuum of observations. The level I diagnosis is made in the field by clinical examination of the affected aquatic animals and their environment (Fig. 11.1). This diagnosis is usually presumptive and may be adequate for endemic diseases with pathognomonic gross lesions. The level II diagnosis is made in the laboratory and includes

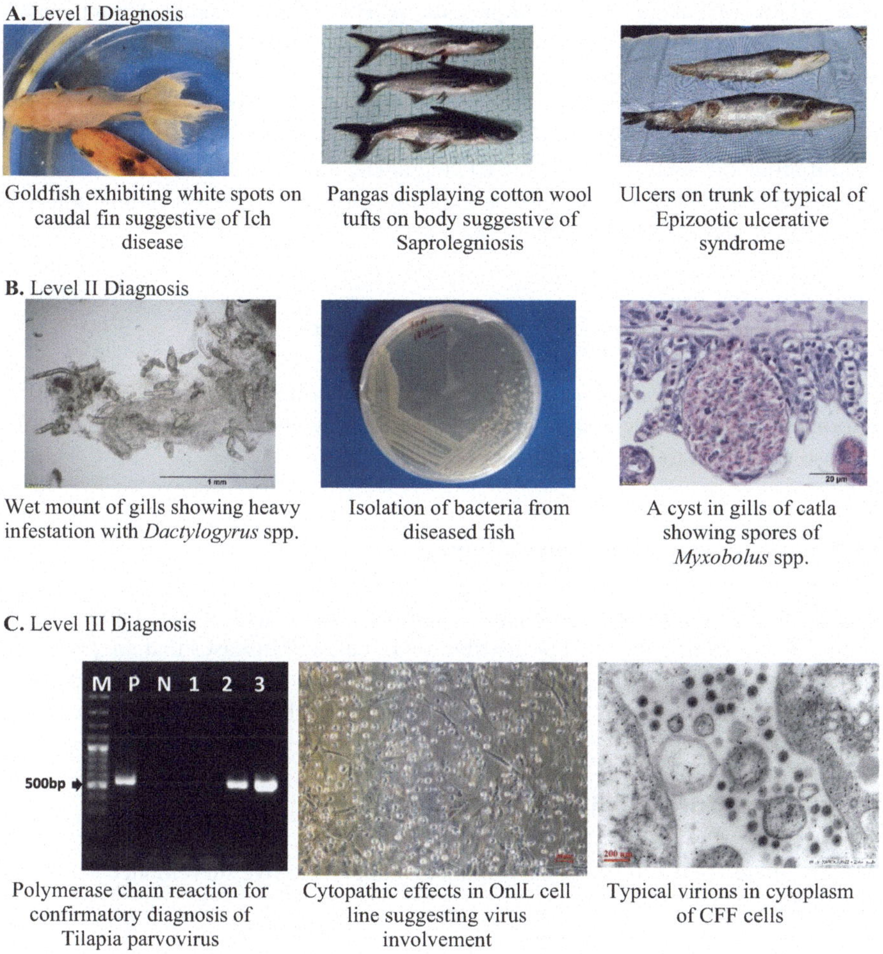

Fig. 11.1 The three levels of disease diagnosis (**a**. Level I diagnosis based on clinical signs; **b**. Techniques used for level II diagnosis; **c**. Techniques used for level III diagnosis)

specializations like parasitology, histopathology, bacteriology and mycology, which require moderate capital and training investment (Fig. 11.1). Level II diagnostics especially histopathology is believed to be the gold standard for new and emerging diseases. Level III diagnostics comprises advanced laboratory techniques including virology, molecular biology, electron microscopy and immunology (Fig. 11.1) which require significant capital and training investment. It is important to mention that even some of the level III diagnostic techniques like virology and electron microscopy reveal the involvement of some pathogens that need to be identified using other techniques. The level III techniques generally target a specific pathogen or a group of pathogens and are vital for fulfilling Koch's postulates. These techniques are also vital for claiming disease freedom for specific diseases. Furthermore, based on level III diagnostic techniques, the World Organization for Animal Health (WOAH) has recognized reference laboratories for WOAH-listed diseases of finfish, crustaceans and molluscs.

Importantly, the three levels of diagnosis build on each other, with each contributing significantly for optimum diagnosis. The level I diagnosis also forms the basis for initiating the level II and level III diagnosis, since results obtained from the higher levels of diagnosis can only be meaningfully interpreted in conjunction with observations and results from the lower levels.

11.3 Aquatic Animal Disease Surveillance

Surveillance is the systematic, ongoing collection, collation and analysis of information related to aquatic animal health and the timely dissemination of information to those who need to know so that action can be taken.

11.3.1 Purposes of the Surveillance

In general, the major purposes of aquatic animal disease surveillance include determining and detecting the level or distribution of cases of diseases that are present in the country and rapidly detecting the incursion of new, emerging or exotic diseases and demonstrating freedom from disease or pathogens that are absent in the country.

11.3.2 Types of Surveillance

Broadly, there are two types of surveillance, namely, passive surveillance and active surveillance.

11.3.2.1 Passive Disease Surveillance

Passive disease surveillance or farmer-based surveillance is a type of surveillance in which the farmers notice a disease and report it to the authorities. Many a times, farmers also send samples to the laboratory for diagnosis of the disease, and the laboratory report makes the disease diagnosis information very accurate. In passive surveillance, the aim is to collect information about all the diseases but not a specific disease, and therefore, it is also known as general surveillance. This type of surveillance is continuous, comprehensive and inexpensive. Importantly, this is the most common form of surveillance in most of the countries and helps in identifying the diseases present in the country. In some countries which have made the reporting mandatory for certain diseases, this type of surveillance is very effective in getting disease reports. One of the major advantages of this type of surveillance is that it helps in the early detection of new, emerging and exotic diseases as the farmers are regularly in direct contact with their fishes in farms and notice clinical signs quickly. The early detection is helpful for managing and controlling diseases, thereby reducing losses. Additionally, passive surveillance helps to meet the disease reporting obligations of member countries to international aquatic animal health organizations. Nevertheless, there are some disadvantages with the passive disease surveillance. Many a times, the farmers may be unable to recognize a disease or may fail to report the disease, which leads to under-reporting and biased reporting. Additionally, as the population size related to the disease reports is unknown, it becomes impossible to calculate useful disease measures like prevalence rates and proportions. Further, this system of surveillance cannot confirm the absence of a disease in an area or provide representative data on disease levels and geographic distribution. Despite these limitations, passive surveillance is considered very useful for detecting new diseases, responding to disease outbreaks, identifying which diseases are present and meeting international reporting obligations. To overcome the limitations of passive disease surveillance, a number of approaches have been developed which include sentinel surveillance system and syndromic surveillance.

11.3.2.1.1 Syndromic Surveillance

A syndrome refers to the collection of signs that suggests a disease situation. The syndromic surveillance is defined as identification of specific signs or groups of signs and analysis of the patterns of these signs in space and time (Cameron et al. 2014). The purpose is to detect abnormal patterns of signs that can be attributed to one of the various diseases. Following detection of abnormal pattern, the disease investigation is undertaken to diagnose the cause of the disease. The syndromic surveillance is considered to be a cost-effective approach for detecting new as well as emerging diseases for which diagnostic tests are not available. This approach can be used to collect three types of data, namely, individual clinical signs; syndromes based on involvement of dominant organ system; and indirect indicators of disease.

11.3.2.1.2 Sentinel Surveillance Systems

In a passive surveillance system, all cases of diseases affecting aquatic animals should be reported in ideal situation. However, in reality, only some farmers report disease cases, and only some diseases are reported, resulting in under-reporting and biased reporting. Therefore, an alternative approach, known as sentinel surveillance system, is used to collect complete information from a few selected farmers, and the selected farmers are assumed to represent the population (Cameron 2002).

In the sentinel system, a small number of farmers work with the fisheries officers to keep a comprehensive record of production (stocking, harvest details, treatments etc.) and disease information (description of disease or unusual mortalities). The farmers are selected to have a wide geographic representation. For example, in each district, two to three sentinel farmers work with district fisheries officer. This kind of surveillance system can provide valuable information about the diseases that are prevalent and the impacts of these diseases on production statistics. However, this is not considered to be useful for detecting new, emerging and exotic diseases.

11.3.2.1.3 Indirect Surveillance

This approach is not based on observation of sick aquatic animals. Herein, the existing data regarding the sale of feed, feed additives, antibiotics etc. is used as an indicator of disease. It is assumed that the sale of feed may decrease, whereas the sale of antibiotics and feed additives like probiotics may increase following the occurrence of disease.

11.3.2.2 Active Surveillance

Active surveillance is a type of surveillance in which authorities make active efforts to gather accurate information about specific disease(s) in a population periodically. This type of surveillance helps in addressing the issues of under-reporting and unrepresentative data, as observed in passive surveillance. Additionally, under active surveillance, survey on a small representative sample of the population helps in collecting better quality information about specific disease(s) which reflects the true disease situation in the population of interest. Further, survey of a part of the population (a sample) rather than every member of the population makes it quicker and less expensive. Here, the sample consists of selected units (i.e. fish, ponds or farmers) from the population, and each unit is examined for disease information. The disease status of the sample provides information from which inferences can be drawn about the entire population. It is assumed that the disease status in the sample represents that of the whole population. Therefore, to ensure that the sample represents the population accurately, it is essential to select a representative sample that is similar to the rest of the population with respect to the disease being studied. Hence, random sampling, in which every unit has an equal chance of being sampled, is considered as the best way to achieve this. The sample size is very important because a larger sample size helps in getting more precise estimates, reduces random errors and gives more reliable results. The precision of the survey results can be measured by confidence intervals, which indicate how close the survey estimate is likely to be to the true population value. The confidence intervals give a range of values in which the real number lies. If the confidence intervals are narrower, the estimate will be more precise. Therefore, all survey estimates should include confidence intervals to show their reliability.

For effective control of diseases, it is important to know how the disease is distributed, how much of the disease is existing, where exactly the disease is present and which group of fishes and age groups are affected. For knowing this, it is important to know the number of affected and unaffected fishes, ponds, farms etc., so that different measures of disease occurrence (i.e. prevalence and incidence rates) can be calculated. The two important measures of disease occurrence are prevalence and incidence. Prevalence is defined as the proportion of animals with the disease at a specific point in time. It is like taking a snapshot to see how the disease is widespread at a particular time. Prevalence can be calculated by dividing the number of diseased fishes by the total number of fishes at risk at that time. On the other hand, the incidence rate is defined as the rate at which the new cases of the disease are occurring over a period of time. It is calculated by dividing the number of new cases by the total time that all fishes were at risk of getting the disease. The incidence rate shows the speed at which the disease is spreading. A disease with a high incidence rate but short duration will have a low prevalence because the disease comes and goes quickly. On the other hand, a disease with a low incidence rate but long duration will have a high prevalence because the disease lasts longer in each affected fish.

In active surveillance, laboratory tests are often used to check if the fishes are infected. However, no laboratory test is perfect, and sometimes these tests give incorrect results. To effectively use these tests, it is important to understand their accuracy and the types of errors they might make. For evaluating a test, it is important to consider its precision and validity, and it is very well described by epidemiological definitions, i.e. sensitivity and specificity. The sensitivity of a test is the proportion of truly diseased animals in the population which are correctly identified by the test as being diseased. Sensitivity also can indicate how exactly the test can identify fishes that are actually diseased. The test with high sensitivity indicates that the test correctly detects most of the animals that are diseased. The specificity of a test is the proportion of truly non-diseased fishes in the population which are correctly identified by the test as being non-diseased. It means how well the test can identify fishes that do not have the disease. A test with high specificity will correctly identify most healthy animals as not having the disease. With the help of sensitivity and specificity, we can judge how reliable a test is for diagnosing diseases in fishes. Importantly, this helps in making informed decisions based on the test results, reducing the chances of missing diseased fishes or wrongly diagnosing healthy ones. Some approaches have been developed to improve the efficiency of active surveillance.

11.3.2.2.1 Risk-Based Surveillance

Risk-based surveillance (RBS) provides an improved strategy particularly for claiming disease freedom. When the disease occurrence is rare, a larger sample size is required to detect positive cases with a defined level of confidence. In such scenario, the utility of random sampling is limited due to financial and operational constraints, and therefore, non-random sampling is considered to be a more efficient option. Herein, the population is divided into higher- and lower-risk strata based on risk factor information and by conducting surveys on populations with a higher risk of being infected. A higher sensitivity can be attained with a lower sample size (Thrusfield 2005). The most important advantage of RBS is the economic criterion. A number of risk factors associated with disease occurrence including spatial factors (e.g. climate, population density, trade), host factors (e.g. age, species), management factors (e.g. biosecurity, antimicrobial usage) and other factors (e.g. history of cases or risky practices) can be applied for RBS (Wells et al. 2009). One of the requirements of RBS is the need of prior information of at least one risk factor that has an association with disease occurrence. Besides, the results of RBS cannot be extrapolated to the general population.

11.4 Surveillance Programme for Aquatic Animal Diseases in India

'The Prevention and Control of Infectious and Contagious Diseases in Animals Act, 2009' and its 2014 amendment provide a legal framework for preventing the spread of diseases in the country. In addition, as a member country of the World Organization of Animal health (WOAH), it is mandatory for the country to meet international disease reporting obligations. Keeping the above in consideration, the Department of Animal Husbandry, Dairying and Fisheries, Ministry of Agriculture, Government of India, started the National Surveillance Programme for Aquatic Animal Diseases (NSPAAD), funded by the National Fisheries Development Board in Hyderabad. In Phase I (April 2013 to March 2022), the programme was initiated in 14 states of aquaculture importance and involved 24 collaborating centres and was coordinated by ICAR-National Bureau of Fish Genetic Resources, Lucknow. In due course of time, the programme was expanded to cover 19 states through 31 collaborating centres. To further strengthen the aquatic animal disease surveillance programme in the country, the Department of Fisheries, Ministry of Fisheries, Animal Husbandry and Dairying, Government of India funded the Phase II of NSPAAD under Pradhan Mantri Matsya Sampada Yojana (PMMSY). The Phase II of NSPAAD has pan-India coverage, and there is involvement of the state fisheries departments and Marine Products Export Development Authority (Fig. 11.2).

In the beginning of the programme, a diagnostic manual was prepared containing the internationally agreed protocols so that all the NSPAAD collaborating centres follow uniform methods for the detection of OIE-/NACA-listed diseases and diseases of national concern. Simultaneously, for strengthening the diagnostic capability for OIE-listed diseases, positive controls were procured from OIE referral laboratories, and the capacity for detection of emerging aquatic animal pathogens is continuously being upgraded. Further, an 'Orientation Training Workshop' was organized involving international resource persons to sensitize the investigators involved in the project. Additionally, several international events are being organized to strengthen the expertise in the area. The major emphasis of the disease surveillance programme has been to strengthen passive disease surveillance in the country so that the diseases are reported and investigated. To strengthen the passive surveillance system in the country, awareness programmes are being organized for fish farmers involving the officers of state fisheries departments. In these programmes, emphasis is given to sensitize the farmers about the clinical signs of important diseases and their management, the importance of timely reporting and the methods for transportation of the diseased fish samples to nearby NSPAAD collaborating centres. Additionally, regular training programmes are organized for the officers of state fisheries departments and research scholars to strengthen their diagnostic capability.

It is important to mention that the strengthening of the surveillance programme is evident by an increase in the number of disease cases being reported by the farmers. Under the programme, the disease cases reported by the farmers are investigated by NSPAAD collaborating centres, and scientific advice for the management

Fig. 11.2 Pan-India implementation of NSPAAD Phase II with NSPAAD collaborating centres and state fisheries departments

of diseases is provided so that losses due to the diseases are minimized. Further to improve the farmer-based reporting of aquatic animal diseases in the country, an android-based mobile application named as 'ReportFishDisease' (RFD) has been developed under NSPAAD Phase II (Fig. 11.3). Using the RFD app, the farmers can report the incidence of disease in finfish, shrimps and molluscs on their farms to field-level officers and fish disease experts as and when required and get scientific advice for the efficient management of disease.

Besides, over the years, 11 diseases have been reported for the first time from the country, and these include goldfish haematopoietic necrosis disease caused by Cyprinid herpesvirus-2 (Sahoo et al. 2016); koi sleepy disease caused by carp oedema virus (Swaminathan et al. 2016); hepatopancreatic microsporidiosis caused

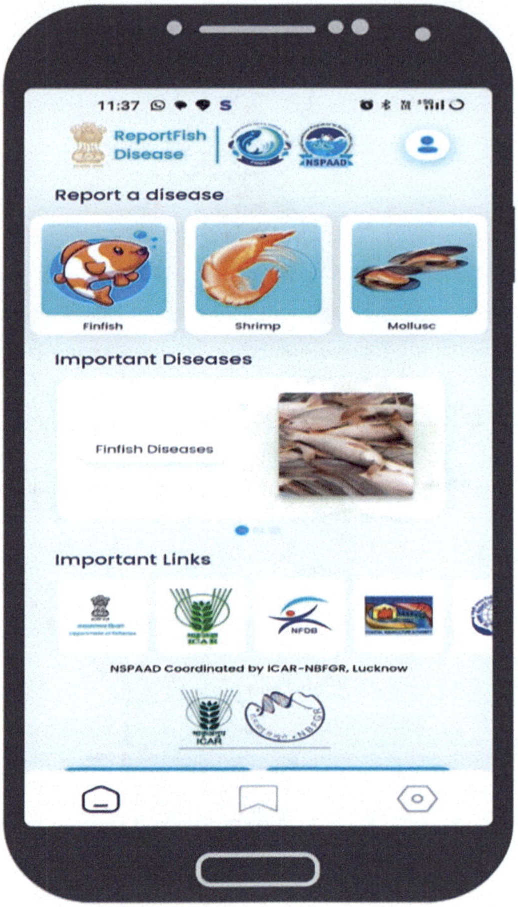

Fig. 11.3 ReportFishDisease app developed under NSPAAD Phase II

by *Enterocytozoon hepatopenaei* (Rajendran et al. 2016); infectious myonecrosis caused by infectious myonecrosis virus (Sahul Hameed et al. 2017); tilapia lake virus disease caused by tilapia lake virus (Behera et al. 2018); epitheliocystis caused by Ca. Actinochlamydia pangasiae sp. nov (Sood et al. 2018); infection with red sea bream iridovirus (Girisha et al. 2020); infection with mud crab reovirus (Ganesan et al. 2023); infection with rhabdovirus (John et al. 2021); and infection with tilapia parvovirus (Rajendran et al. 2023) and Wenzhou shrimp virus 8.

Furthermore, a mechanism has been established for first-time confirmation of exotic and emerging diseases. Following suspicion of a disease for the first time from a country, the same is reported to the NSPAAD coordinating institute, and the samples are shared with two national laboratories. Following confirmation of the disease by three centres, a compiled report is submitted to the competent authority, i.e. the Department of Fisheries, Ministry of Fisheries, Animal Husbandry and Dairying, Government of India. Subsequently, alerts and advisories are sent to the stakeholders. To date, all the first reports of the above disease from the country have been published in peer-reviewed international journals.

11.5 Common Diseases Reported from India

11.5.1 Finfish Diseases

11.5.1.1 Parasitic Diseases

11.5.1.1.1 Myxoboliasis

Myxoboliasis is caused by infestation with myxozoans including *Myxobolus*, *Henneguya* and *Thelohanellus* sp. The disease is common in freshwater fishes. However, the cyprinids in general and Indian major carps in particular are reported to be infected. All age groups are susceptible, but fry and fingerlings are more susceptible. The gills are the main target organ for the disease. In the affected fishes, the gills become pale beginning from the tips and are covered with a heavy coat of slime. In heavy infections, the outer margins of the gills are covered with large, white, opaque parasite cysts which are spherical or irregular in shape (Fig. 11.4). In addition, the diseased fish become lethargic, accompanied by a partial or complete loss of appetite. In severe cases, the growth is stunted. The fish gradually become very weak and emaciated and exhibit increased respiratory movements or float at the surface frequently gulping air.

11.5.1.1.2 Ichthyophthiriasis

Ichthyophthiriasis, also known as 'Ich' disease is caused by *Ichthyophthirius multifiliis*. All Indian major carps, common carp, other cyprinids, murrels and aquarium fishes including goldfish, gourami, guppy, puffer fish, botia, blue betta, clown barb and tetra etc. are highly susceptible host species. Although all age groups of fish are affected, fry and fingerling stages are more vulnerable to the infection. Skin, fins and gills are the target organs. Fish usually develop white spots on the skin (Fig. 11.4). If the parasite is only present on the gills, white spots will not be seen at all, but the mortality rate can be high. Heavy mucus secretion is also an indication of the disease. Fish may show signs of irritation by rubbing at the corners or bottom of the pond or tank, flashing and holding the fins close to the body. More severe infection results in skin sloughing, weakness and lethargy.

11.5.1.1.3 Dactylogyrosis

Dactylogyrosis is due to infestation with monogenean parasites of the genus *Dactylogyrus*. The ectoparasite mostly affects freshwater food fishes including Indian major carps, Chinese carps, other cyprinids, murrels, catfish and ornamental fish like goldfish, guppy, red oscar, angel fish, discus, gourami, swordtail, pufferfish, botia, betta, barbs, tetra etc. All stages of the host are susceptible, but high mortality is observed in the fry and fingerling stages. The gills are the main target organ, but *Dactylogyrus* can be seen on the fins or skin also. Usually there are no clinical signs

Catla showing *Lernaea* infestation

Rohu showing *Argulus* infestation

White cysts on caudal, dorsal and anal fin of rohu due to *Thelohanellus* sp.

White patches in gills of catla due to *Dactylogyrus* infestation

White spots on body of gourami due to infestation with *Ichthyophthirius multifiliis*

White spots in gills of common carp due to *Ichthyophthirius multifiliis*

Fig. 11.4 Common fish disease observed in the country

in fish with few parasites. However, moderate to heavily infected farmed fish show signs of irritation like rubbing at the corners or bottom of the pond/tank, flashing or violent swimming. Heavily infected fish are weak and lethargic and stay in low-current water areas. The high mortality rate among fry and fingerlings is due to respiratory dysfunction.

11.5.1.1.4 Trichodiniasis

Trichodiniasis is caused by saucer-shaped ciliated protozoans. The protozoans are frequently encountered in cultured freshwater fishes; however, infestation has also been reported in marine fishes. The major target organs are the skin and gills. Fingerlings and juveniles are more susceptible to infestation with trichodinids. The affected fish show white patches, mottling of the skin and fins and excessive mucus

Ulcers on trunk of mrigal due to Epizootic ulcerative syndrome

Pangas showing cotton wool growth typical of Saprolegniosis

Protruding eyes and abdominal swelling in tilapia. These non-specific signs are observed in Tilapia lake virus disease.

Motile *Aeromonas* epticaemia in rohu

Fig. 11.4 (continued)

on the skin. The fish may rub their bodies against hard surfaces and exhibit flashing movements. In heavy infection of the gills, the fish show laboured opercular movements.

11.5.1.1.5 Argulosis

Argulosis is caused by infestation with crustacean ectoparasites of the genus *Argulus.* Freshwater carps are mostly affected by this disease, and amongst Indian major carps, rohu is the most susceptible followed by catla and mrigal. Silver carp is considered to be a relatively resistant species among exotic carps. Although all age groups are affected, juveniles and adults are more susceptible, and the skin is the main target organ. Flashing or scratching, violent erratic swimming, loss of appetite, lethargy and denser shoaling are some of the major clinical signs. In addition, fish usually develop red spots or haemorrhagic lesions on the body.

11.5.1.1.6 Lernaeosis

Lernaeosis is caused by infestation with crustacean ectoparasites of the genus *Lernaea*. The disease is most common in cyprinids, including Indian major carps, koi carp, common carp, goldfish etc. Usually, juveniles and young adults are susceptible stages, and the skin, fins, gills, muscle tissues and oral cavity are the target organs. The affected fishes show irritability, emaciation, slow movement and poor appetite, and the parasites can be seen on the body surface (Fig. 11.4). At the site of attachment of the parasite, haemorrhagic lesions occur on the body of the fish, and swelling can be observed around the site of attachment of the parasite.

11.5.1.2 Oomycete Diseases

11.5.1.2.1 Infection with *Aphanomyces invadans*

Infection with *Aphanomyces invadans*, commonly known as epizootic ulcerative syndrome (EUS), is caused by the oomycete *Aphanomyces invadans*. More than 160 fish species are reported to be susceptible to the disease. Usually juvenile and young adults are the susceptible stages. There is no reported infection with *A. invadans* in fry or larval stages. In the affected fish, red spots may be observed on the surface of the body, head, operculum or caudal peduncle at the initial stages of the infection which may progress to large red or grey shallow ulcers, often with a brown necrosis in the later stages (Fig. 11.4).

11.5.1.2.2 Saprolegniosis

Saprolegniosis, commonly known as cotton wool disease, is caused by the oomycete *Saprolegnia parasitica*. Most of the freshwater fish species including carps, catfishes, tilapia, trouts etc. are susceptible. In carps, although all age groups are affected, the disease is more commonly seen in eggs and mature brood fishes. In catfishes particularly in pangas culture, the disease has been reported to cause huge losses in winter months. The affected fishes exhibit cotton wool-like tufts on the integument and gills of fish (Fig. 11.4). In severe cases, it may extend over the entire body surface. The actual cause of death is likely to be associated with impaired osmoregulation and respiratory difficulties, which may happen when infection is associated with the gills.

11.5.1.3 Bacterial Diseases

11.5.1.3.1 Motile *Aeromonas* Septicaemia (MAS)

Motile *Aeromonas* septicaemia (MAS) is caused by any of the three species of the genus *Aeromonas*, namely, *A. hydrophila*, *A. caviae* and *A. veronii*. Many freshwater fishes such as catla, rohu, mrigal, murrels, catfish, goldfish, tilapia, rainbow trout, barbs, few brackish water species and some marine fishes are affected by the disease. All age groups of the fishes are affected, and the skin, muscle, liver and kidney are the target organs. The clinical signs of the affected fish are quite varied and diverse. It includes behavioural (loss of appetite, lethargic and swim lazily at the surface), external (epidermal lesions and exophthalmia) and internal (haemorrhages, pale liver and gill, swollen kidney and spleen) signs, and the fin and tail rot.

11.5.1.3.2 *Edwardsiellosis*

Edwardsiellosis is caused by the Gram-negative intracellular pathogen *Edwardsiella tarda*. Importantly, *E. tarda* isolates from fish have been reclassified into three species, namely, *E. tarda*, *E. piscicida* and *E. anguillarum* of which *E. piscicida* is an important fish pathogen and is associated with huge economic losses in aquaculture.

Many freshwater fishes such as Indian major carps, murrels, catfish, goldfish, tilapia etc. are affected. All age groups of the susceptible fishes are affected by the disease. The kidney, liver and muscle are the target organs. The clinical signs in the affected fishes include cutaneous lesions, abscesses, distended abdomen, swelling and bleeding of the anus, depigmentation and haemorrhages on the ventral parts of the body. Fish shows signs of spiralling movement, and mortality is observed with flared opercula and a gape mouth.

11.5.1.3.3 Columnaris Disease

Columnaris disease is caused by the Gram-negative bacterium *Flavobacterium columnare*. The disease is capable of infecting most freshwater fish species, and usually juveniles and young adults are the susceptible stages of the host. The gills and skin are the main target organs. In scale-less fish, skin lesions begin as areas of discoloration primarily at the base of the dorsal fin and spread to form a pale white band that extends laterally and encircles the body to form a characteristic, pale white 'saddleback' lesion. In fish with scales, gill necrosis is observed as yellowish-white spots on the tips of the primary lamellae of the gills along with fin erosions. The clinical signs of the disease include listlessness, lethargy, inappetence, swimming near the water surface and increased opercular movement.

11.5.1.3.4 *Epitheliocystis*

Epitheliocystis, characterized by the presence of cysts, is caused by intracellular chlamydial infections. More than 90 fish species including rohu and pangas are reported to be affected with the disease. Juveniles of fish are the more susceptible. The skin and gills are the main target organs. The mortalities are mainly due to respiratory distress caused by inflammation, cellular proliferation leading to hyperplasia and necrosis of gill lamellae surrounding the cysts.

11.5.1.3.5 Piscine Mycobacteriosis

Piscine mycobacteriosis, one of the most common bacterial diseases affecting cultured and wild fishes, is primarily a chronic disease caused by species belonging to the genus *Mycobacterium*, commonly referred as non-tuberculous mycobacteria (NTM). The target organs include all tissues including the visceral organs, muscles and gills. Usually, the clinical signs are non-specific and include scale loss, ulceration, spinal defects, emaciation and ascites. Mycobacteria infecting fishes can cause disease in human beings, especially in immunocompromised individuals.

11.5.1.4 Viral Diseases

11.5.1.4.1 Red Sea Bream Iridoviral Disease (RSIVD)

Red sea bream iridoviral disease (RSIVD) is caused by infection with red sea bream iridovirus (RSIV) strain and other genotypes of RSIV including many viruses that are considered to be synonyms of RSIV. The disease is also caused by infectious spleen and kidney necrosis virus (ISKNV), which is one of the viruses related to, but distinct from, RSIV. RSIVD is a significant cause of mortality in farmed red sea bream, and more than 30 other species of cultured marine fish belonging mainly to the orders Perciformes and Pleuronectiformes are susceptible. The susceptibility of juveniles is generally higher than that of adults. The spleen, kidney, heart, intestine and gill are the main target organs. Affected fish become lethargic and exhibit severe anaemia, petechiae in the gills and enlargement of the spleen. Diseased fish swim inactively and show abnormal and conspicuous respiratory exercises caused by anaemia.

11.5.1.4.2 Tilapia Lake Virus Disease

Nile tilapia (*O. niloticus*), grey tilapia (*O. niloticus* × *O. aureus*) and red tilapia (*Oreochromis* sp.), hybrid tilapia (*O. niloticus* × *O. aureus* hybrids) and different species of wild tilapia comprising *Sarotherodon galilaeus*, *Tilapia zillii*, *O. aureus* and *Tristramella simonis intermedia* are susceptible to TiLV. Although all life stages are affected, fingerlings and juveniles are more susceptible. Liver, brain and spleen are the main target organs. The major clinical signs include lethargy, abnormal behaviour such as swimming at the water surface, stop schooling, swirling or loss of balance, loss of appetite, ocular lesions, skin lesions, discoloration and abdominal distension (Fig. 11.4).

11.5.1.4.3 Goldfish Haematopoietic Necrosis Disease

Goldfish haematopoietic necrosis disease is caused by Cyprinid herpesvirus-2 (CyHV-2). Goldfish (*Carassius auratus*), crucian carp (*C. carassius*) and Gibel carp (*C. auratus gibelio*) and hybrids of goldfish are highly susceptible. Although CyHV-2 can infect all the different life stages of goldfish, i.e. egg, fry, fingerling and adult fish, the juveniles are more susceptible. The Gills, kidney, spleen and liver are the main target organs. The clinical signs include yellowish and pale skin, bilateral exophthalmia, fin pustules, decaying white gill filaments, lethargy, anorexia, elevated respiratory efforts, gasping at the surface, erratic swimming, lying at the tank bottom, enophthalmos, necrotic gill patches and mortality across all fish sizes.

11.5.1.4.4 Infection with Tilapia Parvovirus

In the affected fish, the clinical signs include lethargy, loss of scales, haemorrhages on the body, anorexia and change in swimming behaviour, i.e. darting or corkscrew movements. The spleen, kidney, liver, brain, heart and gill are the main target organs. This disease can result in mortality which may range between 60 and 70%.

11.5.1.4.5 Viral Encephalopathy and Retinopathy (VER)

VER is caused by viral nervous necrosis (VNN) virus. To date, the disease has been reported in more than 50 fish species, mainly marine fishes including sea bass, groupers, flatfishes, milk fish, mullets, sole, snapper, cobia, tilapia etc. In addition, few outbreaks have also been documented in freshwater farms. Although the disease mainly affects the larval and juvenile stages, serious mortalities have also been reported in market-size and adult fish. Brain, spinal cord and retina are considered the target organs. The diseased fishes exhibit abnormal swimming behaviour including spiral and/or looping swim pattern, uncoordinated darting, belly-up at rest, swim bladder hyperinflation, anorexia, wasting, dark or pale colour of the skin etc.

11.5.1.4.6 Koi Sleepy Disease

Koi sleepy disease is caused by carp oedema virus. The principal host species are koi and common carp. The gill is the main target organ. The disease is characterized by typical sleepy behaviour, enophthalmia, generalized oedematous condition and gill necrosis, and the affected juvenile carps usually congregate near the surface of the pond or water inlet, whereas the older fish tend to lie on the bottom of the pond and eventually die of anoxia.

11.5.1.4.7 Lymphocystis

Lymphocystis, caused by lymphocystis disease virus, has been reported from over 125 species of fish from both freshwater and marine environments. The disease is characterized by small cream-coloured nodular lesions on the skin and fins of fish. Each nodule is known as lymphocyst, and these nodules may cover the entire body. The diseased fish exhibit reduced growth rate.

11.5.2 Crustacean Diseases

11.5.2.1 Fungal Disease

11.5.2.1.1 Hepatopancreatic Microsporidiosis

Hepatopancreatic microsporidiosis is caused by the microsporidian *Enterocytozoon hepatopenaei* (EHP). The principal host species is *Penaeus vannamei* and *P. monodon*. Although there are no known distinctive clinical signs in shrimps but infection with EHP is known to cause severe growth retardation over time which leads to an increased variability in size. Size variation may not be apparent during the early days of culture (DOC). EHP-infected shrimp typically display soft shells, lethargy, loss of appetite, reduced feed intake, empty midguts and whitish hindgut. The hepatopancreas is the main target organ.

11.5.2.2 Viral Diseases

11.5.2.2.1 White Spot Disease

All decapod crustaceans from marine and brackish or freshwater sources and all life stages from eggs to broodstock are potentially susceptible to white spot syndrome virus (WSSV). The best life stages of crustaceans for detection are the late PL stages, juveniles and adults. Tissues originating from the ectoderm and mesoderm, especially the cuticular epithelium and subcuticular connective tissues, as well as other target tissues (e.g. antennal gland, haematopoietic organ etc.), are the main target tissues for infection with WSSV. Samples from the pleopods, gills, haemolymph, stomach or abdominal muscle are recommended for detection. The most commonly observed clinical sign is white spots embedded within the exoskeleton of shrimps. However, the appearance of white spots is absolutely not a good diagnostic sign of WSSV infection. High degrees of colour variation with a predominance of reddish or pinkish discoloured shrimp are seen in diseased populations. If the affected shrimp appears lethargic, their colour changes to pink or reddish-brown, they gather around the edges of ponds/tanks at the water surface, and there is a rapid reduction in food consumption; then, a very high mortality rate can be expected within a few hours to a few days of the onset of these signs.

11.5.2.2.2 Infectious Hypodermal and Haematopoietic Necrosis

Most of the penaeid species including the principal cultured species, *P. monodon* and *P. vannamei*, are affected by infectious hypodermal and haematopoietic necrosis virus (IHHNV). In *P. vannamei*, all life stages (i.e. eggs, larvae, post-larvae [PL], juveniles and adults) are affected. The affected shrimp exhibit cuticular deformities, specifically a deformed rostrum bent to the left or right, called runt deformity syndrome (RDS), which is pathognomonic for infection with IHHNV. However, in chronic infections, this clinical sign is not always apparent. Populations of juvenile shrimp with RDS display disparate growth with a wide distribution of sizes and many smaller than expected shrimp.

11.5.2.2.3 Infectious Myonecrosis

The principal host species for infectious myonecrosis virus (IMNV) is *P. vannamei*. The disease is most severe in juveniles and subadults of *P. vannamei*. The principal target tissues are the striated, connective tissues, haemocytes and the lymphoid organ tubule parenchymal cells. In chronic infections, the lymphoid organs are the principal target tissue. In the acute phase of infection, the disease can be presumptively diagnosed from clinical signs. Typically, severely affected shrimps become lethargic during or soon after stressful events such as capture by cast-netting, feeding, sudden changes in water temperature or water salinity etc. Shrimp in the acute phase of disease present focal to extensive white necrotic areas in skeletal muscles especially in the distal abdominal segments and tail fan. The paired lymphoid organs may be hypertrophied up to 3–4 times than their normal size.

11.5.2.2.4 Infection with *Macrobrachium rosenbergii* nodavirus

Infection with *Macrobrachium rosenbergii* nodavirus is commonly known as white tail disease (WTD). The aetiological agents are two viral pathogens, namely, *Macrobrachium rosenbergii* nodavirus (*Mr*NV) (primary) and extra small virus (XSV) (associate). MrNV is important in WTD outbreaks in prawns, although the role of XSV in pathogenicity remains unclear. The giant freshwater prawn, *M. rosenbergii*, is highly susceptible. Larvae, PL and early juveniles are susceptible, whereas adults are resistant and act as carriers. The gill tissue, head muscle, heart, abdominal muscle, ovaries, pleopods and tail muscle are the target organs. Infected PL become opaque and develop a whitish appearance, particularly in the abdominal region. The whitish discoloration appears first in the second or third abdominal segment and gradually diffuses both anteriorly and posteriorly. In severe cases, degeneration of telson and uropods may occur. PLs are highly susceptible to WTD.

11.5.3 Molluscan Diseases

11.5.3.1 Infection with *Perkinsus olseni*

Perkinsus olseni has a broad host range including the oysters, pearl oysters, clams and abalone. All the stages after settlement are considered to be susceptible to *P. olseni*. The principal target organs are connective tissue of all organs and haemocytes. The clinical and gross signs are non-specific and may include dead bivalves or gaping bivalves with thin and watery tissue, pale digestive gland and nodules in the mantle and gills. Infected bivalves may be slow to close their valves when disturbed, but these behavioural changes are also non-specific.

11.6 Conclusion

Aquatic animal disease surveillance is important for the sustainable growth of aquaculture sector that contributes significantly to nutritional security and livelihood generation. In spite of the economic and social benefits, the intensification and diversification of aquaculture have increased the threat of disease outbreaks. Effective disease surveillance encompasses passive farmer-based reporting and active surveillance which are essential for early detection, accurate diagnosis and timely intervention, thereby minimizing losses due to diseases. Combining conventional methods of disease diagnosis with advanced techniques can assist in better management of aquatic animal health, reduce economic losses and also fulfil international disease reporting obligations, ultimately promoting a sustainable growth of aquaculture.

NSPAAD has significantly increased the capacity to monitor, diagnose and manage aquatic animal diseases in the country. The success of the programme is evident from increased disease reporting by the farmers, followed by investigation and scientific advice to the fish farmers. The programme has helped in safeguarding the health of aquaculture sector in the country and is also contributing to facilitate the trade in aquatic animals. The continuous efforts in capacity development, farmer awareness and international collaboration will continue to strengthen the programme, ensuring sustainable aquaculture practices and minimizing economic losses due to diseases.

References

Bayliss SC, Verner-Jeffreys DW, Bartie KL et al (2017) The promise of whole genome pathogen sequencing for the molecular epidemiology of emerging aquaculture pathogens. Front Microbiol 8:121

Behera BK, Pradhan PK, Swaminathan TR et al (2018) Emergence of Tilapia Lake Virus associated with mortalities of farmed Nile Tilapia *Oreochromis niloticus* (Linnaeus 1758) in India. Aquaculture 484:168–174

Bondad-Reantaso MG, Subasinghe RP (2010) Aquatic animal diseases and their economic impact health economics. https://thefishsite.com/articles/aquatic-animal-diseases-and-their-economic-impact. Accessed 30 June 2024

Cameron A (2002) Survey toolbox for aquatic animal diseases: a practical manual and software package. Monographs, Australian Centre for International Agricultural Research, number 114075

Cameron A, Njeumi F, Chibeu D et al (2014) Risk-based disease surveillance: a manual for veterinarians on the design and analysis of surveillance for demonstration of freedom from disease. Food and Agriculture Organization of the United Nations, Rome

Chinabut S (1994) EUS in Thailand. In: Roberts RJ, Campbell B, MacRae IH (eds) ODA regional seminar on epizootic ulcerative syndrome at the Aquatic Animal Health Research Institute (pp. 58–60), Bangkok, Thailand, 25–27 January 1994

FAO (2022) The state of world fisheries and aquaculture 2022. Towards blue transformation. FAO, Rome

FAO/NACA (2000) Asia regional technical guidelines on health management for the responsible movement of live aquatic animals and the Beijing consensus and implementation strategy. FAO Fisheries Technical Paper 402. Rome: FAO

Ganesan S, Baskaran B, Raj M et al (2023) Reovirus occurrence in mud crab farming systems and wild-caught brooders located in eastern coastal area of India. Aquac Int 31:739–758

Girisha SK, Puneeth TG, Nithin MS et al (2020) Red sea bream iridovirus disease (RSIVD) outbreak in Asian seabass (*Lates calcarifer*) cultured in open estuarine cages along the west coast of India: first report. Aquaculture 520:734712

Guragain P, Tkachov M, Båtnes AS et al (2021) Principles and methods of counteracting harmful Salmon-arthropod interactions in Salmon farming: addressing possibilities, limitations, and future options. Front Mar Sci 8:701793

Hien NT, Huong NT, Chuong VD et al (2016) Status of acute hepatopancreatic necrosis disease (AHPND) and other emerging diseases of penaeid shrimps in Viet Nam. In: Pakingking RV, de Jesus-Ayson EGT, Acosta BO (eds) Addressing acute hepatopancreatic necrosis disease (AHPND) and other transboundary diseases for improved aquatic animal health in Southeast Asia: proceedings of the ASEAN regional technical consultation on EMS/AHPND and other transboundary diseases for improved aquatic animal health in Southeast Asia (pp 88–95), 22–24 February 2016, Makati City, Philippines

John KR, George MR, Mageshkumar P et al (2021) Isolation and characterisation of a novel rhabdovirus from snakehead fish showing surface ulcerations and mortality from India: first report. Aquaculture 544:737120

Lightner DV, Redman RM, Pantoja CR et al (2012) Historic emergence, impact and current status of shrimp pathogens in the Americas. J Invertebr Pathol 110(2):174–183

Lilley JH, Phillips MJ, Tonguthai K (1992) A review of epizootic ulcerative syndrome (EUS) in Asia. Aquatic Animal Health Research Institute and Network of Aquaculture Centres in Asia-Pacific, Jatujak: Bangkok

Patil PK, Geetha R, Ravisankar T et al (2021) Economic loss due to diseases in Indian shrimp farming with special reference to *Enterocytozoon hepatopenaei* (EHP) and white spot syndrome virus (WSSV). Aquaculture 533:736231

Peterman MA, Posadas BC (2019) Direct economic impact of fish diseases on the East Mississippi Catfish Industry. N Am J Aquac 81(3):222–229

PIB (2023). https://pib.gov.in/PressReleaseIframePage.aspx?PRID=1986155#:~:text=With%20record%20fish%20production%20of,6.724%25%20to%20the%20agricultural%20GVA

Rajendran KV, Shivam S, Ezhil PP et al (2016) Emergence of *Enterocytozoon hepatopenaei* (EHP) in farmed *Penaeus* (*Litopenaeus*) *vannamei* in India. Aquaculture 454:272–280

Rajendran KV, Sood N, Rao BM et al (2023) Widespread occurrence of Tilapia parvovirus in farmed Nile tilapia *Oreochromis niloticus* from India. J Fish Dis. https://doi.org/10.1111/jfd.13871

Sahoo PK, Mohanty J, Garnayak SK et al (2013) Estimation of loss due to Argulosis in carp culture ponds in India. Indian J Fish 60(2):99–102

Sahoo PK, Swaminathan TR, Abraham TJ et al (2016) Detection of goldfish haematopoietic necrosis herpes virus (Cyprinid herpesvirus-2) with multi-drug resistant *Aeromonas hydrophila* infection in goldfish: first evidence of any viral disease outbreak in ornamental freshwater aquaculture farms in India. Acta Trop 161:8–17

Sahul Hameed AS, Abdul Majeed S, Vimal S et al (2017) Studies on the occurrence of infectious myonecrosis virus in pond-reared *Litopenaeus vannamei* (Boone, 1931) in India. J Fish Dis 40:1823–1830

Shinn AP, Pratoomyot J, Paladini G et al (2015) Economic impacts of aquatic parasites on global finfish production. Global Aquaculture Advocate, 82–84. https://www.globalseafood.org/advocate/economic-impacts-of-aquatic-parasites-on-global-finfish-production/#:~:text=Secondary%20losses&text=If%20the%20hatchery%20and%20growout,%241.05%20billion%20to%20%249.58%20billion Accessed 30 May 2024

Shinn AP, Pratoomyot J, Griffiths D et al (2018) Asian shrimp production and the economic costs of diseases. Asian Fish Sci 31S:29–58

Sood N, Pradhan PK, Verma DK et al (2018) *Candidatus* Actinochlamydia *pangasiae* sp. nov. (Chlamydiales, Actinochlamydiaceae), a bacterium associated with epitheliocystis in *Pangasianodon hypophthalmus*. J Fish Dis 41:281–290

Stentiford GD, Neil DM, Peeler EJ et al (2012) Disease will limit future food supply from the global crustacean fishery and aquaculture sectors. J Invertebr Pathol 110(2):141–157

Sunarto A, Rukyani A, Itami T (2005) Indonesian experience on the outbreak of koi herpesvirus in koi and carp (*Cyprinus carpio*). Bull Fish Res Agen Suppl 2:15–21

Swaminathan TR, Kumar R, Dharmaratnam A et al (2016) Emergence of carp edema virus in cultured ornamental koi carp, *Cyprinus carpio koi*, in India. J Gen Virol 97:3392–3399

Tavares-Dias M, Martins ML (2017) An overall estimation of losses caused by diseases in the Brazilian fish farms. J Parasit Dis 41(4):913–918

Thrusfield M (2005) Veterinary epidemiology, 3rd edn. Blackwell Publishing, Oxford

Wei Q (2002) Social and economic impacts of aquatic animal health problems in aquaculture in China. In: Arthur JR, Phillips MJ, Subasinghe RP, Reantaso MB, MacRae IH (eds) Primary aquatic animal health Care in Rural, Small-scale, Aquaculture Development (pp. 55–61). FAO Fish Tech Pap No 406

Wells SJ, Ebel ED, Williams MS et al (2009) Use of epidemiologic information in targeted surveillance for population inference. Prev Vet Med 89:43–50

Whittington RJ, Jones JB, Hyatt AD (2005) Pilchard herpesvirus in Australia 1995-1999. In: Walker P, Lester R, Bondad-Reantaso MG (eds) Diseases in Asian aquaculture V. Fish Health Section, Asian Fisheries Society, Manila, pp 137–140

World Bank (2014) Reducing disease risks in aquaculture. World Bank Report #88257-GLB

Implications of Transboundary Diseases in Wild Finfish Populations: Insights for Conservation

12

Arun Sudhagar, Thangaraj Raja Swaminathan, and Neeraj Sood

Abstract

Transboundary diseases pose significant challenges to the conservation of wild fish populations, with implications for ecosystem health and biodiversity. This chapter explores the dynamics, impacts, and conservation implications of transboundary diseases in wild fish populations, providing insights for effective management strategies. The dynamics of transboundary diseases are shaped by interactions between pathogens, fish hosts, and environmental conditions. Factors such as globalization, climate change, and habitat degradation contribute to the spread and emergence of pathogens, leading to disease outbreaks with profound ecological and socio-economic impacts. Transboundary diseases can have devastating effects on fish populations and ecosystems, including mass mortalities, changes in species composition, and disturbances to ecosystem function. These impacts cascade through aquatic ecosystems, affecting other species, trophic levels, and ecological processes. Furthermore, the socio-economic implications of disease outbreaks, including economic losses, food insecurity, and public health risks, underscore the urgency of effective conservation strategies. Conservation strategies for addressing transboundary diseases in wild fish populations require a multifaceted approach that integrates surveillance and monitoring, habitat restoration, biosecurity measures, collaborative management, and public awareness initiatives. By understanding the complex interactions between pathogens, fish hosts, and their environment, the stakeholders can develop tar-

A. Sudhagar (✉) · N. Sood
ICAR-National Bureau of Fish Genetic Resources, Lucknow, Uttar Pradesh, India
e-mail: Arun.Sudhagar@icar.gov.in; Neeraj.Sood@icar.gov.in

T. R. Swaminathan
Microbiology, Fermentation & Biotechnology Division, ICAR-Central Institute of Fisheries Technology, Kochi, Kerala, India
e-mail: t.swaminathan@icar.gov.in

U. K. Sarkar et al. (eds.), *Sustainable Management of Fish Genetic Resources*,
https://doi.org/10.1007/978-981-97-5250-8_12

geted strategies to mitigate the impacts of transboundary diseases and promote the health and resilience of fish populations and their habitats. Through collaboration, innovation, and a commitment to stewardship, we can work towards sustainable solutions that protect and conserve wild fish populations for posterity. This chapter aims to provide an overview on transboundary diseases as part of broader conservation efforts to safeguard the health and integrity of the aquatic ecosystems.

Keywords

Transboundary diseases · Wild fish populations · Conservation strategies · Ecosystem impacts · Fisheries · Fish diseases

12.1 Introduction

Transboundary diseases in wild fish populations represent a critical and evolving challenge for conservation efforts worldwide (Leaño 2019). As our planet becomes increasingly interconnected through globalization and climate change, the spread of pathogens across the geographical boundaries has become more frequent and impactful. These diseases, caused by a variety of pathogens including viruses, bacteria, parasites, and oomycetes, can have profound implications on the health and sustainability of aquatic ecosystems. In the core, the issue lies in intricate web of interactions between pathogens, fish hosts, and their environment. Pathogens are adept at exploiting vulnerabilities in host populations, whether it be through compromised immune systems, genetic susceptibility, or environmental stressors. Meanwhile, environmental factors, such as water temperature, pH levels, and habitat quality, can influence the transmission and virulence of pathogens, creating a complex interplay of ecological dynamics (Chapman et al. 2021).

The globalization of trade and travel has facilitated the spread of transboundary diseases, with the movement of live fish, contaminated water, and infected equipment (Rodgers et al. 2011). Aquaculture, in particular, has emerged as a hotspot for disease emergence and spread, with intensive farming practices creating ideal conditions for pathogen proliferation. Once introduced into wild populations, these pathogens can establish themselves and spread rapidly, often with devastating consequences for native fish fauna. Climate change further exacerbates the problem, altering ecosystems in ways that favour the proliferation of certain pathogens. Rising water temperatures, altered precipitation patterns, and changing ocean currents can create new opportunities for pathogens to thrive while also stressing fish populations already struggling to adapt to the environmental change (Cascarano et al. 2021). The consequences of these shifts are far-reaching, with potential impacts on biodiversity, ecosystem function, and human well-being. Ballast water from ships is a significant vector for the transmission of fish diseases. As ships take on ballast water in one region and discharge it in another, they inadvertently transport various aquatic organisms, including pathogens. This can introduce non-native diseases to new environments, posing serious threats to local fish populations and ecosystems (Drillet et al. 2018). Similarly, biological vectors such as migratory birds, animals, insects, crustaceans, and molluscs can also serve as potential vectors for disease transmission. The factors influencing the spread of transboundary diseases among the wild fish population are illustrated in Fig. 12.1.

The implications of transboundary diseases in wild fish populations extend beyond ecological concerns to encompass socio-economic and public health considerations. Outbreaks of disease in wild fish populations can devastate these vital resources, resulting in economic losses due to loss of livelihood and food insecurity within impacted communities. In light of these challenges, it is clear that an interdisciplinary approach is needed to address the complex issues of transboundary diseases in wild fish populations. Conservation efforts must be guided by a thorough understanding of the ecological, social, and economic factors driving disease emergence and spread (Young et al. 2017). By integrating scientific research, policy interventions, and community engagement, we can work towards sustainable solutions that protect both the fish populations and the ecosystems they inhabit. In this chapter, we will explore about the dynamics of transboundary diseases in wild fish populations, their impacts on ecosystems and human communities, and the conservation strategies needed to address these pressing challenges. Through collaboration, innovation, and a commitment to stewardship, we can strive to safeguard the health and integrity of our aquatic environments for generations to come.

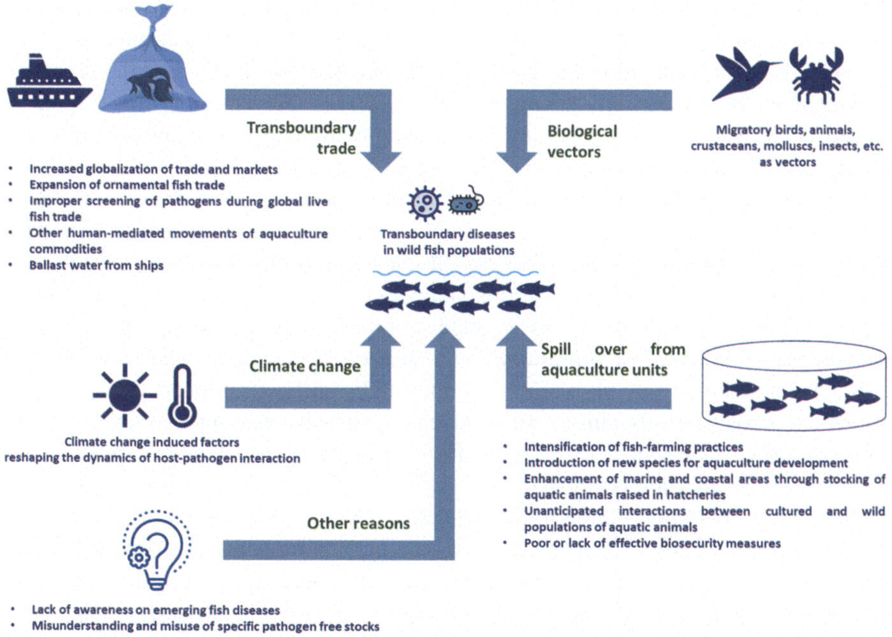

Fig. 12.1 Factors influencing the spread of transboundary diseases among the wild fish population

12.2 The Dynamics of Transboundary Diseases in Wild Fish Populations

Transboundary diseases in wild fish populations are complex phenomena shaped by a myriad of interacting factors, including the ecology of pathogens, the biology of host species, and the environmental conditions of aquatic ecosystems. Understanding the dynamics of these diseases is essential for developing effective conservation strategies aimed at mitigating their impacts and promoting the health and resilience of fish populations. Pathogens, ranging from viruses and bacteria to parasites and fungi, have evolved sophisticated mechanisms for infecting and exploiting their hosts. Some pathogens, such as infectious spleen and kidney necrosis virus (ISKNV) (Swaminathan et al. 2022), Cyprinid herpesvirus 2 (CyHV-2) (Sahoo et al. 2016), and tilapia lake virus (TiLV) (Behera et al. 2018), are highly virulent and can cause mass mortalities in susceptible fish species. Sometimes parasite, such as myxozoan, may not directly kill their hosts but can weaken them, making them more susceptible to predation, competition, or other stressors (Holzer et al. 2021).

The transmission of pathogens among individual fish hosts is influenced by a variety of factors, including host behaviour, population density, and environmental conditions. Close contact between individuals, such as during spawning or feeding aggregations, can facilitate the spread of pathogens through direct contact or the release of infectious particles into the water. High population densities can also increase the likelihood of disease transmission, as pathogens are more likely to encounter susceptible hosts in crowded environments (Krkošek 2017).

Environmental conditions play a crucial role in shaping the transmission dynamics of transboundary diseases in wild fish populations. Water temperature, pH levels, and nutrient concentrations can all influence the survival and infectivity of pathogens in the aquatic environment (Roberts 2012). For example, many pathogens thrive in warm water temperatures, allowing them to replicate and spread more rapidly during the summer months (Sudhagar et al. 2020). Similarly, changes in water chemistry, such as acidification or eutrophication, can alter the susceptibility of fish hosts to infection or create favourable conditions for pathogen growth. The introduction of transboundary pathogens into new environments can have profound consequences for wild fish populations and ecosystems. Invasive pathogens, those introduced from outside their native range, may encounter naive host populations with no prior exposure or immunity to the pathogen. This can lead to explosive outbreaks of disease, with high mortality rates and significant impacts on fish populations and ecosystem dynamics. In some cases, invasive pathogens may also outcompete native pathogens, leading to shifts in disease prevalence or patterns of infection within affected ecosystems.

The human-mediated global trade and travel have paved pathways that are serving as important vectors for disease transmission. The movement of live fish, contaminated water, and infected equipment can all introduce pathogens into new environments, where they may establish themselves and spread rapidly. There is a huge international trade of live ornamental fishes, which potentially facilitates the spread of transboundary diseases (Wood et al. 2022). Aquaculture facilities, in particular, have emerged as hotspots for disease emergence and spread, with intensive farming practices creating ideal conditions for pathogen proliferation and transmission. Many times, the pathogen can spill over from these aquaculture facilities to the wild.

Climate change is also playing an increasingly important role in shaping the dynamics of transboundary diseases in wild fish populations (Combe et al. 2023). The increasing temperatures, shifting precipitation patterns, and evolving ocean currents are not only fostering new environments conducive to pathogen proliferation but also placing additional strain on fish populations already grappling with adapting to environmental shifts. In some cases, climate-driven shifts in habitat suitability or species distributions may facilitate the spread of pathogens into new areas, leading to range expansions or the emergence of novel disease threats.

12.3 Impacts on Fish Populations and Ecosystems

The impacts of transboundary diseases on fish populations and ecosystems are multifaceted and far-reaching, with potential consequences for biodiversity, ecosystem function, and human well-being. Disease outbreaks can have both direct and indirect effects on fish populations, influencing their abundance, distribution, and behaviour. Furthermore, these impacts can cascade through aquatic ecosystems, affecting other species, trophic levels, and ecological processes. One of the most immediate and visible impacts of transboundary diseases on fish populations is the occurrence of mass mortalities (Walker and Winton 2010). Disease outbreaks can lead to rapid declines in fish abundance, with mortality rates sometimes reaching up to 100% in affected populations. This can have significant ecological consequences, particularly in species with small population sizes or low reproductive rates. In extreme cases, mass mortalities can lead to population crashes or local extinctions, disrupting food webs and ecosystem dynamics (Swaminathan et al. 2022). Even in cases where disease outbreaks do not result in mass mortalities, the effects on individual fish can be profound. Infected fish may experience reduced growth rates, impaired reproductive success, or increased susceptibility to predation or other stressors. These sublethal effects can have long-term consequences for population dynamics, influencing recruitment, population structure, and genetic diversity within affected populations.

Transboundary diseases can also alter the composition and structure of fish communities within affected ecosystems. Invasive pathogens, in particular, may selectively target certain species or populations, leading to shifts in species abundance or dominance (Rosa et al. 2022). This can have cascading effects on other species within the ecosystem, including predators, prey, and competitors, as well as symbiotic organisms such as parasites or mutualists. Changes in species composition or diversity can impact ecosystem function, altering nutrient cycling, energy flow, and other ecological processes. Furthermore, the impacts of transboundary diseases on fish populations can extend beyond the aquatic environment to affect terrestrial and marine ecosystems as well. For example, declines in fish populations can have ripple effects through food webs, influencing the abundance and distribution of predators, scavengers, and detritivores (Sumaila and Tai 2020). Changes in fish abundance can also impact nutrient cycling and primary production, with potential consequences for ecosystem productivity and stability.

The loss of fish populations due to transboundary diseases can also have socio-economic implications, particularly for communities that rely on fisheries for food security and livelihoods. Fisheries are a vital source of protein, income, and employment for millions of people worldwide, particularly in developing countries where reliance on fish for nutrition is high (Tidwell and Allan 2001). Disease outbreaks can disrupt these valuable resources, leading to economic losses, food insecurity, and social instability in affected communities. In this section, we explore case scenarios illustrating the impact of a new pathogen on fish populations.

12.3.1 Case 1: Epizootic Ulcerative Syndrome (EUS) in India

Epizootic ulcerative syndrome (EUS) is a formidable disease that has significantly impacted both aquaculture and wild fish populations worldwide. Caused by the oomycete *Aphanomyces invadans*, EUS manifests through invasive infections leading to severe ulcerative lesions and a granulomatous response in the affected fish (WOAH 2023). Its host range includes over 100 fish species, making it a significant threat to aquatic biodiversity, fisheries, and aquaculture. The disease's global spread has been rapid and extensive, affecting numerous countries across various continents. The earliest documented instances of what is now recognized as EUS were cases of mycotic granulomatosis (MG) in Japan around 1971 and red spot disease (RSD) in Australia in 1972. From Japan and Australia, EUS began its transboundary journey through Southeast Asia and the Pacific region. In the late 1970s and early 1980s, EUS was reported in several Southeast Asian countries such as Indonesia, Malaysia, Thailand, Cambodia, and Laos. In the late 1980s, the disease continued its relentless spread, reaching Myanmar, Sri Lanka, and Bangladesh. By July 1988, EUS had made its way to India, first appearing in the Barak Valley region of Assam before spreading to other regions (Kar 2015). The outbreak was linked to the floods from neighbouring Bangladesh, from where the disease is believed to have spread. The rapid dissemination of EUS throughout India led to large-scale fish mortality, severely impacting aquaculture and wild fish populations. Subsequent outbreaks were recorded in Tripura by September 1988, and the disease continued to spread to nearly all Indian states.

The seasonality of EUS outbreaks in India, typically occurring between June and December, is influenced by monsoonal patterns. The disease predominantly affects confined water bodies, such as reservoirs and paddy fields, with a higher incidence rate compared to rivers. Notably, low-salinity areas, such as the brackish-water fisheries of Chilika Lake in Odisha and the backwaters of Kerala, have experienced severe impacts. Environmental factors, including low alkalinity and hardness, have been associated with higher disease intensity.

EUS poses a severe threat to biodiversity, particularly in regions with rich fish diversity. The broad host range has led to widespread mortality, affecting both commercial and subsistence fisheries. This led to severe economic impacts, especially in developing countries where fishing is a primary source of income and nutrition. The economic impact of EUS in India has been profound, leading to the closure of many aquaculture farms and affecting the livelihoods of numerous communities (Jhingran and Das 1990). Even today EUS is a serious threat to fisheries and aquaculture in India. The case of EUS increased awareness and concern among Asian governments have led to improved fish health management practices, benefiting both aquaculture and wild fish populations. This heightened focus on disease management underscores the importance of ongoing research and collaboration to mitigate the impact of transboundary fish diseases.

12.3.2 Case 2: Tilapia Lake Virus (TiLV) Outbreak in Israel

Tilapia are integral to global food security, being the second most farmed fish in the world after carp and a staple protein source in many developing countries. Their farming significantly contributes to local economies, providing livelihoods for millions. The Sea of Galilee in Israel, a major source of potable water and a hub for commercial fishing, saw a severe decline in wild tilapine populations between 2005 and 2010. The production of *Sarotherodon galilaeus*, a predominant species, plummeted from 316 tonnes to as low as 8 tonnes annually. Concurrently, starting in the summer of 2009, Israeli fish farms experienced massive losses of tilapia during the hot seasons (May to October). These outbreaks, characterized by waves of mortality among tilapia of varying weights, spread rapidly from pond to pond. Interestingly, while tilapines suffered high morbidity and mortality, other species such as carp (*Cyprinus carpio*) and grey mullet (*Mugil cephalus*) reared alongside them showed no symptoms, even after long-term cohabitation. Scientific investigations identified the aetiological agent as a novel RNA virus, named tilapia lake virus (TiLV) (Eyngor et al. 2014). Around the same time, TiLV was also reported in Ecuador (Ferguson et al. 2014). Subsequently, the virus spread to about 17 countries worldwide, including India (Behera et al. 2018) leading to large-scale economic losses.

Given the severity of the TiLV outbreak, international organizations have played a crucial role in coordinating efforts to manage and mitigate its impact on global level. In May 2017, the Food and Agriculture Organization of the United Nations (FAO) responded to the escalating threat by releasing a Global Information and Early Warning System (GIEWS) special alert 338 on TiLV. This alert aimed to raise awareness about the virus's spread and its potential economic and ecological consequences, providing essential information to policymakers, researchers, and the aquaculture industry. The World Organization for Animal Health (WOAH) also took significant steps by releasing a technical disease card on TiLV. This document provided detailed information on the virus's characteristics, transmission, clinical signs, and diagnostic methods, serving as a valuable resource for veterinary authorities and aquaculture professionals globally. The dissemination of this technical card facilitated standardized approaches to diagnosis and management, contributing to more effective control measures. Furthermore, TiLV was included in the Network of Aquaculture Centres in Asia-Pacific (NACA) regional Quarterly Aquatic Animal Diseases (QAAD) reporting system for the Asia-Pacific region. This inclusion mandated regular reporting on TiLV cases, enhancing the monitoring, and tracking the virus's spread across different countries. The QAAD system fosters regional collaboration and information sharing, enabling countries to implement timely and coordinated responses to outbreaks. These actions by international organizations underscore the importance of a collaborative global approach to managing transboundary diseases like TiLV.

The TiLV outbreaks severely disrupted local aquatic ecosystems and caused substantial economic losses in the aquaculture industry worldwide. Significant investments in biosecurity, surveillance, and research into resistant strains and vaccines are necessary to manage the virus. Despite these challenges, the outbreak spurred innovation, driving the development of more resilient and sustainable aquaculture practices. The experience from TiLV highlights the importance of vigilance, research, and international collaboration in combating aquatic diseases.

12.3.3 Case 3: Proliferative Kidney Disease Outbreak in the Yellowstone River, Montana, USA

Proliferative kidney disease (PKD) is an emerging disease caused by a myxozoan parasite *Tetracapsuloides bryosalmonae* that affects salmonid group of fishes (Canning et al. 1999). This parasite is responsible for the decline of the wild brown trout population in the Alpine streams of Switzerland and Austria (Wahli et al. 2008; Waldner et al. 2020; Sudhagar et al. 2020). In 2016, an alarming outbreak of PKD struck the Yellowstone River, Montana, USA, igniting concerns among scientists, conservationists, and local communities. The outbreak primarily affects salmonids, particularly mountain whitefish (*Prosopium williamsoni*), and has been known to cause significant mortality in affected populations. Wildlife agencies implemented emergency fishing closures and restrictions on a 300 km stretch of river, limiting public access for a significant period. This subsequently impacted local tourism and the economy (Sage 2016). The outbreak in the Yellowstone River was particularly concerning due to the river's ecological significance and its status as a prized destination for anglers and outdoor enthusiasts. Reports emerged of dead and dying fish lining the riverbanks, sparking immediate action from wildlife officials and researchers to investigate the extent and cause of the outbreak. Initial assessments revealed a high prevalence of PKD among native fish populations, with mortality rates reaching alarming levels in some areas. The parasite's life cycle, which involves an intermediary host, bryozoans, and a definitive host, salmonids, complicated the efforts to contain the outbreak. Scientists hypothesized that environmental stressors, such as low water flows and elevated water temperatures, may have exacerbated the spread of the disease (Hutchins et al. 2021). The outbreak also prompted heightened awareness among the public and policymakers about the threats posed by transboundary diseases to aquatic ecosystems and the importance of proactive conservation measures. While the 2016 outbreak of PKD in the Yellowstone River underscored the vulnerability of wild fish populations to diseases, it also highlighted the importance of adaptive management and interdisciplinary collaboration in responding to emerging threats. By leveraging scientific expertise, community engagement, and proactive conservation strategies, the stakeholders can work together to mitigate the impacts of disease outbreaks and safeguard the health and integrity of aquatic ecosystems for future generations.

12.3.4 Case 4: Infectious Spleen and Kidney Necrosis Virus (ISKNV) Outbreak in Peechi Dam, Kerala, India

Peechi Dam, located within the Peechi-Vazhani wildlife sanctuary in the Western Ghats, Kerala, India, stands as an artificial freshwater tropical lake. A catastrophic fish mortality event unfolded, leading to the death of numerous endemic pearl spot fish (*Etroplus suratensis*) in the Peechi Dam. This led to the complete depletion of pearl spot stock in the affected dam, which exaggerated the livelihood of the local fishers dependent on them. Subsequent clinical investigations led to the identification of ISKNV as the causative agent behind the outbreak (Swaminathan et al. 2022). Interestingly, the presence of ISKNV in India had only been reported recently (Girisha et al. 2020a; Swaminathan et al. 2021). This outbreak assumes critical significance due to the proximity of Peechi Dam to the Western Ghats biodiversity hotspot, a haven for many endemic fish species. Further investigation is needed to know the extent of the spread of ISKNV among the endemic fish population in the region. This can help to develop effective management strategies to prevent the spread of the virus.

12.4 Conservation Implications and Strategies

The conservation implications of transboundary diseases in wild fish populations are profound, requiring a comprehensive and multidimensional approach to address the complex challenges they pose. Effective conservation strategies must consider the ecological and socio-economic dimensions of disease management while also recognizing the interconnectedness of aquatic ecosystems and the need for collaborative action at local, regional, national, and global scales. One of the key conservation implications of transboundary diseases is the need for enhanced surveillance and monitoring of wild fish populations. Regular monitoring programmes can provide early warning of disease outbreaks, allowing for rapid response and targeted intervention measures. This may include the implementation of biosecurity protocols, quarantine, screening of inbound fish stock, movement restrictions, or other control measures aimed at preventing the spread of pathogens to new areas. By tracking disease trends and identifying emerging threats, the surveillance efforts can help to inform conservation priorities and allocate resources effectively. Aquatic animal disease surveillance is frequently overlooked or neglected. In India the National Surveillance Programme for Aquatic Animal Diseases (NSPAAD) is a collaborative initiative designed to oversee and address the diseases impacting fish and shellfish culture across the country (Sood et al. 2021). Within the scope of the NSPAAD project, numerous transboundary diseases have been documented in India, including instances where the wild fish populations have been affected. Notable examples include the identification of pathogens such as ISKNV and TiLV in the wild fish. Such regionally focused surveillance initiatives play a crucial role in the early detection of transboundary diseases, enabling the development of contingency plans to address aquatic animal emergencies proactively. Important transboundary and emerging finfish pathogens documented from India are listed in Table 12.1.

Biosecurity measures are essential for preventing the introduction and spread of transboundary diseases in wild fish populations (Subasinghe et al. 2023). This includes the implementation of stringent biosecurity protocols in aquaculture facilities, fisheries, and other areas where human-mediated pathways of disease transmission are likely to occur. Measures such as quarantine procedures, pathogen screening, and disinfection protocols can help to minimize the risk of disease introduction and spread, protecting both wild fish populations and the aquaculture industry from the devastating impacts of disease outbreaks. Moreover, implementing rigorous biosecurity measures can effectively mitigate the risk of pathogen transmission from aquaculture facilities to the natural environment.

Habitat protection and restoration are also important components of conservation strategies aimed at addressing transboundary diseases in wild fish populations. Healthy aquatic ecosystems are better able to support diverse fish communities and buffer against the impacts of disease outbreaks. By protecting and restoring critical habitats such as spawning grounds, nursery areas, and migratory corridors, the conservation efforts can help to strengthen the resilience of fish populations to disease and other stressors. This may include initiatives such as habitat restoration projects, pollution reduction measures, or the establishment of protected areas to safeguard important fish habitats from anthropogenic threats.

Collaborative management approaches are also critical for effective conservation of wild fish populations affected by transboundary diseases. Collaboration between governments, conservation organizations, researchers, and local communities can help to leverage expertise, resources, and knowledge to develop coordinated strategies for disease prevention and control. This may involve the establishment of multi-stakeholder partnerships, the sharing of data and information, and the development of joint research initiatives aimed at understanding the drivers of disease emergence and spread. By working together, the stakeholders can develop holistic and science-based approaches to disease management that address the complex challenges posed by transboundary diseases in wild fish populations.

Public awareness and education are other important components of conservation strategies aimed to address transboundary diseases in wild fish populations. By raising awareness about the importance of healthy fish populations and the threats posed by transboundary diseases, the education programmes can empower

Table 12.1 Important transboundary and emerging finfish pathogens documented from India

Sl. no	Pathogen	Host	Status in captive and wild fish in India	Reference
1.	*Aphanomyces invadans* (epizootic ulcerative syndrome)	Broad host range	Reported from both captive and wild	Jhingran and Das (1990); Das et al. 2021; WOAH 2023)
2.	Cyprinid herpesvirus 2 (CyHV-2)	Goldfish (*Carassius auratus*)	Reported from captive goldfish	Sahoo et al. (2016)
3.	Carp oedema virus (CEV)	Koi carp (*Cyprinus carpio* var. *koi*)	Reported from captive koi carp	Swaminathan et al. (2016)
4.	Tilapia lake virus (TiLV)	Tilapia (*Oreochromis niloticus*) (*Oreochromis mossambicus*)	Reported from both captive and wild tilapia	Behera et al. (2018); Uma et al. (2019)
5.	Infectious spleen and kidney necrosis virus (ISKNV)	Giant gourami (*Osphronemus goramy*), pearl spot (*Etroplus suratensis*), and various ornamental fishes (broad host range)	Reported from various captive ornamental fishes and from wild pearl spot	Girisha et al. (2020a); Swaminathan et al. (2021, 2022)
6.	Red sea bream iridovirus (RSIV)	Asian sea bass (*Lates calcarifer*)	Reported from captive Asian sea bass	Girisha et al. (2020b); Sumithra et al. (2022)
7.	Koi ranavirus (KIRV)	Koi carp (*Cyprinus carpio* var. *koi*)	Reported from captive koi carp	George et al. (2015)
8.	Similar damselfish virus (SRDV)	Similar damselfish (*Pomacentrus similis*)	Reported from captive similar damselfish	Sivasankar et al. (2017)
9.	Lymphocystis disease virus (LCDV)	Grass carp (*Ctenopharyngodon idella*)	Reported from captive grass carp	Shahi et al. (2020)
10.	Viral nervous necrosis (VNN)	Asian sea bass (*Lates calcarifer*), flathead grey mullet (*Mugil cephalus*), milk fish (*Chanos chanos*), and many other important fish species of marine, brackish, and freshwater origin (broad host range)	Reported from both captive and wild fish	Jithendran et al. (2011); John et al. (2023)
11.	Snakehead rhabdovirus (SHRV)	Striped snakehead (*Channa striata*)	Reported from captive striped snakehead	John et al. (2021)
12.	Tilapia parvo virus (TiPV)	Tilapia (*Oreochromis niloticus*)	Reported from captive tilapia	Rajendran et al. (2023)

communities to act to protect their local ecosystems. This may involve outreach initiatives such as community workshops, school programmes, or public awareness campaigns aimed at promoting sustainable fishing practices, responsible aquaculture, and the conservation of aquatic habitats. By engaging and empowering local communities, conservation efforts can foster a sense of stewardship and ownership over natural resources, leading to more effective and sustainable conservation outcomes.

12.5 Conclusion

The transboundary diseases present multiple challenges in wild fish populations with far-reaching implications for ecosystems, economies, and public health. Addressing these challenges requires a holistic approach that integrates ecological and socio-economic considerations. By implementing targeted conservation strategies, such as surveillance and monitoring, habitat restoration, biosecurity measures, collaborative management, and public awareness initiatives, the stakeholders can work together to mitigate the impacts of transboundary diseases and promote the health and resilience of fish populations and their habitats. Through collaboration, innovation, and a commitment to stewardship, we can attempt to protect and conserve wild fish populations, ensuring their sustainability and the well-being of ecosystems as well as human communities for generations to come.

References

Behera BK, Pradhan PK, Swaminathan TR et al (2018) Emergence of Tilapia Lake Virus associated with mortalities of farmed Nile Tilapia *Oreochromis niloticus* (Linnaeus 1758) in India. Aquaculture 484:168–174

Canning EU, Curry A, Feist SW et al (1999) *Tetracapsula bryosalmonae* n.sp. for PKX organism, the cause of PKD in Salmonid fish. Bull Eur Assoc Fish Pathol 19(5):203–206

Cascarano MC, Stavrakidis-Zachou O, Mladineo I et al (2021) Mediterranean aquaculture in a changing climate: temperature effects on pathogens and diseases of three farmed fish species. PathoGenetics 10(9):1205

Chapman JM, Kelly LA, Teffer AK et al (2021) Disease ecology of wild fish: opportunities and challenges for linking infection metrics with behaviour, condition, and survival. Can J Fish Aquat Sci 78:995–1007

Combe M, Reverter M, Caruso D et al (2023) Impact of global warming on the severity of viral diseases: a potentially alarming threat to sustainable aquaculture worldwide. Microorganisms 11(4):1049

Das BC, Haridas DV, Jacob N et al (2021) Outbreaks of epizootic ulcerative syndrome in Kerala, India following episodes of flooding. Dis Aquat Org 143:189–193

Drillet G, Juhel G, Trottet A et al (2018) Aquaculture biosecurity challenges in the light of the ballast water management convention. Asian Fish Sci 31S:168–181

Eyngor M, Zamostiano R, Kembou Tsofack JE et al (2014) Identification of a novel RNA virus lethal to tilapia. J Clin Microbiol 52(12):4137–4146

Ferguson HW, Kabuusu R, Beltran S et al (2014) Syncytial hepatitis of farmed tilapia, *Oreochromis niloticus* (L.): a case report. J Fish Dis 37(6):583–589

George MR, John KR, Mansoor MM et al (2015) Isolation and characterization of a ranavirus from koi, *Cyprinus carpio* L., experiencing mass mortalities in India. J Fish Dis 38:389–403

Girisha SK, Kushala KB, Nithin MS et al (2020a) First report of the infectious spleen and kidney necrosis virus (ISKNV) infection in ornamental fishes in India. Transbound Emerg Dis 68(2):964–972

Girisha SK, Puneeth TG, Nithin MS et al (2020b) Red sea bream iridovirus disease (RSIVD) outbreak in Asian seabass (*Lates calcarifer*) cultured in open estuarine cages along the west coast of India: first report. Aquaculture 520:734712

Holzer AS, Piazzon MC, Barrett D et al (2021) To react or not to react: the dilemma of fish immune systems facing myxozoan infections. Front Immunol 12:734238

Hutchins PR, Sepulveda AJ, Hartikainen H et al (2021) Exploration of the 2016 Yellowstone River fish kill and proliferative kidney disease in wild fish populations. Ecosphere 12(3):e03436

Jhingran AG, Das MK (1990) Epizootic ulcerative syndrome in fishes. CIFRI 65:14

Jithendran KP, Shekhar MS, Kannappan S et al (2011) Nodavirus infection in freshwater ornamental fishes in India: diagnostic histopathology and nested RT-PCR. Asian Fish Sci 24:12–19

John KR, George MR, Mageshkumar P et al (2021) Isolation and characterisation of a novel rhabdovirus from snakehead fish showing surface ulcerations and mortality from India: first report. Aquaculture 544:737120

John KR, Sivasankar P, George MR (2023) Status of emerging and reemerging fish viral infections in India. In: Saxena SK (ed) Viral outbreaks—global impact and newer horizons. IntechOpen. https://doi.org/10.5772/intechopen.109012

Kar D (2015) Epizootic ulcerative fish disease syndrome. Elsevier

Krkošek M (2017) Population biology of infectious diseases shared by wild and farmed fish. Can J Fish Aquat Sci 74:620–628

Leaño EM (2019) Transboundary aquatic animal diseases: history and impacts in ASEAN aquaculture. In: Tendencia EA, de la Peña LD, de la Cruz JMV (eds) Aquatic emergency preparedness and response systems for effective management of transboundary disease outbreaks in Southeast Asia: proceedings of Asian Regional Technical Consultation. (pp. 72–79) Aquaculture Department, Southeast Asian Fisheries Development Center. http://hdl.handle.net/10862/3466

Rajendran KV, Sood N, Rao BM et al (2023) Widespread occurrence of tilapia parvovirus in farmed Nile tilapia *Oreochromis niloticus* from India. J Fish Dis. (in press). https://doi.org/10.1111/jfd.13871

Roberts RJ (2012) The aquatic environment. In: Roberts RJ (ed) Fish pathology, 4th edn. Wiley, pp 1–16

Rodgers CJ, Mohan CV, Peeler EJ (2011) The spread of pathogens through trade in aquatic animals and their products. Rev Sci Tech 30(1):241–256

Rosa GM, Ayala Botto G, Mitra AT et al (2022) Invasive fish disrupt host-pathogen dynamics leading to amphibian declines. Biol Conserv 276:109785

Sage JL (2016) Economic contributions of the Yellowstone River to Park County, Montana, vol 346. Institute for Tourism and Recreation Research Publications. https://scholarworks.umt.edu/itrr_pubs/346

Sahoo PK, Swaminathan TR, Abraham TJ et al (2016) Detection of goldfish haematopoietic necrosis herpes virus (Cyprinid herpesvirus-2) with multi-drug resistant *Aeromonas hydrophila* infection in goldfish: first evidence of any viral disease outbreak in ornamental freshwater aquaculture farms in India. Acta Trop 161:8–17

Shahi N, Mallik SK, Kala K et al (2020) Seasonal emergence of benign epidermal tumor in farm-reared adult grass carp (*Ctenopharyngodon idella*) caused by lymphocystis disease virus at Uttarakhand, India. Aquaculture 526:735408

Sivasankar P, John KR, George MR et al (2017) Characterization of a virulent ranavirus isolated from marine ornamental fish in India. Virus Disease 28(4):373–382

Sood N, Pradhan PK, Swaminathan TR et al (2021) National surveillance programme for aquatic animal diseases—a stepping stone for establishing disease governance system in India. Curr Sci 120(2):273–277

Subasinghe R, Alday-Sanz V, Bondad-Reantaso MG et al (2023) Biosecurity: reducing the burden of disease. J World Aquac Soc 54(2):397–426

Sudhagar A, Kumar G, El-Matbouli M (2020) The malacosporean myxozoan parasite *Tetracapsuloides bryosalmonae*: a threat to wild salmonids. PathoGenetics 9(1):16

Sumaila UR, Tai TC (2020) End overfishing and increase the resilience of the ocean to climate change. Front Mar Sci 7:523

Sumithra TG, Krupesha Sharma SR, Neelima L et al (2022) Red sea bream iridovirus infection in cage farmed Asian sea bass (*Lates calcarifer*): insights into the pathology, epizootiology, and genetic diversity. Aquaculture 548(1):737571

Swaminathan TR, Kumar R, Dharmaratnam A et al (2016) Emergence of carp edema virus in cultured ornamental koi carp, *Cyprinus carpio koi*, in India. J General Virol 97(2):3392–3399

Swaminathan TR, Raj NS, Preena PG et al (2021) Infectious spleen and kidney necrosis virus-associated large-scale mortality in farmed giant gourami, *Osphronemus goramy*, in India. J Fish Dis 44(12):2043–2053

Swaminathan TR, Johny TK, Nithianantham SR et al (2022) A natural outbreak of infectious spleen and kidney necrosis virus threatens wild pearlspot, *Etroplus suratensis* in Peechi Dam in the Western Ghats biodiversity hotspot, India. Transbound Emerg Dis 69(5):e1595–e1605

Tidwell JH, Allan GL (2001) Fish as food: aquaculture's contribution. EMBO Rep 2(11):958–963

Uma A, Rebecca G, Karthik P, Gangatharan S, Ganesh Babu S, Anitha R (2019) Prevalence of tilapia lake virus (TiLV) in farmed and wild fishes of north Tamil Nadu, India. In: Asian-Pacific Aquaculture 2019, organized by the World Aquaculture Society during June 19–21 at Chennai, pp 10

Wahli T, Bernet D, Segner H et al (2008) Role of altitude and water temperature as regulating factors for the geographical distribution of *Tetracapsuloides bryosalmonae* infected fishes in Switzerland. J Fish Biol 73(9):2184–2197

Waldner K, Bechter T, Auer S, Borgwardt F, El-Matbouli M (2020) A brown trout (*Salmo trutta*) population faces devastating consequences due to proliferative kidney disease and temperature increase: a case study from Austria. Ecol Freshw Fish 29(3):465–476

Walker PJ, Winton JR (2010) Emerging viral diseases of fish and shrimp. Vet Res 41(6):51

WOAH (2023) Infection with *Aphanomyces invadans* (Epizootic Ulcerative Syndrome). In: Manual of diagnostic tests for aquatic animals, 10th edn. World Organisation for Animal Health

Wood LE, Guilder J, Brennan ML et al (2022) Biosecurity and the ornamental fish trade: a stakeholder perspective in England. J Fish Biol 100(2):352–365

Young HS, Wood CL, Kilpatrick AM et al (2017) Conservation, biodiversity and infectious disease: scientific evidence and policy implications. Philos Trans R Soc Lond Ser B Biol Sci 372(1722):20160124

Conservation of Genetic Resources: Restoration of Fish Stocks

13

V. S. Basheer, Charan Ravi, and Uttam Kumar Sarkar

Abstract

India is one of the 34 biodiversity hotspots of the world. It harbours more than 3500 fish species and more than 800 freshwater fish species. Different ecosystems harbour diverse fish species, which provide livelihoods to millions of people and serve as an important source of nutrition. As these ecosystems face multiple threats due to anthropogenic activities, pollution, overfishing, and the introduction of exotic fishes, there are chances of decline of fish stocks. Stock enhancement involves the release of cultured or captive-bred individuals into natural habitats to supplement wild populations. In India, this practice holds immense potential for replenishing declining fish stocks, conserving endangered species, and enhancing ecosystem resilience. By augmenting natural recruitment, stock enhancement can bolster fishery yields, thus ensuring food security and supporting the livelihoods of millions dependent on inland fisheries. Many efforts are being made by different research organizations and governments for the restoration of fishes, like cryopreservation of milt, breeding of the fishes, and ranching of the fishes, so that fisheries will be sustainable. Many projects were formulated for the genetic characterization and conservation of fishes. Coupling with these efforts, stock enhancement programmes are being carried out. ICAR-CMFRI is continuously doing the shrimp stock enhancement using hatchery-produced *Penaeus semisulcatus* in Mandapam to improve the fishery from reservoirs. A few examples of stock enhancement programmes of fishes carried

V. S. Basheer (✉) · C. Ravi
Centre for Peninsular Aquatic Genetic Resources, ICAR-National Bureau of Fish Genetic Resources, CMFRI Campus, Kochi, Kerala, India
e-mail: VS.Basheer@icar.gov.in; Ravi.Charan@icar.gov.in

U. K. Sarkar
ICAR-National Bureau of Fish Genetic Resources, Lucknow, Uttar Pradesh, India
e-mail: Uttam.Sarkar@icar.gov.in

U. K. Sarkar et al. (eds.), *Sustainable Management of Fish Genetic Resources*,
https://doi.org/10.1007/978-981-97-5250-8_13

out in different parts of the world and some of the efforts taken by the ICAR-NBFGR in collaboration with other research organizations are explained in the chapter.

Keywords

Freshwater fishes · Stock enhancement · Conservation · Genetic characterization · Cryopreservation · Exotic fishes

13.1 Introduction

India harbours a rich diversity of freshwater fish species, with more than 800 fish species recorded in our water bodies (Ponniah and Gopalakrishnan 2000). Not only does this fish contribute significantly to the country's biodiversity, but it also plays a crucial role in supporting the livelihoods of people, especially in rural areas where fishing is the main occupation. India's freshwater ecosystems are under threat as a result of rampant habitat destruction, pollution, overfishing, and the introduction of exotic species, which have resulted in the decline of fish stocks and threatened the sustainability of fisheries. Overexploitation of the resources coupled with habitat alteration and other anthropogenic activities not only led to a reduction in fish production in most of the natural inland waters over the years but also pushed several endemic species to become threatened/endangered. The conservation of natural resources in aquatic ecosystems is crucial because most genetic resources used for food still come from the wild because of the low level of domestication in the fisheries sector. Aquaculture is expanding rapidly and now accounts more than 50 percent of the fish production that are directly consumed by humans (Bartley et al. 2007). Aquaculture production has almost tripled as compared to the capture fisheries, according to recent statistics. There are different threats to biodiversity in general and fish diversity in particular due to several reasons. There are a wide range of environmental threats which include overexploitation, habitat alterations, reduction of natural habitat area, construction of dams, diversion or reclamation of river beds for urbanization, and the introduction of exotic species, for improving fish production. Conservation of India's vast and diverse fish genetic resources is essential to maintain ecological as well as socio-economic equilibrium. This will provide a sustainable harvest of these resources either from the wild or via artificial propagation. Indian fish genetic resources, consisting of over 11 percent of the world's 20,000 known species of fish, are among the richest in the world. However, overexploitation, habitat degradation, and pollution have threatened the populations of many freshwater species. In response, freshwater species stock enhancement has emerged as a crucial strategy to restore and maintain fish diversity while supporting fisheries-dependent communities. Other factors impacting fish stocks are pollution, habitat degradation (Molony and Bird 2002; Molony et al. 2003), and climate change (Eaton and Scheller 1996; Morgan et al. 2001).

13.2 Challenges Facing Freshwater Fish Species

Freshwater fish species are confronted with numerous challenges due to anthropogenic activities and environmental changes, including climate change, habitat destruction, water pollution, the introduction of exotic fish species, overfishing, and disease. Freshwater ecosystems are being altered by climate change in various ways, such as changes in water temperature, precipitation patterns, and water flow. Disruption in the distribution and abundance of fish species, alteration in their reproductive cycles, and an increase in the frequency and severity of extreme weather events like droughts and floods can be caused by these changes. Loss of habitat reduces the breeding grounds, disrupts migration patterns, and diminishes the availability of food sources. The destruction and alteration of freshwater habitats through activities such as dam construction, urbanization, agriculture, and deforestation pose a significant threat to freshwater fish. Pollution from various sources, including industrial waste, agricultural runoff, sewage discharge, and improper waste disposal, contaminates freshwater ecosystems. Fish tissues can accumulate chemical pollutants, such as heavy metals, pesticides, and pharmaceutical materials, which can cause physiological and reproductive problems.

Commercial fishing may lead to overexploitation of freshwater fish populations, which can cause population declines and even extinction of species (Mustafa 2003). Native fish populations can be devastated by the introduction of non-native species into freshwater habitats. Exotic invasive species often outcompete native species for resources, prey on them, or bring diseases, to which native species have no resistance (Singh and Lakra 2011a). Diseases and parasites, which can propagate rapidly under certain conditions, such as overcrowding or compromised water quality, can be a threat to fish populations in freshwater ecosystems.

13.3 Strategies for Restoration

Stock enhancement of indigenous freshwater fish species in India is an important aspect of conservation and sustainable fisheries management. There are a number of steps to be taken for restoration or stock enhancement of fish species, such as habitat restoration, breeding and stocking of the desired species, integrated aquaculture, community-based stock enhancement, and implementing regulatory measures. The survival of freshwater fish species depends on protecting and restoring their natural habitats. This includes conserving wetlands, rivers, lakes, and other aquatic ecosystems. Implementing captive breeding programmes for indigenous freshwater fish species will help to supplement the natural populations. Such programmes involve breeding fish in controlled environments and releasing them into suitable habitats. Promoting integrated aquaculture practices that incorporate indigenous fish species alongside other aquaculture activities can also contribute to enhance the fish stock. Monitoring and managin in different periods is required for balancing the population of stocked fishes (Mustafa 2003). Controlling invasive species that compete with or prey upon indigenous fish species is essential for maintaining the ecological balance of freshwater ecosystems (Singh and Lakra 2011a, b). Educating the public, policymakers, and stakeholders about the importance of conserving indigenous freshwater fish species and their habitats is key to garnering support for conservation initiatives. Collaboration among government agencies, NGOs, research institutions, and local communities is essential for the success of freshwater fish conservation initiatives. Working together enables the sharing of resources, expertise, and best practices to achieve common conservation goals (Bartley and Leber 2004).

13.4 Restoration of Depleted Stock

Restoration of severely depleted stock due to overfishing or any other perilous consequences can be achieved by habitat restoration and releasing the cultured juveniles into the natural habitats for improving the stock and exploitation in the future. It is to reconstruct conservation hatcheries for the implementation of possible conservation strategies to lend a hand to restoring endangered or threatened species in the wild. The genetic reliability of a population is greatly affected by the large-scale restocking undertaken to improve the depleted stocks of commercially important species. The term 'stock enhancement' is often used to describe most forms of stocking with an ultimate goal of enhancement practice to increase the stock size of the species which occurred naturally to optimize its harvest or increase catch rates. Stock enhancement and restocking are potential measures that could either reduce the time needed to rebuild certain capture fisheries to a more productive level or increase the productivity of fisheries (Bell et al. 2006). Culture-based fisheries include stocking of small water bodies undertaken on a regular basis, and the activity is the only means of satisfying the fishery in the form of stock enhancement practices. Over the past decades, marine stock enhancement has begun to be treated scientifically, and strides are now being on the way in gaining a scientific understanding of the stock-replenishment potential by hatchery releases (CMFRI 2019). Restocking and stock enhancement must be positioned within a fishery management scheme to facilitate which integrates releases with appropriate control of fishing effort and habitat protection. By implementing these strategies and initiatives, India can enhance the stocks of indigenous freshwater fish species, conserve biodiversity, and promote sustainable fisheries management for future generations.

13.5 Fish Stock Enhancement Studies Worldwide

Fishery enhancement based on culture-based fisheries is in practice in natural environments in many countries. Worldwide, many countries have demonstrated the potential for releasing cultured juveniles to augment fisheries in natural water bodies (Anderson 2002). Aquaculture-based enhancement has been practiced on a large scale since the last century, but seed transportation might have a much longer history. Serious thought on research on stock enhancement issues might have started in the late nineteenth century by the Norwegian cod stocking programme (Schwach 1998). Initially some enhancement efforts have yielded good results, generated economic and social benefits, and helped create better fisheries management, but there were many studies that did not give a positive result, and only a few such success stories have been documented (Pinkerton 1994). Drummond (2004) reported stock enhancement programmes of Japanese and New Zealand scallop enhancements (Uki 2006), Alaska salmon enhancement was demonstrated by Heard (2003), and De Silva (2003) reported Asian culture-based lake fisheries in reservoirs. Many countries enhanced the fishery through culture-based enhancement fisheries systems (Bell et al. 2006, 2008). Interest in the release of cultured juveniles as a possible management tool for coastal fisheries has become easy after methods for producing juveniles in captivity and at reasonable cost. Dedicated scientific conferences and reviews have also helped advance the science of releasing a wide range of valuable coastal species (Mustafa 2003; Bartley and Leber 2004). The fisheries enhancement programme by releasing seeds was mainly aimed at a production imperative rather than from a fisheries management perspective (Leber 2004; Lorenzen 2008). As a result, few releases of cultured juveniles have helped to optimize production from coastal fisheries resources. The most celebrated exception is the stock enhancement of scallops in Japan (Ventilla 1982), although similar success stories were also reported from China for shrimp (Wang et al. 2006), scallops, and sea cucumbers (Lovetelli et al., 2004). De Silva and Funge-Smith (2005) reviewed the work done on stock enhancement of different rivers and reservoirs of Asia. Their review was mainly focused on the resources and practices of management and enhancement of inland fisheries and how these resources can be enhanced to continue to provide food and income to the common man. The review also provides suggestions and recommendations on how to improve enhancement practices and practical issues related to improving fisheries. The effect of enhancement on wild fish stocks and the implications for hatchery management of stocks and stocking strategies are also covered. Ibrahim and Ilias (2006) reported stock enhancement programmes in Malaysia for improving the fishery stocks. On the other hand, enhancement initiatives that are poorly integrated into fisheries systems may contribute to management failure by encouraging or compensating for counterproductive changes in fishing patterns or for habitat degradation (Meffe 1992; Taylor 1999). In Asian countries, so far, the economic viability of stock enhancement of large lacustrine water bodies and rivers is lacking, the fisheries of such water bodies being dependent mostly on natural recruitment (De Silva and Funge-Smith 2005). Overall, fish stock enhancement programme became a big success story especially in freshwater reservoirs (FAO 1998).

13.6 Case Studies of Indigenous Fishes in Kerala

Habitat protection and management are important parameters for fishery enhancement. A suitable environment will give good productivity, and such areas need to be protected. Many areas in important water bodies were declared protected area (PA) for conserving the fish and other aquatic life. *Etroplus suratensis* (also known as the green chromide) is the state fish of Kerala, and it is a regional delicacy (Padmakumar et al. 2011). All over Kerala, this species is exploited in huge numbers for marketing, since it is a high-value fish. Stock enhancement programmes for this species are very important for restoring populations and supporting fisheries in Kerala. Regional Agricultural Research Station, Kerala Agricultural University, Kumarakom, had taken up a project for enhancing the fish stocks in the Vembanad Lake, the biggest lake in Kerala. To enhance, conserve, and rehabilitate the declining fisheries of pearlspot, utilizing the unique biological characteristics of the species, a 'fish sanctuary' for pearlspot (Karimeen) was established in the open Vembanad Lake near Kumarakom, by establishing a protected area within the Vembanad Lake. A variety of artificial substrates such as concrete tetrapods, laterite rocks, coconut shells, and coconut piles were planted to facilitate the natural breeding of the fish. The substrates installed on the lake floor provided nesting surfaces for fish, as the fish are known to attach their eggs on underwater substrates. The habitat requirements and spawning behaviours of this species were taken into consideration while developing this system, and simultaneously mangrove canopies were also developed around the margin of the sanctuary. Although the major objective of the fish reserve has been the protection of a minimum spawning stock to ensure recruitment to fished areas, the spillover effects of the sanctuary by adult fish movement from the fish reserve to the adjacent waters were perceptible in the fishing zones close to the sanctuary. Population census by visual fish counts and sighting frequency of pearlspot, the target species in the designated sanctuary, indicated that the fish utilize the sanctuary habitats and deposited substrate materials abundantly for breeding. The increased fishery recruitment and yield of fish in the sanctuary zone during the succeeding seasons indicate that these engineered breeding habitats are viable strategies for the conservation of endemic fish species. Considering the social benefit of the system, the Kumarakom Grama Panchayat, the local administration, has come forward to support the maintenance of the system in partnership with fishermen communities. Given the critical level of overexploitation of lake fisheries, the protective sanctuaries and engineered spawning habitats have been found to be a viable management option to maintain spawning stock and sustain fisheries. Although habitat management is recognized as the cornerstone of fishery enhancement and is widely adopted in developed countries, where annual fish surveys are organized to monitor the stock, it is for the first time that such a model has been tested in our country. Since stock enhancement is not a programme to be started and left abandoned, a wider programme of fishery management by the establishment of a network of sanctuaries of the kind is suggested for the conservation of such endemic species, subject to heavy exploitation. Seeds of *Etroplus* are produced regularly in hatcheries in captivity, and regular stocking was done in the Vembanad Lake for improving the fishery. Under this programme, fishermen reported more than a 10% increase in production from this area by stocking regularly over 3 years (Table 13.1) (Padmakumar et al. 2011).

Efforts were also made to ranch the other indigenous fishes in the Vembanad Lake. The species selected were *Labeo dussumieri* and *Horabagrus brachysoma*.

Labeo dussumieri, commonly called as Malabar Labeo and *Pullan* in local parlance, is an indigenous fish of Kerala. The ICAR-NBFGR had developed the breeding protocol, milt cryopreservation of this species and rearing protocol of young ones (NATP 2004). In a joined effort with RARS, KAU, Kumarakom, the young ones of *L. dussumieri* were stocked in the Vembanad Lake consecutively for 3 years, and the fishermen reported more than an 11% increase in the capture of this species from the Vembanad Lake. Genetic stock characterization work was also complete for this species (Gopalakrishnan et al. 2009). Similarly, yellow catfish, *Horabagrus brachysoma*, a delicacy of Kerala, was also stocked in the Vembanad Lake consecutively for 3 years, and an increase of 9% in production was reported by the fishermen. Conservation measures such as cryopreservation of milt and induced breeding of the yellow catfish were standardized by the ICAR-NBFGR earlier (NATP 2004). Detailed genetic stock characterization was studied by Abdul Muneer et al. (2009). All these three species were stocked for enhancement of production under the NATP project. Subsequently, the Department of Fisheries, Kerala, has taken the enhancement of these species in the reservoirs of Kerala.

The mahseer is a large, iconic freshwater fish species found in rivers across India. The population of this iconic species is declining due to many reasons, including dynamiting. Efforts are being carried out by different agencies, to conserve and enhance the stock of mahseer populations in various states. There was a lot of confusion about the mahseer species, and many agencies are working on the conservation and stock enhancement of this species. Conservation organizations like the Wildlife Association of South India (WASI), Bengaluru, in collaboration with the ICAR-NBFGR, and other different government agencies have initiated projects to protect habitat restoration, establishment of protected areas, and community-based conservation initiatives for mahseer conservation. Though TATA hatchery at Lonavala, Maharashtra, is breeding mahseer, since proper identification was not followed, hybrid mahseers were produced which in turn reduces the productivity, if it ranched in the natural waters.

Sea ranching involves the production of seeds, release of seeds, monitoring, and harvest. The release of young ones into the sea for stock enhancement is called sea ranching (Uno 1984). ICAR-Central Marine Fisheries Research Institute, Mandapam

Table 13.1 Percentage of fish production by repeated stocking and habitat restoration

Species stocked	2002	2003	2004	Increase in production (%) 2005
Labeo dussumieri	1,50,000	2,00,000	2,00,000	11.0
Horabagrus brachysoma	1,00,000	1,50,000	1,00,000	9.0
Etroplus suratensis	15,000 & Habitat restoration and protection of breeding ground	15,000 & Habitat restoration and protection of breeding ground	15,000 & Habitat restoration and protection of breeding ground	10.0

Centre, initiated the sea ranching programme of green tiger shrimp *Penaeus semisulcatus* from 1985 onwards and continued for 3 years. Ranching improved the fishery of the *Penaeus semisulcatus* from the Mandapam area, and the fishery was sustained. ICAR-CMFRI is continuously ranching *Penaeus semisulcatus* in the Mandapam waters, which yields a sustainable production of shrimp from that area.

13.7 Conservation of Genetic Diversity for Stock Enhancement

Conservation of genetic resources is important to ensure the future adaptability of natural populations; to preserve life history and behavioural and morphological traits that ensure sustainable fisheries; and to conserve genetic diversity for cultural reasons. When a species is of special concern, in situ conservation could prove inadequate. Gene banks can hold live animals or cryopreserved gametes, and they can be considered as a last line of defence against species extinction. A live gene bank (LGB) is an ex situ facility where endangered/endemic species are maintained in near-natural conditions. Such gene banks can contribute to the recovery and utilization of genetic diversity and can be used in conservation programmes and genetic enhancement. The developments in biotechnology have made it possible for cryopreservation of fish spermatozoa in sperm banks, and work on embryo cells and pluripotent embryonic stem cell (ESC) preservation is also being attempted. There are two aspects of conservation: the first is managing the declining population, and the second is managing endangered species. The most important aspect to be considered in both cases is that if there are differences in gene frequencies among populations (as observed in many freshwater species among catchments), the populations should be stocked separately in a live gene bank to maintain the genetic diversity. In such cases, the genetic structure of the original wild population should be assured in restocking programmes, before any new fish are released into the waters.

13.8 Efforts of the ICAR-NBFGR in Conservation and Enhancement of Fish Genetic Resources of the Western Ghats

The Western Ghats in India, one of the 34 globally recognized biodiversity 'hotspots', is known for their high levels of endemism as well as rich and highly varied species diversity, including the freshwater fishes (Ponniah and Gopalakrishnan 2000). A concerted effort made by the National Bureau of Fish Genetic Resources (NBFGR) has described genetic variation and population structure for seven prioritized fish species from their major range of natural distribution in the region. These species include *Labeo dussumieri*, *Puntius denisonii*, *P. chalakkudiensis*, *Horabagrus brachysoma*, *H. nigricollaris*, *Gonoproktopterus curmuca*, and *Etroplus suratensis*. Distinct population structure was observed in many of these species indicating that propagation-assisted restoration programmes must be stock-specific to replenish declining populations (Gopalakrishnan et al. 2009; Abdul Muneer et al. 2009). Considering the increased popularity of ornamental fish at household level and to curb the indiscriminate exploitation from wild, captive seed production and rearing technology of 15 indigenous species with export potential such as *Pristolepis marginata*, *Horabagrus nigricollaris*, *Chela fasciata*, *Danio malabaricus*, *Puntius filamentosus*, *P. fasciatus*, and *Mesonoemacheilus triangularis* have been standardized by the College of Fisheries, Kochi, in a joint programme with NBFGR. A handbook on the ornamental fishes of the Western Ghats with descriptions, coloured images, and other relevant information on nearly 150 native ornamental fishes has also been prepared (Anna Mercy et al. 2007). The ICAR-NBFGR in collaboration with the RARS, Kumarakom, has made significant achievements in the conservation of several commercially important and endangered freshwater species of regional importance (Padmakumar et al. 2011). Stock-specific, breeding-assisted river ranching of two food fishes, *Horabagrus brachysoma* and *Labeo dussumieri*, has been successfully carried out in Kerala; the landings of *H. brachysoma* after 2 years increased from 1.8% to 11%, and that of *L. dussumieri* showed an increase from 0.68% to 9% of the total landings from the Vembanad Lake and adjacent rivers in the state. To strengthen such activities further, the establishment of live gene banks (LGB) in the region can have a significant bearing on the protection and conservation of the dwindling population of several endemic fish species.

13.9 Conservation and Stock Enhancement: Way Forward

The fish stocks need to be conserved first for future generations. In comparison to the cryopreservation of gametes, stock enhancement is currently the most practical to maintain live gene banks of aquatic animals. The limits to the number of individuals of each species that can be maintained in LGBs, the varied environmental requirements of different species, the difficulties in propagating individuals in a way that maintains genetic diversity, and the high cost of holding facilities are the main impediments at present. The maintenance of aquatic species in aquatic protected areas can help to overcome these problems. In well-located aquatic protected areas, whole communities of species can be maintained for the cost of effective enforcement of the protected area. Another advantage is that individuals are not removed from their roles in production on-site. Rather, aquatic protected areas allow the number of individuals of exploited species to increase and to attain a greater mean body size (Bohnsack 1993). In the case of most marine fisheries species, this has resulted in an exponential increase in egg production, dispersal of greater numbers of 'seed' to areas outside the protected area and replenishment of the proportion of the population vulnerable to fishing (Harvey et al. 1998). The benefits of establishing aquatic protected areas as gene banks are not limited to the populations on-site. Representative groups from populations that are at risk elsewhere can be transferred to aquatic protected areas, provided that the genetic stocks are the same and that the health and bio-safety requirements are met. The concept of genetic diversity conservation may have an apparent simplistic look, but the measures to be taken for conservation require certain creative approaches. For ex situ operations, gene banking will focus largely on the target species, without much diversification into the banking of the food and other beneficial organisms. For the effective implementation of conservation and management programmes, the participation of the public, voluntary organizations, government agencies, scientists, planners, and administrators is required. It is clear that commitment to a live gene bank will have to be long lasting with total dedication to the programme that involves breeding and maintaining multiple generations without losing genetic diversity.

Invest in scientific research and monitoring programmes to assess the effectiveness of restoration efforts, track changes in fish populations, and adapt management strategies accordingly. Provide training and capacity-building programmes for fishery managers, conservation practitioners, and local communities to enhance their knowledge and skills in sustainable fisheries management and restoration techniques. Advocate for the development and implementation of policies and regulations that support fish stock restoration, habitat conservation, and sustainable fisheries management at the local, national, and international levels. Foster collaboration and partnerships between government agencies, non-governmental organizations, research institutions, and the private sector to leverage resources, expertise, and support for fish stock restoration initiatives. By adopting a holistic approach that integrates ecological conservation, sustainable management, community involvement, and policy support, India can effectively restore and conserve its fish stocks for future generations.

References

Abdul Muneer PM, Gopalakrishnan A, Musammilu KK et al (2009) Genetic variation and population structure of endemic yellow catfish, *Horabagrus brachysoma* (Bagridae) among three populations of Western Ghat region using RAPD and microsatellite markers. Mol Biol Rep 36:1779–1791

Anderson JL (2002) Aquaculture and the future: why fisheries economists should care. Mar Resour Econ 17(2):133–115

Anna Mercy TV, Gopalakrishnan A, Kapoor D et al (2007) Ornamental fishes of the Western Ghats of India. National Bureau of Fish Genetic Resources, Lucknow, India

Bartley DM, Leber KM (eds) (2004) Marine ecology—biotic and abiotic interactions. Marine Ranching. FAO Fisheries Technical Paper 493, Rome, 213

Bartley DM, Harvey BJ, Pullin RSV (eds) (2007) Workshop on status and trends in aquatic genetic resources: a basis for international policy. 8–10 May 2006, Victoria, British Columbia, Canada. FAO Fisheries Proceedings. No. 5. Rome, 179

Bell JD, Bartley DM, Lorenzen K et al (2006) Restocking and stock enhancement of coastal fisheries: potential, problems and progress. Fish Res 80:1–8

Bell JD, Leber KM, Blankenship HL et al (2008) A new era for restocking, stock enhancement and sea ranching of coastal fisheries resources. Rev Fish Sci 16:1–8

Bohnsack JA (1993) Marine reserves-they enhance fisheries, reduce conflicts and protect resources. Oceanus 36:63–72

CMFRI (2019) Sea ranching of hatchery produced green tiger shrimp for the sustainable shrimp productivity. Aquac Spectr 2(2):56

De Silva S (2003) Culture-based fisheries: an underutilised opportunity in aquaculture development. Aquaculture 221:221–243

De Silva S, Funge-Smith S (2005) A review of stock enhancement practices in the inland water fisheries of Asia. Asia-Pacific Fishery Commission, Bangkok, Thailand. RAP Publication No. 2005/12, 93 p

Drummond K (2004) The role of stock enhancement in the management framework for New Zealand's southern scallop fishery. In: Leber KM, Kitada S, Blankenship HL et al (eds) Stock enhancement and sea ranching: developments, pitfalls and opportunities second edition, pp 397–411

Eaton JG, Scheller RM (1996) Effects of climate warming on fish thermal habitat in streams of the United States. Limnol Oceanogr 41(5):1109–1115

FAO (1998) Marine ranching: global perspectives with emphasis on the Japanese experience. FAO Fisheries Circular 943, Rome, 259p

Gopalakrishnan A, Musammilu KK, Basheer VS et al (2009) Low genetic differentiation in the populations of the Malabar carp. *Labeo dussumieri* as revealed by allozymes, RAPD and microsatellites. Asian Fish Sci 22(2):359–391

Harvey B, Ross C, Greer D et al (1998) Action before extinction: an international conference on conservation of fish genetic diversity. Proceedings of a conference held Feb. 16–18, 1998, Vancouver, B.C., Canada. World Fisheries Trust, Victoria, B.C., Canada, p. 259

Heard WR (2003) Alaska salmon enhancement: a successful program for hatchery and wild stocks. In: Nakamura Y, McVey JP, Leber K et al (eds) Ecology of aquaculture species and enhancement of stocks proceedings of the 30th U.S.-Japan Meeting on Aquaculture. Sarasota, Florida, December, 3–4, 2001. UJNR Technical Report No. 30. Mote Marine Laboratory, Sarasota

Ibrahim K, Ilias Z (2006) Fishery stock enhancement in Malaysia. In: Primavera JH, Quinitio ET, Eguia MRR (eds) Proceedings of the Regional Technical Consultation on Stock Enhancement for Threatened Species of International Concern. Iloilo City, Philippines

Leber KM, Kitada S, Blankenship L, Svasand T. (Eds.) (2004) Stock Enhancement and Sea Ranching: Developments, Pitfalls and Opportunities. 2nd. Oxford: Blackwell Publishing; 562 p

Lorenzen K (2008) Understanding and managing enhancement fisheries systems. Rev Fish Sci 16(1–3):10–23. https://doi.org/10.1080/10641260701790291

Lovetelli et al., (2004) Advances in sea cucumber aquaculture and management. FAO Fisheries Technical Paper 463, Rome, p. 440

Meffe GK (1992) Techno-arrogance and halfway technologies: salmon hatcheries on the Pacific coast of North America. Conserv Biol 6:350–354

Molony BW, Bird C (2002) Annual report on the monitoring of the recreational marron fishery in 2000, with an analysis of changes in the long term data from this fishery. Fish Res Rep West Aust 137:1–44

Molony BRC, Lenanton GJ, Norriss J (2003) Stock enhancement as a fisheries management tool. Reviews in Fish Biology and Fisheries 13(4):409–432. https://doi.org/10.1007/s11160-005-1886-7

Morgan GM, Pitelka LF, Shevliakova E (2001) Elicitation of expert judgments of climate change impacts on forest ecosystems. Clim Chang 49:279–307

Mustafa S (2003) Stock enhancement and sea ranching: objectives and potential. Rev Fish Biol Fish 13:141–149

NATP (2004) Project report on Germpalsm and inventory and Live Gene Banking of freshwater fishes. Final report of ICAR=NATP funded project. 400 p

Padmakumar KG, Bindu L, Sreerekha PS et al (2011) Breeding of endemic catfish, *Horabagrus brachysoma* in captive conditions. Curr Sci 100(8):1232–1236

Pinkerton E (1994) Economic and management benefits from the coordination of capture and culture fisheries: the case of Prince William sound pink salmon. N Am J Fish Manag 14:262–277

Ponniah AG, Gopalakrishnan A (eds) (2000) Endemic fish diversity of the Western Ghats. National Bureau of Fish Genetic Resources, Lucknow, India, p 347

Schwach V (1998) The impact of artificial hatching of cod on marine research. In: Lenz W, Streicher S (eds) Historisch-Meereskundliches Jahrbuch, vol 5. Deutsches Meeresmuseum, Stralsund, pp 27–47

Singh AK, Lakra WS (2011a) Risk and benefit assessment of alien fish species of the aquarium trade into India. Rev Aquac 3(1):3–18

Singh AK, Lakra WS (2011b) Ecological impacts of exotic fish species in India. Aquac Asia 16(2):23–25

Taylor EB (1999) Species pairs of north temperate freshwater fishes: evolution, taxonomy, and conservation. Rev Fish Biol Fish 9:299–324

Uki N (2006) Stock enhancement of the Japanese scallop *Patinopecten yessoensis* in Hokkaido. Fish Res 80:62–66

Uno Y (1984) An ecological approach to mariculture of shrimp: shrimp ranching fisheries. In: Proceedings of the First International Conference on the Culture of Penaeid Prawns/Shrimps, Iloilo City, Philippines, 1984. SEAFDEC Aquaculture Department

Ventilla RF (1982) The scallop industry in Japan. Adv Mar Biol 20:309–382

Wang Q, Zhuang Z, Deng J et al (2006) Stock enhancement and translocation of the shrimp *Penaeus chinensis* in China. Fish Res 80:67–79

Artificial Intelligence-Based Modeling for Sustainable Management of Fish Genetic Resources: Status and Opportunities

14

Achal Singh, Tanwy Dasmandal, Rejani Chandran, Ganesan Kantharajan, and Uttam Kumar Sarkar

Abstract

Artificial intelligence (AI) is fundamentally reshaping the management of fish genetic resources and conservation biology, ushering in unprecedented opportunities for sustainable practices. Across the globe, AI is being applied to various facets of fisheries, from modeling fish abundance and distribution to predicting the effects of environmental changes. In India, efforts to embrace technological advancements have seen notable progress in applying AI to sectors like fisheries in recent years. While a range of machine learning algorithms have been explored for addressing diverse challenges in fisheries resource management, artificial neural networks stand out due to their ability to capture non-linear relationships among variables. Despite these advancements, significant hurdles remain. Challenges such as poor underwater image quality, the black-box nature of deep learning algorithms, and the high cost of sensor technologies for data collection, particularly in developing countries, continue to impede progress. Addressing these limitations is crucial to fully harnessing the potential of AI-based modeling for sustainable fishery management. This chapter discusses the potential applications of AI-based models in fish genetic resource management with case studies and future scope for sustainable fisheries development.

Keywords

Artificial intelligence · Artificial neural network · Aquatic genetic resources · Fish conservation · Machine learning

A. Singh (✉) · T. Dasmandal · R. Chandran · G. Kantharajan · U. K. Sarkar
ICAR-National Bureau of Fish Genetic Resources, Lucknow, Uttar Pradesh, India
e-mail: Achal.Singh@icar.gov.in; tanwy.dasmandal@icar.gov.in; rejani.chandran@icar.gov.in; kantharajan.g@icar.gov.in; Uttam.Sarkar@icar.gov.in

U. K. Sarkar et al. (eds.), *Sustainable Management of Fish Genetic Resources*,
https://doi.org/10.1007/978-981-97-5250-8_14

14.1 Introduction

Artificial intelligence (AI) is a rapidly evolving field that aims to create systems and machines capable of performing tasks that typically require human intelligence, such as learning, problem-solving, and decision-making. AI models are the core components of these intelligent systems, designed to process and analyse data, recognize patterns, and make predictions or decisions (Xu et al. 2021). Modeling in the context of AI refers to the process of developing mathematical or computational representations of complex systems or phenomena. These models can be used to simulate, predict, and optimize the behaviour of various systems, from natural processes to human-made technologies. One of the key aspects of AI modeling is the use of machine learning algorithms, which enable systems to learn and improve from experience without being explicitly programmed. These algorithms can discover patterns and relationships within large datasets, allowing AI models to make accurate predictions and informed decisions. Beyond machine learning, several other important aspects are crucial to the development and application of AI systems like deep learning, computer vision and image processing, robotics, expert systems, and so on (Mukhamediev et al. 2022). The utility of AI is well recognized and catalyses the progress of activities towards achieving Sustainable Development Goals (SDGs) by 2030. The application of AI modeling spans a wide range of domains, including agriculture, healthcare, finance, transportation, and scientific research. As the field continues to advance, the potential for AI-driven innovation and problem-solving is vast, with the promise of transforming the way we interact with technology and understand the world around us. A similar area of priority which requires the involvement of AI is the fisheries sector for its sustainable management and utilization.

India possesses a rich diversity of fish genetic resources, encompassing a wide range of freshwater, brackish water, and marine species. Globally, more than 36,775 fish species have been reported (Fricke et al. 2024), and India contributes nearly 3183 fish species, out of which 917 are purely freshwater inhabitants; 1548 are marine species, and 394 are brackish water species (Sarkar et al. 2024). India also harbours several endemic fish species, of which 659 are freshwater fish species and 81 are marine fish species (India Endemic Freshwater Fish Checklist (https://lntreasures.com/indiaff.html). With its rich fish genetic resources, the fisheries sector has highly contributed to the country's economy, with an export of about 6679 million US dollars from species like shrimp, squid, dried fish, etc. in 2019–2020 (Department of Fisheries 2020). However, with the rapid increase in India's population, the pressure on agricultural food production systems including fisheries is also increasing at a faster rate. A sizeable portion of India's total fish production originates from the natural waterbodies, which emphasize their role in ensuring the food and nutritional security of rural populations through subsistence and small-scale fisheries. With the continuous and rapid growth witnessed in this sector, the world is gaining access to unprecedented amounts of data on fisheries and aquaculture. Further, the generation of remote sensing datasets through satellites and other unmanned aerial vehicles (UAVs) provides the scope to understand the complex interaction between fish and the environment. This necessitates the adoption of advanced computational technologies like artificial intelligence, which is based on innovative data systems and analytics, to improve economic, social, and environmental sustainability. For the rapid adoption of AI and ML in India, NITI Aayog has drafted a nationwide strategy for the drive, and the national AI portal named INDIAai (https://negd.gov.in/national-ai-portal-indiaai/) has been launched in 2020 to catalyse the innovations in the field and bolster its applications in various sectors. In the farming sector, the adoption of AI has been increasing steadily in recent years, driven by advancements in technology, growing demand for sustainable farming practices, and the need to address challenges such as climate change, resource scarcity, and food security. The main areas in agriculture where AI has found its application are in addressing complex challenges like monitoring crop health, optimizing irrigation and fertilization, and managing weeds, pests, and disease outbreaks that enable farmers to make data-driven decisions, maximize yields, and minimize inputs, leading to improved efficiency and sustainability (Javaid et al. 2022).

In fisheries, the fish production in general and the shape morphometry in particular are influenced by multiple sets of parameters and the relationship within these parameters. These parameters can be largely grouped into physical habitat, climatic, physicochemical, and hydrological parameters in an interrelated frame. At the community level, factors include stream width, water depth, flow status, physicochemical water quality parameters, and climatic factors that influence the fish assemblage structure in aquatic ecosystems (Sarkar et al. 2020). At the population level, the influence on phenotypic characteristics observed in *Chitala chitala* from various river systems in India is attributed to the stream hydrodynamics (Chandran et al. 2022). Likewise, the morphometric differences reported in the *Megalaspis cordyla* populations along the Indian coast revealed the association with coral reef features in the Gulf of Mannar region (Sajina et al. 2011). However, the occurrence, distribution, and abundance of fish species are correlated with optimal habitat conditions in its natural environment. Likewise, the optimal farming conditions in the aquaculture systems are expected to yield higher production than the stressed environment. This relationship makes the basis for the use of AI in prediction modeling to derive meaningful conclusions from complex datasets. Attempts have been made to integrate AI in solving problems like fish abundance and distribution modeling, biomass estimation, fish age prediction, species classification, and predicting the effect of environmental changes to fish community, as well as disease identification. Unlike the agriculture sector, which has witnessed rapid integration of AI-driven technologies, the fisheries sector faces distinctive challenges stemming from the dynamic nature of aquatic ecosystems, limited data availability, regulatory complexities, and economic constraints. However, recently, several institutes across the world like the University of Milan, Italy; the Institute of Marine Research, Europe; and several others as well as Indian institutes like ICAR-Central Inland Fisheries Research Institute, CSIR-Indian Institute of Toxicology Research, ICAR-Central Institute of Fisheries Education, and Indian Institutes of Technology are addressing the challenges and are actively engaged in research and development related to AI applications in fisheries management, aquaculture, and marine conservation.

The present compilation discusses the different areas of the fisheries sector where AI has been implemented and the key features that are used for modeling and also demonstrates artificial neural network model structure in a few aspects of fisheries management. This chapter also highlights the limitations existing in the application of AI in fisheries as well as the way forward.

14.2 Global and Indian Scenario of AI in Fisheries

AI has become essential for solving complex problems in fisheries through the use of advanced technologies and systems. In the field of aquaculture, AI techniques offer opportunities for real-time monitoring, data analytics, predictive modeling, and decision support systems, which can significantly enhance the understanding and management of fish growth and health (Mustapha et al. 2021). By analysing historical growth data, AI models can learn and adapt to optimize feed management, water quality control, and environmental monitoring, leading to improved growth performance and resource efficiency (Bagheri et al. 2019). AI-based techniques and tools, including disease detection and diagnosis systems, early warning systems, and behavioural analysis, enable rapid and accurate disease identification for timely interventions and minimizing the spread of infections (MacIntyre et al. 2023).

The potential and capability of this technology at a global level can be understood through the number of related publications in AI, which has grown by 34.5% from 2019 to 2020, a much higher growth than from 2018 to 2019 (19.6%). In the last 6 years, the number of AI-related publications on ARXIV grew by more than sixfold, from 5478 in 2015 to 34,736 in 2020 (Zhang et al. 2021). Further, AI publications represented 3.8% of all peer-reviewed scientific publications worldwide in 2019, up from 1.3% in 2011. The popularization of AI in solving complex problems can be judged from the tremendous growth in peer-reviewed AI publications during the period 2000–2019. In 2019, peer-reviewed AI publications grew to >1,20,000 as compared to 2000 (<20,000). The country-wise analysis explains that in the top 15 countries, the United States has contributed >7500 publications, followed by the European Union and China with >3000 papers. Also, 9 countries along with India have contributed <1000 publications in the pool during the period 2000–2019 (Zhang et al. 2021).

The application of AI in different areas of fisheries has gained significant attention worldwide since 2016. The utility of AI in making informed decisions and improving operational practices is made possible through advanced algorithms, big data analytics, machine learning, and robotics. This has leveraged the monitoring, stock assessment, and fish behaviour, growth, and health assessment of aquatic ecosystems which ensures the long-term viability of such habitats (Fig. 14.1). However, the potential applications of AI-based models in fish and fisheries are not explored adequately in comparison with other sectors which offer scope for the sustainable development of the sector. The keyword-based assessment of papers published on the use of AI in traceability in the fisheries sector revealed that <20 papers were published during 2000–2022 (Fernandes-Salvador et al. 2022).

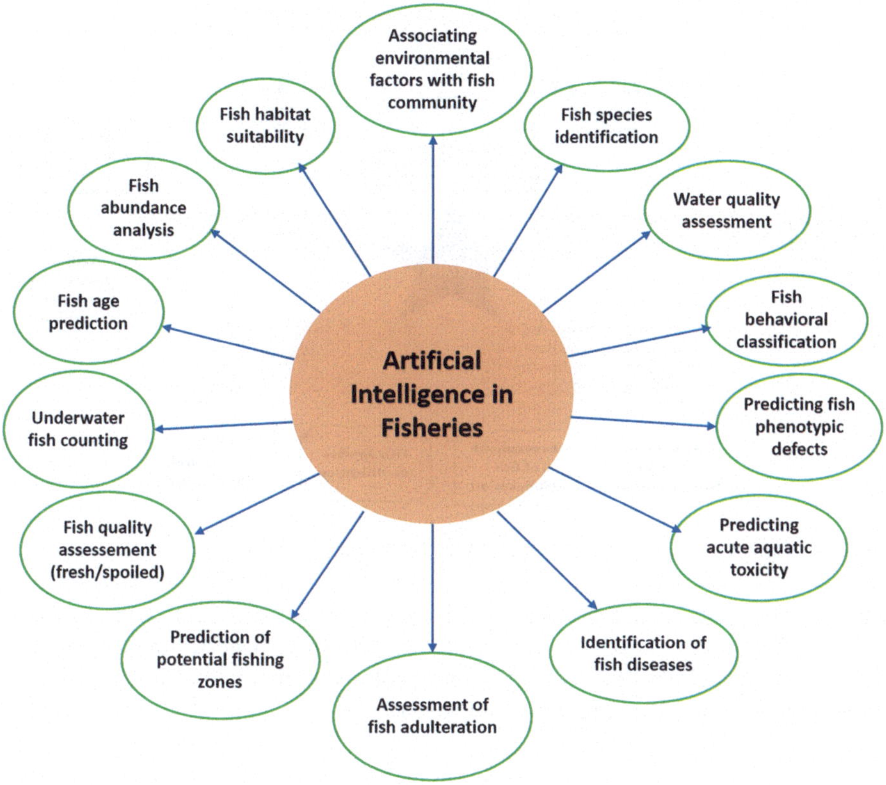

Fig. 14.1 Application of artificial intelligence in fisheries

An artificial neural network (ANN) is a subset of AI that follows the networking of the neurons in the human brain and makes decisions like a human being in the computing system. The ANN is being widely applied in the field of fisheries including the prediction of the weight of crabs (Tureli et al. 2011), the influence of climatic factors on yellowfin tuna (Singh et al. 2015), the influence of weather conditions on pelagic fish landings (Ho et al. 2013), and the automated monitoring and quantification of fish discards onboard in fishing trawlers (French et al. 2020). It is also worth noting that a comparison between the performance of the AI method and the traditional method in predicting mussel weight observed that the AI method performed better than the traditional method (Uba 2020).

In recent years, India has experienced a notable surge in the adoption of AI across diverse sectors, with the fisheries industry being no different. With a coastline of over 8118 km and a rich aquatic biodiversity, India's fisheries sector plays a crucial role in providing livelihoods, ensuring food security, and contributing to the nation's economy. Against the backdrop of evolving challenges, such as overfishing, habitat degradation, and climate change, the integration of AI technologies presents immense opportunities to enhance the efficiency, sustainability, and resilience of

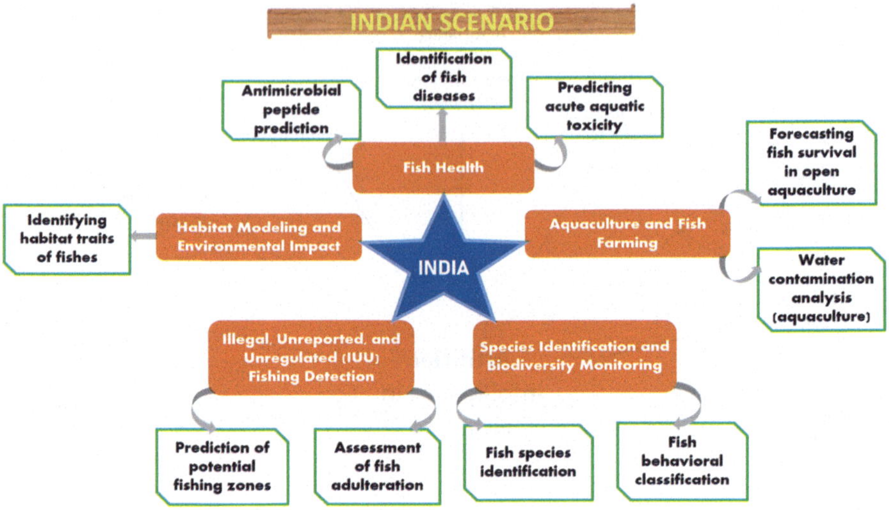

Fig. 14.2 The major areas explored through AI in Indian fisheries

fisheries. The fisheries stakeholders in India leverage the use of AI to address pressing challenges and unlock new avenues for growth from fisheries management to optimizing aquaculture practices. The areas explored in the field of Indian fisheries through the lens of AI are presented in Fig. 14.2.

14.3 Features for AI-Based Models in Fish Genetic Resource Management

The use of AI in different fields, including fisheries, has increased dramatically over the past few decades. With the increasing volume of data being generated, AI models are progressively becoming more dependable for tasks involving classification and prediction. The accuracy of AI models is not solely contingent on the volume of available data but also varies based on the features extracted, selected, and utilized for modeling, which serve as crucial determinants for the efficacy of AI models. Feature extraction is inherently integrated into the layers of deep learning models like convolutional neural networks, necessitating no manual input and demand extraction through machine learning models. Therefore, a comprehensive understanding of the types of features fed into the models for diverse problems is essential. The most common models/algorithms used in developing AI-based relationship networks in the fisheries sector are provided in Table 14.1. In the fisheries sector, a plethora of data types such as fish size, species composition, fish image, and environmental data can be integrated and analysed to furnish valuable insights spanning all facets of fisheries management. Below is a concise summary of the features applicable to diverse studies across various sectors of fisheries resource management.

Table 14.1 Models/algorithms used in developing AI-based relationship networks in the fisheries sector

Sl. no.	Name of the model/ algorithm	Features	Purposes
1.	Artificial neural network or neural network model	• Sigmoid function (N TanH), input, hidden, output layer (diagram) • Development of model in training sub-dataset and validation in another sub-dataset • Optimized NN model provides a prediction profiler	Prediction of fish stock association of variables influencing a system
2.	Partial least square model	• Generate covariance matrix and determination of latent factors (or PLS factor) • Analysis of factors provides control of variations patterns in predictors and response data matrix	Factor-based prediction modeling; associate fish growth with climate parameters
3.	Linear regression model	Model response with many predictors; assumes normality feature	Weight & length, on a logarithmic scale, measure growth over location
4.	Gaussian process regression	A probabilistic (Bayesian) approach also used in AI	Response prediction of associated variables influencing a system under normality assumption
5.	Support vector regression	Most suitable for classification between two groups and maximum margins from boundary line which separate the groups	Classification among groups based on a criterion (e.g. fish with positive allometric growth ($b > 3$) and negative allometric growth ($b < 3$))
6.	Random forest	Produce multiple decision trees with multiple data samples and final prediction based on average values	Prediction of fish stock and its growth; association of biological traits change over time
7.	Extreme learning machine	Simple and single layer, feedforward neural network; low evaluation speed	Neural network of predictors producing response variable as species growth; in normality case, this is similar to linear regression model approach

(continued)

Table 14.1 (continued)

Sl. no.	Name of the model/ algorithm	Features	Purposes
8.	Boosting trees (BoTs)	• Helps in transforming weak decision trees that cause overfitting problems to strong decision trees and performs well with numerical data • Capture the non-linearity feature of predictors and response • Weak performance in dealing with sparse features (or predictors)	Prediction of fish stocks and growth parameters, association of multiple traits changing over generation tree
9.	Bagging trees (BaTs)	• Prediction for trees' out-of-bag observations • Variance reduction procedure for statistical learning method • Multiple trees trained on different samples observed as correlated	
10.	Kernel ridge regression	Non-linear based regression model used in prediction modeling, based on squared error	Fish stock prediction with predictors; better substitute for linear regression model

14.3.1 Modeling the Spatial Distribution and Abundance of Fish Populations

Spatial distribution and abundance estimation modeling are crucial for conservation and management decision-making, as well as for understanding ecosystem health and monitoring outbreaks of nuisance species, which primarily relies on environmental variables data. Traditionally, habitat and environmental factors such as distance from the river bank, depth, local slope of the bottom, drainage density, mean water depth at baseflow, velocity, channel width, vegetation cover, substrate composition, presence of terrestrial leaf litter, and woody material cover were utilized to study the abundance and spatial occupancy of fish species (Mastrorillo et al. 1997; Brosse et al. 1999, 2001). More recently, additional environmental parameters including water level, water temperature, air temperature, precipitation, pH, dissolved oxygen, water conductivity, nutrient levels, alkalinity, suspended solids, and chemical oxygen demand have been integrated into models predicting fish species richness and abundance (Ngor et al. 2023). However, with advancements in image processing and AI, fish images have emerged as another valuable resource for abundance estimation. Image features such as geometric features (bounding box dimensions, convex hull parameters) and texture features (histogram shape index, grey standard deviation, entropy) are now utilized for this purpose (Marini et al. 2018).

14.3.2 Fish Biomass Estimation

Estimating biomass in fish is essential for optimizing daily feeding, controlling stocking densities, and determining the optimal time for harvesting. Quantitative estimation of fish biomass forms the basis of scientific fishery management and conservation strategies aimed at sustainable fish production. Key predictive variables for biomass estimation include fish size parameters such as length, area, height, perimeter, area-to-length ratio, area-to-perimeter ratio, minimum width, and maximum width, which can be derived from fish image data. Certain features extracted from image texture, size, shape, and gradients have demonstrated a high correlation with the biomass target variable (Pache et al. 2022). These features include the major and minor axis (Suwannakhun and Daungmala 2018), area and perimeter (Suzuki 1985), extreme points with Euclidean distance (Liberti et al. 2014), K-curvature (Bakar et al. 2015; Vinh and Tri 2015), and Hu-moments (Huang and Leng 2010).

14.3.3 Fish Age Prediction

Predicting fish age is a critical aspect of fish stock management, aiding in the analysis of growth patterns, maturity, recruitment and mortality rates, and assessment of population status regarding exploitation levels. It also plays a pivotal role in developing stock-specific sustainable management strategies. The predictive variables used in modeling through artificial neural network, random forest, decision tree, and other AI algorithms for fish age prediction primarily include length, weight, and sex (Benzer et al. 2022). However, additional parameters such as geometrical characteristics (e.g. external shape of otoliths, scales, and vertebrae), otolith images (shape, size, inner structure), and CpG methylation also hold promise for fish age prediction (Politikos et al. 2022; Anastasiadi and Piferrer 2023).

14.3.4 Fish Identification and Classification

Traditionally, biologists identify fish from their ichthyologic characteristics such as meristic and morphometric characters, scale morphology, and so on which is laborious, time-consuming, and expensive. So, automating the system using imagery and artificial intelligence is gaining popularity. Some important features that can be extracted from the fish images for modeling species classification include texture features (image quality features like average or mean value, standard deviation, contrast, dissimilarity, homogeneity, and energy), fish shape features (23 anchor points determined, distance and angle measurements, eye-end mouth angle, caudal fin angle, fish head angle, radius of fish eye, length of pectoral fin), RGB and HSV colour features, statistical and wavelet-based texture features, body length, anal fin length, caudal fin length, dorsal fin length, pelvic fin length, and pectoral fin length (Hu et al. 2012; Ogunlana et al. 2015). Further, fish species classification in the natural environment based on AI models helps in estimating the relative abundance of fish species, monitoring changes in their populations as well as health status, and disease identification.

14.3.5 Fish Abundance Estimation

Fish abundance estimation is an essential component of stock management, and with the advent of image processing technologies, it can now be achieved solely from underwater image data. For modeling purposes, several image features are to be extracted which include geometric features of the fish image (area, perimeter, and convex area), pixel features (area and perimeter of the fish-connected areas, perimeter-area ratio, and complexity), texture features (energy, entropy, contrast, and correlation in four directions, i.e. 0°, 45°, 90°, and 135°, of the grey level co-occurrence matrix), coordinate features, and edge features (edge length) (Zhang et al. 2020). The AI models produce more accurate and consistent results for fish abundance estimation using underwater video footage and imageries. A study by Ditria et al. (2020) demonstrated the better performance of AI models in determining the abundance of luderick (*Girella tricuspidata*) in seagrass meadows in Queensland, Australia. Such AI-based deep learning models offer a rapid, economical, and more precise alternative to manual data analysis methods currently used to assess fish abundance.

14.3.6 Effect of Environmental Changes on Fish Community

Climate change is causing shifts in weather patterns, sea levels, and ocean currents, which can have far-reaching consequences for fish communities, impacting their abundance, distribution, behaviour, and overall health. Leveraging AI technologies can provide potent tools for investigating these effects, thereby enhancing our comprehension of ecosystem dynamics, facilitating evidence-based decision-making, and bolstering adaptive management strategies amidst environmental uncertainties. When employing AI to model the effects of climate change on fish communities, primarily, river flow and water quality variables including temperature, pH, electric conductivity, turbidity, dissolved oxygen, suspended solids, biochemical oxygen demand, ammonia nitrogen, nitrate-nitrogen, and total phosphorus can be employed (Hu et al. 2020).

14.3.7 Fish Disease Identification

Timely identification of fish disease is an important step in curbing the disease spread and for initiating management measures. Fish disease prediction through AI requires features that can be extracted from image data like contrast, correlation, energy, entropy, homogeneity, and statistical features like mean (greyscale colour intensity of pixels), SD, variance, kurtosis, and skewness. However, to increase the accuracy of prediction, some non-image features like environmental parameters (water temperature, water quality, and the growth phase of the fish), as well as symptoms in the head, abdomen, gills, scales, fins, muscle, and body, can also be used. With all the requisite features available, various artificial intelligence algorithms such as regression models, support vector machines (SVM), random forest (RF), and artificial neural networks (ANN) can be trained. The choice of algorithm depends on several factors, including the nature of the data, the specific task at hand, and computational resources. Since many variables used as features, such as environmental variables and fish composition data, exhibit non-linear relationships, regression models often struggle to capture the nuances within the data effectively. For capturing non-linear relationships, SVM, RF, and ANN are commonly preferred. The SVM models were applied for the identification of disease-infected salmon fish from healthy individuals by employing the trained models using the images of infected and fresh fish, which produced the results with >90% accuracy (Ahmed et al. 2022).

14.4 Artificial Neural Network-Based Models in Fisheries

SVM and RF face limitations with highly non-linear data, rely on predefined features that may not always capture the most relevant information, can be sensitive to noise, and are primarily designed for classification tasks. For these reasons, ANN is frequently used and often yields superior performance in fisheries management studies (Mastrorillo et al. 1997; Brosse et al. 1999, 2001; Benzer et al. 2022). The ANN model will be applied to a response matrix and a predictor matrix with an input layer, a hidden layer, and a processing centre (or node) at each hidden layer. The k-fold method can be used for dataset partitioning, learning rate (or boosting), and the output layer. The processing centres (or nodes or neuron) use the activation function.

The neural network works in the following manner: each input layer consists of several neuron centres (or nodes or processing centres), and at each node (or neuron), the residual is estimated as the difference between the estimated value and the scaled-up estimated value for the response variable. The scaled-up estimated value is calculated from the transfer function or activation function. This residual variable is then used as input for the second node (or neuron or processing centre) of the same layer, and again, the same process of residual computation is performed on the second node and the second residual (at the second node, the difference between the estimated value to scaled-up estimated value is calculated), which will serve as the input for the third node (or neuron). This process is continued till the last node of the last layer.

For optimization of the ANN model, the validation method can be on several folds, i.e. k-fold whole data divided into k-fold and the hidden layer structure was the transfer function for the first and second layers. Each layer is composed of many such nodes, and these values were substituted as input values in the ANN model structure for the training and validation of the ANN model, and this was iterated several times to obtain an optimized neural network of the ANN model. The iteration process of the training and validation of artificial neural network model was stopped only at the stage (iteration) when the performance value of the root mean square error (RMSE) in the validation set started continuously decreasing even as the performance on the training set started increasing. The performance of the ANN model on the coefficient of determination (R^2), RMSE, mean absolute deviation (MAD), and minus-log-likelihood value on the training and validation sub-dataset were recorded for each iteration.

14.4.1 Influence of Climate Variables on Fish Production

The ANN model on fish production (or response) matrix vs climate parameters (or predictor variables) matrix can be applied to study the network of relationship (or neural net diagram), the simultaneous relationship of parameters influencing production (prediction profiler), and the goodness-of-fit test on the model. The ANN models were used in studying the relationship between the capture and aquaculture production of fish and the climate variables (minimum and maximum temperature (°C) and rainfall (mm)) for time series data from 1951 to 2016. The relationship pattern was utilized in the training and validation of the ANN model and may be used for projecting the capture and aquaculture production (Tureli et al. 2011; Singh et al. 2015). The ANN model shown in Fig. 14.3 consists of three input layers, viz. minimum temperature (Mnt), maximum temperature (Mxt), and rainfall (Rfn), each layer consisting of three neurons (or processing centres or nodes) and one hidden layer. The validation method used was k-fold, and the whole dataset was divided into two folds with one hidden layer serving as a transfer function (sigmoid-TanH) for the three nodes at the hidden layer. The prediction profiler obtained through further analysis may be used for identifying the parameters influencing fish production (Uba 2020).

14.4.2 Fish Shape Morphometry Using Image Analysis

The morphology and physiology of fishes in aquatic environments are influenced by many parameters including physical habitat and water quality. The impact of such parameters on the morphology of fish species may be observed by analysing the images of fish species. The digital image of fish specimens is used for generating coordinates (x, y) for the identified landmarks. The body shape structure of a specimen is characterized by a variety of body landmarks. The body shape variables can be defined by a line (or distance between landmarks) connecting any two landmarks. The body shape variation over landmarks can be studied by the supervised use of the lines (or shape variables) as body shape descriptors. The supervised

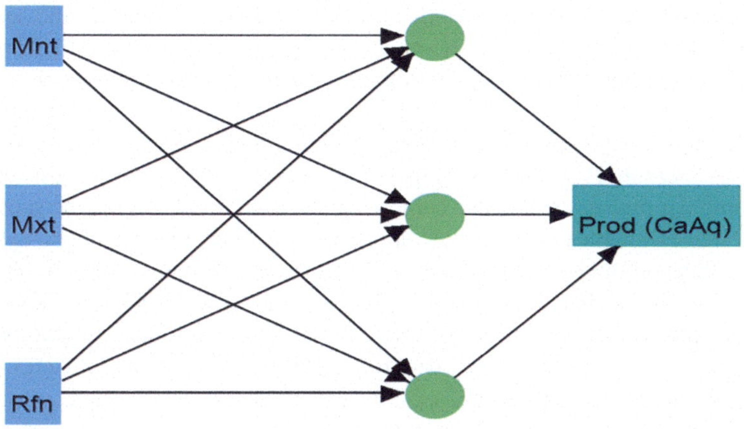

Fig. 14.3 Artificial neural networks model constructed for predicting the influence of climate variables on fish production (input layers: minimum temperature (Mnt), maximum temperature (Mxt), rainfall (Rfn), one hidden layer with 3 neurons (green colour), and predicted output layer (Prod (CaAq)))

learning of lines may be based on certain criteria or rules that can be derived from the shape variables describing the body structure of the fish species. The images of specimens collected through statistical frames are used in the Procrustes system which provides the Procrustes shape (or mean shape) of all specimens. The mean shape (or Procrustes shape) is obtained by applying Procrustes transformation (2D) on the coordinate (x, y) data matrix. The deviation in coordinate value from the mean coordinate measures the changes in shape over landmarks for fish species. The change in body shape over the landmark is measured through the deviation in coordinates, and thus a prediction profiler may be derived for obtaining information on body shape changes at a specific landmark or the role of landmarks in changing the body shape structure. The prediction profiler may be derived from ANN models of body shape measures (or Procrustes principal components) as the response matrix and the Procrustes distance as the predictor matrix. This can be applied to study the network of relationships between different parameters, the simultaneous relationship of predictors influencing the body shape structure (using the prediction profiler), and any two predictors affecting the body shape measurements (using the surface profiler).

AI-based supervised learning detects diseases better than human expertise in certain situations as it considers many body parameters in combination which simultaneously influences the response for body shape structure. The diagnostic test may be derived using the shape variables as descriptors of fish species. The efficacy of a diagnostic test is determined by higher sensitivity vs (1-specificity), i.e. detection of a true case when a disease is present vs detection of a false case when the disease is absent (false). The level of detection of a diagnostic test for a disease depends on the sensitivity vs (1-specificity) curve known as the receiver operating characteristic (ROC) curve. A diagnostic test performs better if the sensitive value is close to unity in the ROC curve, and for a satisfactory performance, the sensitive value should be greater than 0.5. Therefore, the disease symptoms or effects may be observed through the role of shape variables in the body shape composition of fish species.

The fish specimens are selected using a minimum and maximum score over the body shape measure Procrustes principal components 1 and 2. The analysis of body shape composition over Procrustes components is used for observing the sensitivity value of shape variables in the ROC curve analysis. Thus, the sensitivity values of shape variables in the ROC curve may be used in developing a criterion of a diagnostic test for body shape composition. The area under an ROC curve provides a measure of discrimination or classification and allows investigators to compare the performance of two or more diagnostic tests. The ROC curves are a useful tool in the assessment of the performance of a diagnostic test over the range of possible values of a predictor variable (Chandran et al. 2023).

14.4.3 Relationship of Environmental Parameters in the River Landscape

In the era of climate change, and with increasing human intervention, the river landscape is changing faster and may affect the associated biota in multiple ways. The changes in the river landscape with respect to the physicochemical parameters on a temporal scale can be assessed by constructing the metadata matrix. The environmental data available from the published sources may be supplemented with remote sensing datasets that can be used to understand the complex relationship between the variables along the river stretch. The metadata matrix constructed for 3 landscapes, viz. Allahabad, Kanpur, and Varanasi, and 13 physicochemical parameters (temperature, pH, total dissolved solids, electrical conductivity, sulphate, phosphate, alkalinity, chloride, total hardness, nitrate, dissolved oxygen, biological oxygen demand, chemical oxygen demand) collected from published literature revealed that the first 2 principal components accounted for 30.45%, 21.35%, and cumulatively 51.80% variation. The scores on the principal components were then used to characterize the three landscapes of Allahabad, Kanpur, and Varanasi during the period 1960–2019.

14.5 Limitations of Using AI in Fisheries and Future Scope

- The majority of the machine learning approaches in fisheries like fish disease detection, live health assessment of fishes, species classification, fish counting, etc. use underwater images as the input data. However, the major challenge of this approach is the poor quality of images due to water turbidity, lighting conditions, and equipment limitations. In future research along with the improvement of underwater image acquisition platforms, machine learning algorithms can also be integrated with acoustic technology for more accurate and reliable results.

- Automated fish counting and sizing from images can be inaccurate due to variations in fish appearance and occlusions. In future, stereo imaging or multi-view systems can be used to improve depth perception, integrating complementary sensing modalities, such as acoustics and LiDAR, and developing robust algorithms that account for variations in fish size and orientation.
- Black-box nature of AI: Some AI techniques, such as deep learning, can be opaque and difficult to interpret, making it challenging to understand the reasoning behind the model's outputs. In aquaculture, where decisions have significant implications for fish welfare and economic outcomes, interpretability is crucial for gaining trust and understanding the rationale behind AI-driven recommendations. Combining AI techniques with other interpretable methods, such as rule-based systems, expert knowledge, or analytical models, can enhance transparency and interpretability. For example, a deep learning model could be used for data processing and feature extraction, while the final decision-making is handled by an interpretable rule-based system or decision support tool.
- Implementing AI technologies in aquaculture may involve upfront costs for sensors, data collection systems, computational resources, and skilled personnel. Small-scale fish farmers or those with limited resources may find it challenging to adopt AI due to the associated costs and infrastructure requirements. To deal with this cost issue, we need to primarily identify high-impact areas where AI can provide significant monetary benefits, such as early disease detection, feed optimization, or water quality monitoring, rather than attempting a complete overhaul of the entire fisheries operations and then gradually scaling up as benefits are realized and resources become available.

14.6 Conclusion

The application of artificial intelligence (AI)-based modeling in fisheries management and conservation holds significant potential for promoting sustainable practices and ensuring the long-term viability of aquatic ecosystems and fisheries resources. Artificial neural network models can be applied to derive neural net diagrams, prediction profilers, and surface profilers for the management of aquatic and fish genetic resources. The ANN model with a suitable activation function at the processing centre (node) and many hidden layers as per the complexity of the data can be applied to gain deeper knowledge about data structures. Likewise, different AI-based models can enhance our understanding of complex marine ecosystems, fish population dynamics, and the impacts of various environmental and anthropogenic factors. Leveraging AI models such as machine learning, deep learning, and predictive modeling techniques can provide valuable insights for decision-making processes related to stock assessments, catch quotas, habitat conservation, and ecosystem-based management. However, the use of AI-based models in fisheries faces multiple challenges, which should be addressed through cost-effective and reasonable approaches. By leveraging the strengths of AI while addressing its limitations, integrating human expertise, and fostering collaboration among stakeholders, we can harness the full potential of these technologies to promote sustainable and resilient fisheries for generations to come.

Acknowledgements The author express gratitude to the Director, ICAR-National Bureau of Fish Genetic Resources, Lucknow, for the necessary support and guidance.

References

Ahmed MS, Aurpa TT, Azad MAK (2022) Fish disease detection using image based machine learning technique in aquaculture. J King Saud Univ Comput Inf Sci 34(8):5170–5182

Anastasiadi D, Piferrer F (2023) Bioinformatic analysis for age prediction using epigenetic clocks: application to fisheries management and conservation biology. Front Mar Sci 10:1096909

Bagheri M, Akbari A, Mirbagheri SA (2019) Advanced control of membrane fouling in filtration systems using artificial intelligence and machine learning techniques: a critical review. Process Saf Environ Prot 123:229–252

Bakar MZA, Samad R, Pebrianti D et al (2015) Finger application using K-curvature method and Kinect sensor in real-time. In: 2015 international symposium on technology management and emerging technologies (ISTMET). IEEE, pp 218–222

Benzer S, Garabaghi FH, Benzer R et al (2022) Investigation of some machine learning algorithms in fish age classification. Fish Res 245:106151

Brosse S, Guegan JF, Tourenq JN et al (1999) The use of artificial neural networks to assess fish abundance and spatial occupancy in the littoral zone of a mesotrophic lake. Ecol Model 120(2–3):299–311

Brosse S, Lek S, Townsend CR (2001) Abundance, diversity, and structure of freshwater invertebrates and fish communities: an artificial neural network approach. N Z J Mar Freshw Res 35(1):135–145

Chandran R, Singh A, Singh RK et al (2022) Phenotypic variation of *Chitala chitala* (Hamilton, 1822) from Indian rivers using truss network and geometric morphometrics. PeerJ 10:e13290

Chandran R, Singh KR, Singh A et al (2023) Evaluating the influence of environmental variables on the length-weight relationship and prediction modelling in flathead grey mullet, *Mugil cephalus* Linnaeus, 1758. PeerJ 11:e14884

Department of Fisheries (2020) Handbook of fisheries statistics 2020. Ministry of fisheries, animal husbandry and dairying, Government of India. (Final_Book.pdf (dof.gov.in))

Ditria EM, Lopez-Marcano S, Sievers M et al (2020) Automating the analysis of fish abundance using object detection: optimizing animal ecology with deep learning. Front Mar Sci 7:429

Fernandes-Salvador JA, Oanta GA, Olivert-Amado A et al (2022) Research for PECH Committee—Artificial Intelligence and the fisheries sector, European Parliament, Policy Department for Structural and Cohesion Policies, Brussels

French G, Mackiewicz M, Fisher M et al (2020) Deep neural networks for analysis of fisheries surveillance video and automated monitoring of fish discards. ICES J Mar Sci 77(4):1340–1353

Fricke R, Eschmeyer WN, van der Laan R (eds) (2024) Eschmeyer's catalog of fishes: genera, species, references. http://researcharchive.calacademy.org/research/ichthyology/catalog/fishcatmain.asp. Accessed 27 Apr 2024

Ho DJ, Siti MD, Jafar-Sidik M et al (2013) Influence of weather condition on pelagic fish landings in Kota Kinabalu, Sabah, Malaysia. J Trop Biol Conserv 10:11–21

Hu J, Li D, Duan Q et al (2012) Fish species classification by color, texture and multi-class support vector machine using computer vision. Comput Electron Agric 88:133–140

Hu JH, Tsai WP, Cheng ST et al (2020) Explore the relationship between fish community and environmental factors by machine learning techniques. Environ Res 184:109262

Huang Z, Leng J (2010) Analysis of Hu's moment invariants on image scaling and rotation. In: 2010 2nd international conference on computer engineering and technology, vol 7. IEEE, pp V7–V476

Javaid M, Haleem A, Singh RP et al (2022) Enhancing smart farming through the applications of Agriculture 4.0 technologies. Int J Intell Netw 3:150–164

Liberti L, Lavor C, Maculan N et al (2014) Euclidean distance geometry and applications. SIAM Rev 56(1):3–69

MacIntyre CR, Chen X, Kunasekaran M et al (2023) Artificial intelligence in public health: the potential of epidemic early warning systems. J Int Med Res 51(3):03000605231159335

Marini S, Fanelli E, Sbragaglia V et al (2018) Tracking fish abundance by underwater image recognition. Sci Rep 8(1):13748

Mastrorillo S, Lek S, Dauba F et al (1997) The use of artificial neural networks to predict the presence of small-bodied fish in a river. Freshw Biol 38(2):237–246

Mukhamediev RI, Popova Y, Kuchin Y et al (2022) Review of artificial intelligence and machine learning technologies: classification, restrictions, opportunities and challenges. Mathematics 10(15):2552

Mustapha UF, Alhassan AW, Jiang DN et al (2021) Sustainable aquaculture development: a review on the roles of cloud computing, internet of things and artificial intelligence (CIA). Rev Aquac 13(4):2076–2091

Ngor PB, Uy S, Sor R et al (2023) Predicting fish species richness and abundance in the lower mekong basin. Front Ecol Evol 11:1131142

Ogunlana SO, Olabode O, Oluwadare SAA et al (2015) Fish classification using support vector machine. Afr J Comp ICT 8(2):75–82

Pache MCB, Sant'Ana DA, Rozales JVA et al (2022) Prediction of fingerling biomass with deep learning. Ecol Inform 71:101785

Politikos DV, Sykiniotis N, Petasis G et al (2022) DeepOtolith v1. 0: an open-source AI platform for automating fish age reading from otolith or scale images. Aust Fish 7(3):121

Sajina AM, Chakraborty SK, Jaiswar AK et al (2011) Stock structure analysis of *Megalaspis cordyla* (Linnaeus, 1758) along the Indian coast based on truss network analysis. Fish Res 108(1):100–105

Sarkar UK, Bakshi S, Lianthuamluaia L, Mishal P, Das Ghosh B, Saha S, Karnatak G (2020) Understanding enviro-climatological impact on fish biodiversity of the tropical floodplain wetlands for their sustainable management. Sustainable Water Resour Manage 6:1–12

Sarkar UK, Lalramliana, Chandran R et al (2024) Fish species identification by morphological characters. In: Singh M, Sahu A, Sarkar UK (eds) DNA barcoding of fishes. International books and periodical supply service, pp 21–32

Singh A, Suzuki N, Sakuramoto K (2015) Influence of climatic conditions on the time series fluctuation of Yellowfin Tuna *Thunnus albacares* in the South Pacific Ocean. Open J Mar Sci 5:247–264

Suwannakhun S, Daungmala P (2018) Estimating pig weight with digital image processing using deep learning. In: 2018 14th international conference on Signal-Image Technology & Internet-Based Systems (SITIS). IEEE, pp 320–326

Suzuki S (1985) Topological structural analysis of digitized binary images by border following. Comput Gr Image Process 30(1):32–46

Tureli BC, Kokcu P, Ibrikci T (2011) Application of artificial neural networks (ANNs) for weight predictions of Blue Crabs (*Callinectes sapidus* Rathbun, 1896) using predictor variables. Mediterr Mar Sci 12:439–446

Uba KIN (2020) Development of a model for predicting mussel weight: a comparison of traditional and artificial intelligent methods. J Fish 8:837–842

Vinh TQ, Tri NT (2015) Hand gesture recognition based on depth image using kinect sensor. In: 2015 2nd National Foundation for Science and Technology Development Conference on Information And Computer Science (NICS). IEEE, pp 34–39

Xu Y, Liu X, Cao X et al (2021) Artificial intelligence: a powerful paradigm for scientific research. Innovation 2(4):100179

Zhang L, Li W, Liu C et al (2020) Automatic fish counting method using image density grading and local regression. Comput Electron Agric 179:105844

Zhang D, Mishra S, Brynjolfsson E et al (2021) The AI Index 2021 Annual Report, AI Index Steering Committee, Human-Centered AI Institute, Stanford University, Stanford, CA

National Legislations: Best Practices and Challenges in India for Sustainable Management of Fish Genetic Resources

15

Tarachand Kumawat and Lalit Kumar Tyagi

Abstract

The chapter examines India's legislative frameworks and policies that support the conservation and sustainable use of fish genetic resources. Emphasizing the critical role of biodiversity in food security and ecosystem resilience, the present assessment details the evolution of fisheries management in India, highlighting key legislations. The chapter also discusses the importance of recent amendments and the integration of international guidelines to address emerging challenges like invasive species and habitat degradation. By providing an overview of India's diverse aquatic ecosystems, the chapter underscores the significance of national and state-specific legislations in safeguarding fish biodiversity. Key themes include the balance between economic development and conservation goals, the role of community involvement in fisheries management, and the application of advanced technologies and interdisciplinary research in informing policy decisions. The authors also present case studies demonstrating successful legislative implementations and explore the socio-economic and ethical considerations impacting local fishing communities. The chapter concludes with recommendations for strengthening legislative frameworks, promoting interdisciplinary collaboration, enhancing community participation, and investing in capacity building and technological advancements to ensure the long-term sustainability of India's fish genetic resources. Through a holistic approach, integrating ecological, socio-economic, and institutional aspects, India can advance towards effective conservation while supporting the livelihoods of its fishing communities.

T. Kumawat · L. K. Tyagi (✉)
ICAR-National Bureau of Fish Genetic Resources, Lucknow, Uttar Pradesh, India
e-mail: tarachand.kumawat@icar.gov.in; Lalit.Tyagi@icar.gov.in

U. K. Sarkar et al. (eds.), *Sustainable Management of Fish Genetic Resources*,
https://doi.org/10.1007/978-981-97-5250-8_15

Keywords

National legislations · Sustainable fisheries management · Fish genetic resources · Biodiversity conservation · Community involvement

15.1 Introduction

15.1.1 Importance of Sustainable Management of Fish Genetic Resources

Sustainable management of fish genetic resources is crucial for maintaining biodiversity, ensuring food security, and supporting livelihoods, especially in countries like India with significant reliance on fisheries. Fish genetic diversity contributes to ecosystem resilience, enabling species to adapt to environmental changes and resist diseases (FAO 2020a, b). Effective management practices help prevent the depletion of fish stocks, support conservation efforts, and promote the sustainable use of aquatic resources, aligning with global sustainability goals (World Bank 2019).

In India, the sustainable management of fish genetic resources is vital due to the country's extensive dependence on fisheries for economic and nutritional needs. The fisheries sector employs millions of people and is a major source of animal protein for the population. Ensuring the sustainability of fish genetic resources helps safeguard these livelihoods and food sources against the impacts of overfishing, habitat destruction, and climate change (National Fisheries Development Board 2021).

15.1.2 Role of National Legislations in Promoting Sustainability

National legislations play a vital role in promoting the sustainability of fish genetic resources by providing a legal framework for conservation, management, and sustainable use. These legislations establish guidelines and regulations that help protect aquatic habitats, regulate fishing practices, and promote the conservation of endangered species. They also facilitate the implementation of international agreements and conventions at the national level, ensuring compliance with global standards and best practices (Government of India 2022).

In India, several key legislations contribute to the sustainable management of fish genetic resources. The Indian Fisheries Act (1897) provides the foundational framework for fisheries regulation, while the Wildlife Protection Act (1972) includes provisions for the protection of aquatic species and habitats. The Environment Protection Act (1986) and the National Biodiversity Act (2002) further strengthen the legal framework by addressing broader environmental and biodiversity conservation issues (Central Marine Fisheries Research Institute 2021).

The recent amendments to these Acts, such as the Wildlife Protection Amendment Act (2022), have introduced new provisions to address emerging challenges, including the management of invasive species and the protection of critical habitats

(Government of India 2022). These legislative measures, combined with recent policies such as Draft National Fisheries Policy 2020 reflect a comprehensive approach to sustainable fisheries management in India.

15.1.3 Overview of India's Diverse Aquatic Ecosystems and Fish Biodiversity

India is home to a vast array of aquatic ecosystems, ranging from rivers, lakes, and reservoirs to coastal and marine environments. These diverse ecosystems support a rich biodiversity of fish species, making India one of the global hotspots for fish genetic resources. The country's freshwater systems, including the Ganges, Brahmaputra, and Godavari river basins, are known for their unique fish fauna, which includes species like the Indian major carps and mahseers (Central Inland Fisheries Research Institute 2020).

Coastal and marine ecosystems, such as the Sundarbans, Gulf of Mannar, and Andaman and Nicobar Islands, harbor a wide variety of marine fish species, many of which are of significant commercial and ecological importance. Coral reefs, mangroves, and seagrass beds in these regions provide critical habitats for numerous fish species, contributing to the overall biodiversity and productivity of India's marine environment (Central Marine Fisheries Research Institute 2019).

The sustainable management of these diverse fish genetic resources is essential for maintaining ecosystem health, supporting fisheries productivity, and preserving the ecological balance. National legislations and policies aimed at protecting aquatic habitats, regulating fishing practices, and promoting conservation efforts play a crucial role in ensuring the long-term sustainability of India's fish genetic resources (National Fisheries Development Board 2021).

15.2 Historical Context

15.2.1 Evolution of Fisheries Management in India

The evolution of fisheries management in India has been a complex process shaped by the country's vast and diverse aquatic ecosystems. Historically, traditional fishing practices were predominant, with local communities relying on indigenous knowledge and sustainable practices to manage fish stocks. However, with the advent of colonial rule, there was a significant shift towards commercial exploitation and the introduction of modern fishing techniques.

In the early twentieth century, the colonial administration began to introduce regulations aimed at controlling fishing activities. The Indian Fisheries Act of 1897 marked one of the earliest attempts to regulate fisheries, focusing primarily on inland waters. This Act provided the legal framework for the protection of fish during their breeding season and prohibited the use of certain types of fishing gear that were deemed destructive.

Post-independence, the Indian government recognized the need for a more comprehensive approach to fisheries management. The establishment of the Central Inland Fisheries Research Institute (CIFRI) and the Central Marine Fisheries Research Institute (CMFRI) in 1947 was pivotal in advancing fisheries science and management in the country. These institutes played crucial roles in conducting research, providing scientific advice, and developing management strategies for both inland and marine fisheries.

15.2.2 Global and Indian Reviews

Globally, the principles and guidelines for sustainable fisheries management have been significantly influenced by the United Nations Food and Agriculture Organization (FAO) through instruments such as the Code of Conduct for Responsible Fisheries (FAO 1995). This code provides a comprehensive fisheries and aquaculture management framework, emphasizing sustainability, ecosystem-based management, and a precautionary approach.

In the Indian context, reviews and research articles have highlighted the unique challenges and opportunities in managing the country's diverse fisheries resources. Studies such as those by Sathiadhas and Panikkar (2015) and Ghosh (2019a, b) have provided valuable insights into the evolution of fisheries management and the impact of various legislations on the sustainability of fish genetic resources.

15.3 Key National Legislations

The development of national legislation related to fisheries and aquatic resources in India has been marked by several key milestones:

15.3.1 The Indian Fisheries Act (1897)

15.3.1.1 Historical Significance and Primary Objectives

The Indian Fisheries Act of 1897 is one of the first formal legislative efforts to regulate fishing activities in the country. It aimed to protect fish during breeding seasons and regulate the use of destructive fishing methods. This Act laid the foundation for future fisheries legislation in India (Sathiadhas and Panikkar 2015). Its primary objective was to provide a legal framework for the regulation and conservation of fisheries resources. The Act aimed to prevent destructive fishing practices, ensure sustainable fishery resources, and regulate fishing methods and seasons to promote conservation. Historically, it served as a foundational document for the management of fisheries, setting a precedent for subsequent laws and policies in the sector (Pillai 2012).

15.3.1.2 Amendments and Current Relevance

Over the years, the Act has undergone several amendments to address emerging challenges in fisheries management. Despite its age, the Indian Fisheries Act remains relevant as it provides a legal basis for the regulation of inland and marine fisheries. The Act has been supplemented by various state-specific fisheries legislations that cater to local ecological and socio-economic conditions. These amendments have been crucial in adapting the Act to contemporary conservation needs and sustainable practices (Sinha and Katiha 2002).

15.3.2 The Wildlife Protection Act (1972) and its Amendments

15.3.2.1 Provisions Related to Aquatic Species

The Wildlife Protection Act of 1972 is a comprehensive legislation aimed at the protection of wildlife in India. Although the Wildlife Protection Act of 1972 primarily focused on terrestrial wildlife, it included provisions for the protection of aquatic species. The Act includes provisions for the protection of endangered species of flora and fauna, including aquatic species. It classifies wildlife into schedules and provides varying degrees of protection to species listed in these schedules. Aquatic species, such as certain fish and amphibians, are included in these schedules, affording them legal protection against exploitation and habitat destruction (Jeyabaskaran 2013). The Act was amended several times, with significant changes in 2022, to address invasive species and strengthen habitat protection (Government of India 2022). These amendments underscore the growing recognition of the importance of aquatic biodiversity within the broader conservation framework.

15.3.2.2 Impact of the 2022 Amendment Act on Invasive Species and Aquatic Habitats

The 2022 Amendment to the Wildlife Protection Act introduced significant changes, particularly concerning invasive species. It defined invasive alien species and provided for their regulation, thereby addressing a major threat to native aquatic habitats and species. This amendment empowers authorities to control the introduction and spread of invasive species, enhancing the protection of aquatic ecosystems (Government of India 2022).

15.3.2.3 Role in Protecting Fish Habitats and Genetic Resources

The Wildlife Protection Act plays a crucial role in protecting fish habitats by regulating activities that could lead to habitat destruction or pollution. The inclusion of fish habitats in protected areas and the regulation of fishing practices contribute to the conservation of fish genetic resources. By preventing overfishing and habitat degradation, the Act helps maintain the genetic diversity of fish populations (Sarkar et al. 2008).

15.3.3 The Merchant Shipping Act (1958)

This Act aims to promote and regulate the development and ensure the efficient maintenance of Indian mercantile shipping in a manner best suited to serve the national interests. Under this Act the National Shipping Board was established to provide for the registration of Indian ships and generally to amend and consolidate the law relating to merchant shipping. The Act was amended in 1983 to provide for the registration and control of Indian fishing boats. The Act enables defining a fishing vessel and prescribes a registration procedure besides provision for data collection. It applies to fishing boats of 20 m and above. The Board has played a very distinguished role in the maritime development of the country, as its deliberations and recommendations have contributed to the evolution of a sound and pragmatic National Maritime Policy over the years.

15.3.4 The Territorial Waters, Continental Shelf, EEZ and Other Maritime Zones Act, 1976

The Territorial Waters, Continental Shelf, EEZ and other Maritime Zones Act, 1976, recognizes the sovereign rights to conservation and management of living resources in the Indian EEZ, in addition to their exploration and exploitation. It further gives powers to the central government to make rules, inter alia, for conservation and management of the living resources of the EEZ and for the protection of the marine environment.

15.3.5 The Maritime Zones of India (Regulation of Fishing by Foreign Vessel) Act, 1981

The Maritime Zones of India (Regulation of Fishing by Foreign Vessel) Act, 1981, aims to regulate the fishing by foreign vessels in the EEZ of India. The Act deals with the prohibition of fishing in the maritime zones of India by foreign vessels, grant of licenses, prohibition of fishing by Indian citizens using foreign vessels, and cancellation or suspension of licenses or permits. It also deals with foreign vessels entering the maritime zones of India without license or permit to stow gear and the power for seizure and contravention. The Act permits fishing for scientific research, investigations, etc. (Edwin 2022).

15.3.6 The Environment Protection Act (1986)

15.3.6.1 Framework for Environmental Regulation

The Environment Protection Act of 1986 provided a comprehensive legal framework for the protection and improvement of the environment, including aquatic ecosystems. This Act empowered the government to take necessary measures to prevent and control environmental pollution, which indirectly benefited the conservation of fish genetic resources (Ghosh 2019a, b). This Act is instrumental in addressing environmental issues that affect aquatic ecosystems, such as water pollution and industrial discharges (Kumar and Chopra 2012).

15.3.6.2 Impact on Aquatic Ecosystems and Fish Genetic Resources

The Act's stringent regulations on pollutants and its emphasis on environmental quality standards have a direct impact on aquatic ecosystems. By controlling pollutants that enter water bodies, the Act helps maintain the health of aquatic habitats, which is essential for the survival and genetic diversity of fish species. The Act also promotes the restoration of degraded aquatic ecosystems, thereby supporting fish genetic resources (Gupta and Asher 2013).

15.3.7 The National Biodiversity Act (2002)

15.3.7.1 Focus on Conservation and Sustainable Use of Biological Resources

The National Biodiversity Act of 2002 is a landmark legislation aimed at conserving biological diversity, promoting sustainable use of its components, and ensuring fair and equitable sharing of benefits arising from the use of genetic resources. This Act has specific provisions for the conservation of fish genetic diversity and the regulation of access to these resources (Kumar and Tripathi 2020). It emphasizes the conservation of biodiversity, including aquatic species, through the establishment of Biodiversity Management Committees (BMCs) and People's Biodiversity Registers (PBRs) (National Biodiversity Authority 2012).

15.3.7.2 Mechanisms for Protecting Fish Genetic Diversity

The Act includes provisions for the protection of fish genetic diversity by regulating access to biological resources and promoting sustainable practices. It mandates that any use of biological resources, including fish, should be done in a manner that does not threaten their survival or genetic diversity. The Act also facilitates the involvement of local communities in conservation efforts, recognizing their role in managing and protecting biodiversity (Gadgil et al. 2000).

15.3.8 The Coastal Aquaculture Authority Act (2005)

The Coastal Aquaculture Authority Act of 2005 was enacted to regulate coastal aquaculture activities to ensure sustainable development and minimize environmental impact. This Act established the Coastal Aquaculture Authority, which grants permissions, ensures compliance with environmental standards, and promotes best practices in coastal aquaculture (Mohan 2018).

15.3.9 National Fisheries Policy of 2020

In 2020, the National Fisheries Policy of 2020 was drafted by merging existing policies such as the National Policy on Marine Fisheries of 2017, Draft National Inland Fisheries and Aquaculture Policy of 2020, and Draft National Mariculture Policy of 2019. The Inland Fisheries and Aquaculture Policy of 2020 aims to promote sustainable aquaculture practices and enhance the productivity of inland water bodies. This policy emphasizes the need for scientific management, community participation, and the use of technology to achieve sustainable fisheries development (Sinha and Ahmed 2020). It aims to enhance fish production and productivity through sustainable practices. It also emphasizes the modernization of inland fisheries and aquaculture sectors, the adoption of advanced technologies, and the improvement of infrastructure. The policy seeks to create a conducive environment for the sustainable development of the fisheries sector (Department of Fisheries 2020). The merged policy document has envisioned export growth and enhanced fishers' income and better choices for consumers through sustainable development of the fisheries sector. The policy lays down a clear road map for the unhindered growth of the sector by the optimal harnessing of fisheries resources in sync with the best national and international standards. It seeks to consolidate the sectoral gains and ensure sustainable growth through policy support to enable and accelerate fisheries development responsibly and inclusively through multipronged strategies and focused interventions to provide the requisite impetus for realizing the full potential across the country. It will also create a conducive environment for attracting investment into the fisheries sector. The policy intends to strengthen the government's initiative to double fishers and fish farmers' incomes and double exports for improving the lives and livelihoods of fishers and their families. The policy focuses on managing the fisheries resources sustainably and responsibly by following the Ecosystem Approach to Fisheries (EAF) management and modern, scientific, and diverse fishing practices in the ocean (Salim Shyam and Anuja 2022). The policy places a strong emphasis on sustainable aquaculture practices to ensure long-term productivity and environmental health. It advocates for the adoption of eco-friendly techniques, the efficient use of resources, and the conservation of aquatic biodiversity. By promoting sustainable aquaculture, the policy aims to safeguard fish genetic resources and enhance the resilience of aquatic ecosystems (Beveridge et al. 2020).

15.3.10 State Fisheries Acts

15.3.10.1 State Inland Fisheries Acts

India's diverse aquatic ecosystems necessitate state-specific legislation for effective fisheries management. Various states have enacted their own Inland Fisheries Acts to regulate the sustainable use and conservation of inland fish resources. These legislations generally cover aspects such as fishing rights, licensing, habitat protection, and measures to control overfishing.

Specific case studies from major states:

West Bengal: West Bengal Inland Fisheries Act (1984)—This Act emphasizes the regulation of fishing activities in the state's water bodies, including rivers, lakes, and reservoirs. It includes provisions for licensing, restrictions on certain types of fishing gear, and the conservation of fish habitats. The Act has been instrumental in managing the state's rich fish biodiversity and supporting local fishing communities.

Case study: Implementation of community-based fishery management in the Sundarbans region has shown significant success in balancing conservation and livelihood goals (Barman et al. 2018).

Kerala: Kerala Inland Fisheries and Aquaculture Act, 2010—Kerala's legislation focuses on promoting sustainable aquaculture practices and managing inland fisheries resources. The Act includes regulations for fish farming, protection of aquatic environments, and measures to prevent the introduction of invasive species.

Case study: The Vembanad Lake fisheries management plan has demonstrated effective governance through stakeholder participation and scientific monitoring (Ramasamy et al. 2021).

15.3.10.2 Challenges in Uniform Implementation across States

Despite the presence of state-specific legislations, several challenges hinder uniform implementation:

Regulatory disparities: Differences in regulations and enforcement mechanisms across states lead to inconsistencies in fisheries management.

Resource constraints: Limited financial and human resources at the state level often impede the effective enforcement of regulations.

Stakeholder conflicts: Balancing the interests of different stakeholders, such as commercial fishers, local communities, and conservationists, remains a significant challenge.

Lack of coordination: Insufficient coordination between state and central authorities can lead to fragmented management efforts.

15.3.11 State Marine Fisheries Acts

15.3.11.1 Regulations for Marine Fisheries

Fisheries within the territorial waters, which extend up to a distance of 12 nautical miles from shore, is a state subject, and the primary responsibility for its development rests with the state government. The central government is responsible for all fishing activity that takes place beyond this limit—the deep sea or EEZ (exclusive economic zone) that extends up to a distance of 200 nautical miles. State Marine Fisheries Acts are designed to regulate fishing activities within the territorial waters of coastal states. These regulations aim to ensure sustainable exploitation of marine resources, protect marine ecosystems, and support the livelihoods of coastal communities.

Tamil Nadu: Tamil Nadu Marine Fishing Regulation Act (1983)—This Act governs marine fishing activities within the state's territorial waters. It includes provisions for the licensing of fishing vessels, restrictions on the use of certain types of fishing gear, and the establishment of closed seasons to allow fish stocks to replenish.

Example: The implementation of the Act has led to improved compliance with sustainable fishing practices and reduced conflicts between traditional and mechanized fishers (Srinivasan and Ravikumar 2017).

Gujarat: Gujarat Fisheries Act (2003)—The Act provides a comprehensive framework for the management of marine fisheries in Gujarat. It focuses on the conservation of marine resources, regulation of fishing methods, and protection of endangered marine species.

Example: Successful community-led initiatives in the Gulf of Kutch have demonstrated the benefits of collaborative management in conserving marine biodiversity (Patel and Mehta 2020).

15.3.12 Balancing State and Central Policies for Marine Conservation

Achieving a balance between state and central policies is crucial for effective marine conservation:

Central policies: The central government, through acts like the Wildlife Protection Act and the Environment Protection Act, provides a broad framework for marine conservation. The Coastal Aquaculture Authority Act (2005) further regulates aquaculture activities along the coast.

State adaptation: States need to adapt these central guidelines to local contexts, ensuring that regional ecological and socio-economic conditions are considered.

Collaborative governance: Effective marine conservation requires collaborative governance involving both state and central authorities. Initiatives like the Marine Fisheries Regulation and Management (MFRM) Bill, 2020, aim to harmonize state and central efforts in regulating marine fisheries (Srinivasan 2020).

15.4 National Fisheries Regulations and Supportive Policies

15.4.1 National Fisheries Code

15.4.1.1 Alignment with International Guidelines

India's National Fisheries Code aligns closely with the principles outlined in the UN Code of Conduct for Responsible Fisheries (CCRF). The CCRF provides a comprehensive framework to ensure that fisheries are managed responsibly, with an emphasis on sustainability, ecosystem health, and the protection of fish genetic resources (FAO 1995). The Code advocates conservation and management measures based on the best scientific evidence available. The CCRF's guidelines serve as a benchmark for national policies, promoting best practices in fisheries management, including the preservation of biodiversity and the sustainable use of aquatic resources.

15.4.1.2 Principles and Objectives for National Fisheries Management

The National Fisheries Code in India is designed to achieve the following objectives:

- Sustainable use: Ensuring the long-term sustainability of fish stocks through scientifically informed management practices.
- Biodiversity conservation: Protecting the genetic diversity of fish species by regulating fishing activities and promoting habitat conservation.
- Community participation: Involving local communities in the decision-making process to enhance compliance and ensure equitable benefits.
- Ecosystem approach: Adopting an ecosystem-based approach to fisheries management that considers the interdependence between species and their habitats.
- Regulatory framework: Establishing a robust regulatory framework that includes monitoring, control, and surveillance measures to combat illegal, unreported, and unregulated (IUU) fishing (Pomeroy and Berkes 1997).

15.4.2 Marine/Mariculture Policy

15.4.2.1 Framework for Sustainable Mariculture Practices

India's Marine/Mariculture Policy provides a structured approach to promoting sustainable mariculture. This policy is crucial for enhancing food security, improving livelihoods, and conserving marine biodiversity. Key elements include:

- Environmental impact assessments: Mandatory evaluations to ensure mariculture projects do not harm the marine environment.
- Best management practices: Guidelines for site selection, species selection, and farming techniques that minimize ecological footprints.
- Research and development: Encouraging innovation in mariculture practices through dedicated research programs (Soto et al. 2008).

15.4.2.2 Policy Guidelines and Implementation Strategies

The policy outlines specific strategies for implementation:

Zoning and spatial planning: Identifying suitable areas for mariculture to reduce conflicts with other marine activities.

Capacity building: Training programs for stakeholders to adopt sustainable practices.

Financial support: Providing incentives and subsidies to promote environmentally friendly mariculture operations (FAO 2011).

15.4.3 Pollution Acts

15.4.3.1 The Water (Prevention and Control of Pollution) Act, 1974

The Water (Prevention and Control of Pollution) Act of 1974 is a pivotal piece of legislation aimed at preventing and controlling water pollution. It establishes water quality standards and regulates the discharge of pollutants into water bodies. The Act is enforced by the Central and State Pollution Control Boards, which monitor water quality and ensure compliance (CPCB 2019).

15.4.3.2 The Air (Prevention and Control of Pollution) Act, 1981

Similarly, the Air (Prevention and Control of Pollution) Act of 1981 focuses on controlling air pollution by regulating emissions from industrial and vehicular sources. This Act helps protect aquatic ecosystems from airborne pollutants that can settle into water bodies, thereby safeguarding fish habitats (CPCB 2019).

15.4.3.3 Impact on Aquatic Ecosystems and Measures for Pollution Control

Both Acts have significantly contributed to reducing pollution levels in India's water bodies, thus aiding in the conservation of aquatic ecosystems. Measures include:

Regulatory compliance: Industries are required to install pollution control devices and adhere to discharge limits.

Monitoring and surveillance: Regular monitoring of water and air quality by pollution control boards.

Public awareness campaigns: Initiatives to educate the public and industries about the importance of pollution control (Nathanson 2013).

15.4.4 Coastal Aquaculture Authority Act (2005)

15.4.4.1 Regulation and Management of Coastal Aquaculture

The Coastal Aquaculture Authority Act (2005) regulates coastal aquaculture activities to ensure sustainable development. It provides guidelines for site selection, farming practices, and waste management to minimize environmental impacts. The Act mandates registration of all coastal aquaculture farms and periodic inspections to ensure compliance with environmental standards (CAA 2020).

15.4.4.2 Role in Protecting Coastal and Marine Fish Genetic Resources

The Act plays a crucial role in protecting coastal and marine fish genetic resources by:

Preventing overexploitation: Regulating the intensity and methods of aquaculture to avoid the depletion of native fish stocks.

Habitat protection: Ensuring that aquaculture practices do not destroy critical habitats such as mangroves and coral reefs.

Biosecurity measures: Implementing measures to prevent the introduction and spread of invasive species and diseases (Naylor et al. 2000).

15.5 Fisheries Governance in India

Fisheries governance in India operates through a comprehensive and multitiered framework involving both the central and state governments. The Department of Fisheries, one of the two departments under the Ministry of Fisheries, Animal Husbandry and Dairying, is pivotal in this governance structure. It is responsible for formulating policies and schemes related to developing inland, marine, and coastal fisheries and overseeing various fishery institutes across the country. These institutes include the Fishery Survey of India in Mumbai, the Central Institute of Fisheries Nautical and Engineering Training (CIFNET) in Kerala, the Central Institute of Coastal Engineering for Fishery (CICEF) in Bangalore, the National Institute of Fisheries Post Harvest Technology and Training (NIFPHATT) in Kochi, the National Fisheries Development Board (NFDB) in Hyderabad, and the Coastal Aquaculture Authority in Chennai.

In addition to the Department of Fisheries, the Fisheries Science Division under the Indian Council of Agricultural Research (ICAR) plays a crucial role in fisheries and aquaculture research and academic aspects. It coordinates and monitors programs through five resource-specific fisheries research institutes: ICAR-Central Marine Fisheries Research Institute (CMFRI), ICAR-Central Inland Fisheries Research Institute (CIFRI), ICAR-Central Institute of Fisheries Technology (CIFT), ICAR-Central Institute of Freshwater Aquaculture (CIFA), and ICAR-Central Institute of Brackishwater Aquaculture (CIBA). Additionally, it oversees a deemed university, the ICAR-Central Institute of Fisheries Education (CIFE); a bureau, the ICAR-National Bureau of Fish Genetic Resources (NBFGR); and a directorate, the ICAR-Directorate of Coldwater Fisheries Research (DCFR). A network of regional, research, and field centers supports these institutions. These institutions and organizations collectively contribute to the development, management, and conservation of fisheries resources in India, ensuring sustainable practices and the well-being of fishing communities. The list of fisheries-related institutions/organizations in India and their major activities is provided in Appendix.

15.6 Best Practices in Legislative Framework

15.6.1 Integrated Approach to Fisheries Management

An integrated approach to fisheries management in India involves the harmonization of various policies and regulations across national and state levels, addressing the ecological, economic, and social dimensions of fisheries. This approach is essential for ensuring the sustainable use of fish genetic resources.

The integrated management framework incorporates multiple aspects, including habitat conservation, pollution control, and sustainable aquaculture practices. For instance, the Coastal Regulation Zone (CRZ) Notification (2011) helps regulate activities in coastal areas to protect fragile coastal ecosystems (Ministry of Environment, Forest and Climate Change 2011). Additionally, the National Fisheries Policy emphasizes ecosystem-based management to balance resource use and conservation.

15.6.2 Community Involvement and Co-management Strategies

Community involvement and co-management strategies are crucial for effective fisheries management. Co-management is a collaborative arrangement where the responsibility of resource management is shared between the government and local communities. This approach empowers communities, ensures compliance with regulations, and promotes sustainable practices.

In India, several community-based management initiatives have shown success. For example, the involvement of local communities in the management of Chilika Lake in Odisha has significantly improved the health of the ecosystem and the livelihoods of local fishermen (Jalal 2017). Similarly, the Maharashtra State Cooperative Fisheries Federation (MSCOF) promotes cooperative management of fisheries resources, enhancing community participation and resource sustainability (Patil and Mude 2017). Similarly, studies conducted in the selected reservoirs of Madhya Pradesh, Himachal Pradesh, and Odisha have shown that where fishing cooperative societies were active and functioning effectively, their performance contributed significantly towards conservation and sustainable management of fishery resources (Tyagi et al. 2013, 2014, 2015). The studies give enough pointers to view cooperative institutions of fisherfolks as social organizations having dormant capabilities (beyond production motives), which could be sincerely and effectively identified, promoted, and utilized for resource conservation and management goals in fishery resources. However, the mere existence of cooperative institutions is no guarantee that these will serve educational and managerial purposes among the fisherfolks. Their effectiveness and vibrancy may be determined by the dynamics of several location-specific socio-economic and political factors (Tyagi et al. 2015). The studies also revealed that the involvement of resource users through their organizations and institutions not only supplements the managerial efforts of the state agencies but also improves the prospects of sustainable utilization of the resource due to the dynamic impact of users' organizations on the resource-use behavior of their members and a sense of ownership of the resource among the users (Tyagi et al. 2015).

15.6.3 Successful Case Studies of Legislative Implementation

Successful implementation of fisheries legislation can be seen in various case studies across India:

Chilika Lake, Odisha: The Chilika Development Authority (CDA) was established to manage and restore Chilika Lake. Through effective legislation and community involvement, the CDA has improved fish production and biodiversity conservation (Jalal 2017).

Gujarat Fisheries Act: Gujarat's proactive fisheries management policies have led to significant improvements in marine fisheries. The state has implemented strict fishing regulations, including seasonal bans and gear restrictions, to sustain fish stocks (Shyam 2018).

Sustainable aquaculture in Andhra Pradesh: The state has promoted sustainable aquaculture practices through legislative support and technical guidance, leading to increased fish production and economic benefits for local communities (Rao and Ravichandran 2018).

15.6.4 Role of Technology and Scientific Research in Informing Policy Decisions

The role of technology and scientific research in fisheries management cannot be overstated. Advances in technology and research provide valuable data and tools for informed decision-making and effective management.

Remote sensing and GIS: These technologies are used for mapping and monitoring aquatic ecosystems, helping to identify critical habitats and assess environmental changes. For example, remote sensing has been employed to monitor the health of coral reefs and mangroves along India's coast (Ramachandran et al. 2018).

Genetic research: Genetic studies help in understanding the population structure and diversity of fish species. This information is vital for the conservation of genetic resources. Recent studies on the genetic diversity of Indian major carps have provided insights into breeding programs and stock management (Reddy et al. 2019).

Fishery stock assessment models: These models are used to estimate fish population sizes and predict the impact of fishing on stock sustainability. The use of such models has improved the management of marine fisheries in India, ensuring sustainable harvests (Vivekanandan et al. 2017).

15.6.5 Global and Indian Reviews

Global perspective: The FAO's Code of Conduct for Responsible Fisheries (CCRF) provides comprehensive guidelines for sustainable fisheries management globally (FAO 1995). The CCRF emphasizes the importance of an ecosystem approach, community participation, and the use of scientific research to inform policy decisions.

Indian perspective: In India, the National Fisheries Policy of 2020 aligns with global best practices, focusing on sustainable development, resource conservation, and community welfare (Department of Fisheries 2017).

15.7 Challenges in Implementation

15.7.1 Enforcement Issues

15.7.1.1 Gaps in Monitoring and Enforcement of Regulations

The effectiveness of fisheries regulations in India is often hampered by significant gaps in monitoring and enforcement. Despite the existence of comprehensive laws, the implementation on the ground remains weak due to insufficient surveillance infrastructure and limited enforcement capabilities. For instance, the enforcement of the Indian Fisheries Act (1897) and the Wildlife Protection Act (1972) is often inconsistent, with many instances of illegal fishing and habitat destruction going unpunished. Additionally, the decentralized nature of fisheries management, with different states having their own regulations, further complicates enforcement efforts (Bavinck et al. 2013).

15.7.1.2 Strategies to Strengthen Enforcement Mechanisms

To address these gaps, several strategies can be employed:

Enhanced surveillance technology: Implementing advanced surveillance technologies, such as satellite monitoring and unmanned aerial vehicles (UAVs), can improve real-time tracking of fishing activities and ensure compliance with regulations (Pramod et al. 2014).

Capacity building: Training and equipping local enforcement agencies with the necessary skills and tools is crucial. This includes regular training programs and workshops for fisheries officers and local law enforcement (Salagrama 2006).

Community-based monitoring: Engaging local communities in monitoring efforts can enhance compliance. Community members can act as watchdogs and report illegal activities, fostering a sense of ownership and responsibility towards resource management (Pomeroy et al. 2007).

15.7.2 Stakeholder Conflicts

15.7.2.1 Balancing the Interests of Different Stakeholders

Balancing the diverse interests of stakeholders such as fishermen, conservationists, and industry players is a persistent challenge in fisheries management. Fishermen rely on the resource for their livelihoods, while conservationists prioritize the preservation of biodiversity. Industry players often seek to maximize profits, sometimes at the expense of sustainability (Chuenpagdee et al. 2013). These conflicting interests can lead to disputes and hinder the implementation of effective management strategies.

15.7.2.2 Conflict Resolution Frameworks

Effective conflict resolution frameworks are essential to harmonize these interests:

Inclusive decision-making: Establishing platforms for inclusive decision-making where all stakeholders can voice their concerns and contribute to policy formulation is crucial. Participatory governance models have been successful in various regions, promoting transparency and mutual understanding (Jentoft and McCay 1995).

Mediation and negotiation: Employing professional mediators to facilitate negotiations between conflicting parties can help in reaching consensus. Structured negotiation processes can address grievances and find middle-ground solutions (Pomeroy and Rivera-Guieb 2005).

Collaborative management: Co-management approaches, where management responsibilities are shared between the government and local communities, have shown promise in reducing conflicts and enhancing compliance with regulations (Berkes 2009).

15.7.3 Resource Constraints

15.7.3.1 Financial and Human Resource Limitations

Financial and human resource constraints significantly impact the implementation of fisheries management policies in India. Limited funding restricts the ability to conduct necessary research, maintain surveillance infrastructure, and enforce regulations effectively (FAO 2020a, b). Additionally, the shortage of trained personnel further exacerbates these challenges, leading to inadequate monitoring and enforcement.

15.7.3.2 Addressing Capacity-Building Needs

To overcome these limitations, targeted capacity-building initiatives are essential:

Funding for research and infrastructure: Increasing investment in fisheries research and infrastructure development is critical. This includes funding for scientific research, the development of monitoring technologies, and the maintenance of enforcement equipment (Blyth et al. 2002).

Training programs: Establishing regular training programs for fisheries management personnel can enhance their capabilities. These programs should focus on the latest research findings, technological advancements, and best practices in enforcement (Cochrane and Garcia 2009).

15.7.4 Awareness and Education

15.7.4.1 Enhancing Awareness Among Local Communities and Stakeholders

Raising awareness about the importance of sustainable fisheries management among local communities and stakeholders is vital for the success of conservation efforts. Many fishermen and community members may not fully understand the long-term benefits of sustainable practices, leading to resistance against regulations (Salagrama 2006).

15.7.4.2 Role of Education in Promoting Sustainable Practices

Education plays a crucial role in promoting sustainable practices:

Community outreach programs: Implementing community outreach programs that educate local communities about the ecological and economic benefits of sustainable fisheries can foster a culture of conservation (Pomeroy et al. 2007).

School curricula: Integrating fisheries and environmental education into school curricula can instill an early understanding of sustainability principles among the youth, ensuring long-term attitudinal change (Cinner et al. 2009).

Public awareness campaigns: Conducting public awareness campaigns using various media channels can reach a broader audience, emphasizing the importance of sustainable fisheries and encouraging compliance with regulations (FAO 2020a, b).

15.8 Socio-Economic and Ethical Considerations

15.8.1 Impact of Legislations on Local Fishing Communities

The implementation of national legislations for the sustainable management of fish genetic resources in India has had significant socio-economic impacts on local fishing communities. These communities, often reliant on traditional fishing practices, face challenges adapting to new regulatory frameworks designed to conserve fish populations and habitats.

Legislations such as the Wildlife Protection Act (1972) and its amendments, including the 2022 Amendment, have introduced stricter measures to protect aquatic species and their habitats. While these measures are crucial for conservation, they can limit access to traditional fishing grounds and resources, affecting the livelihoods of small-scale fishers. For instance, the prohibition of fishing in certain protected areas can lead to reduced income and economic instability for those who depend on these areas for their daily catch (Sinha et al. 2023).

Furthermore, the Coastal Aquaculture Authority Act (2005) regulates coastal aquaculture practices to ensure environmental sustainability. However, compliance with these regulations can be costly and complex, posing additional burdens on small-scale aquaculture farmers. These regulations often require significant investments in infrastructure and technology, which may not be feasible for all community members (Ramachandran et al. 2022).

15.8.2 Balancing Economic Development with Conservation Goals

Balancing economic development with conservation goals presents a complex challenge in the context of sustainable fisheries management. While the conservation of fish genetic resources is essential for long-term ecological balance and biodiversity, it must be harmonized with the economic needs of local communities.

Policies such as the National Fisheries Development Board (NFDB) initiatives aim to promote sustainable fishing practices while enhancing the economic well-being of fishing communities. Programs supporting the adoption of sustainable aquaculture techniques and providing financial assistance for technological upgrades help mitigate the economic impact of conservation regulations (Kumar et al. 2021).

Moreover, integrating community-based management approaches can foster a sense of ownership and stewardship among local fishers. Collaborative management strategies, where local communities are involved in decision-making processes, can ensure that conservation measures are more effectively implemented and respected (Chuenpagdee and Jentoft 2018). This approach not only supports ecological sustainability but also strengthens the socio-economic resilience of fishing communities.

15.8.3 Ethical Considerations in Fish Genetic Resource Management

Ethical considerations play a vital role in the management of fish genetic resources. Ensuring that conservation efforts do not disproportionately disadvantage vulnerable communities is a key ethical concern. The principle of equity must be upheld, ensuring that all stakeholders, including marginalized and small-scale fishers, have fair access to resources and opportunities (FAO 2020a, b).

Another ethical dimension involves the responsible use of biotechnology and genetic research in fisheries management. While advances in genetic technologies can significantly enhance conservation efforts, they also raise ethical questions regarding the manipulation of genetic materials and the potential consequences for natural populations. Ethical frameworks must guide the application of these technologies to ensure they are used responsibly and do not harm the natural genetic diversity of fish populations (Waples et al. 2019).

Additionally, the conservation of fish genetic resources must respect the cultural and traditional practices of local fishing communities. Traditional ecological knowledge is invaluable for sustainable resource management and should be integrated with scientific approaches to create comprehensive conservation strategies. Recognizing and preserving the cultural heritage of fishing communities is crucial for maintaining the social fabric and ensuring the long-term success of conservation initiatives (Govan et al. 2009).

15.9 International Perspectives and India's Role

15.9.1 Comparison with Global Best Practices

Global best practices in the sustainable management of fish genetic resources are informed by international guidelines and conventions, such as the UN Code of Conduct for Responsible Fisheries (UN CCRF) and the Convention on Biological

Diversity (CBD). These frameworks emphasize ecosystem-based management, conservation of biodiversity, and sustainable use of aquatic resources.

15.9.1.1 Ecosystem-Based Management (EBM)

EBM is a holistic approach that considers entire ecosystems, including human impacts, rather than focusing on single species. Countries like Canada and Norway have effectively implemented EBM in their fisheries management. For instance, Norway's Fisheries Management Plan incorporates scientific assessments and stakeholder input to maintain fish stocks at sustainable levels (Smith et al. 2022).

15.9.1.2 Marine Protected Areas (MPAs)

Establishing MPAs is a widely adopted strategy to conserve marine biodiversity. Australia's Great Barrier Reef Marine Park is an exemplary model, balancing conservation and sustainable use through zoned areas that regulate activities like fishing and tourism (Jones et al. 2021).

15.9.1.3 Community-Based Management

Community involvement in fisheries management is crucial for sustainable practices. The Philippines and Indonesia have successfully implemented community-based fisheries management (CBFM) systems, empowering local communities to manage and protect their resources (Pomeroy et al. 2020).

India has begun integrating these global best practices into its national policies. The Indian Marine Fisheries Bill, 2021, aims to align with the UN CCRF by promoting responsible fishing practices and sustainable resource management (Government of India 2021).

15.9.2 India's Participation in International Agreements and Conventions

India is a signatory to several international agreements and conventions that guide the conservation and sustainable management of fish genetic resources:

15.9.2.1 Convention on Biological Diversity (CBD)

India ratified the CBD in 1994, committing to the conservation of biological diversity, sustainable use of its components, and fair sharing of benefits arising from genetic resources. The National Biodiversity Action Plan (NBAP) reflects these commitments by incorporating strategies to conserve fish genetic resources (CBD 2020).

15.9.2.2 United Nations Convention on the Law of the Sea (UNCLOS)

India is a party to UNCLOS, which provides a legal framework for marine and coastal resource management. This includes provisions for the sustainable use of marine resources and the protection of the marine environment (UN 1982).

15.9.2.3 International Convention for the Conservation of Atlantic Tunas (ICCAT)

Though not in the Atlantic, India's alignment with ICCAT's principles for responsible fisheries management and conservation of fish stocks reflects its commitment to international standards (ICCAT 2021).

15.9.2.4 Agreement on Port State Measures (PSMA)

India's participation in the PSMA demonstrates its commitment to preventing, deterring, and eliminating illegal, unreported, and unregulated (IUU) fishing through stringent port controls (FAO 2016).

15.9.3 Lessons Learned from Other Countries

15.9.3.1 Integrated Approaches and Technological Innovations

Countries like Japan and New Zealand have integrated traditional knowledge with modern scientific techniques to manage their fisheries sustainably. Japan's satoumi concept emphasizes harmonious coexistence between humans and marine ecosystems, integrating local ecological knowledge with scientific management (Makino 2018).

15.9.3.2 Policy and Governance Frameworks

Norway's success in sustainable fisheries is attributed to its strong legal and institutional frameworks that ensure compliance and enforcement. The Norwegian Fisheries Directorate plays a crucial role in implementing fisheries policies, supported by robust scientific research and data collection (FAO 2020a, b).

15.9.3.3 Community Engagement and Co-management

The Philippines has shown that involving local communities in decision-making processes leads to better compliance and sustainability outcomes. The implementation of Marine Protected Areas (MPAs) with community participation has significantly improved fish stocks and biodiversity (Alcala and Russ 2018).

India can learn from these examples by enhancing its legal frameworks, integrating traditional and scientific knowledge, and fostering community participation in fisheries management. Strengthening institutions like the National Fisheries Development Board (NFDB) and the Marine Products Export Development Authority (MPEDA) can also enhance the sustainable management of fish genetic resources.

India's efforts to align with international best practices and agreements demonstrate its commitment to the sustainable management of fish genetic resources. By learning from global examples and strengthening domestic policies and institutions, India can achieve its conservation goals while supporting the livelihoods of its fishing communities.

15.10 Recommendations for the Future

15.10.1 Strengthening Legislative Frameworks

To enhance the sustainable management of fish genetic resources in India, it is crucial to strengthen the existing legislative frameworks. Current laws need to be periodically reviewed and updated to reflect the latest scientific advancements and socio-economic realities. For instance, the amendments in the Wildlife Protection Act, 1972, such as those introduced by the Wildlife Protection (Amendment) Act, 2022, have included provisions for managing invasive alien species, which is a step in the right direction (Government of India 2022). Similar updates should be considered for other relevant acts and policies to ensure comprehensive protection and management of fish genetic resources. Additionally, integrating fish genetic resource management into broader environmental legislation can create a more cohesive and effective regulatory environment.

15.10.2 Promoting Interdisciplinary Research and Collaboration

Interdisciplinary research is vital for addressing the complex challenges associated with the conservation and sustainable use of fish genetic resources. Collaboration between ecologists, geneticists, economists, and social scientists can lead to a more holistic understanding and innovative solutions. For example, recent studies have highlighted the importance of integrating genetic diversity assessments with ecological and socio-economic research to develop effective conservation strategies (Meek et al. 2021). In India, institutions like the Central Marine Fisheries Research Institute (CMFRI) and the National Bureau of Fish Genetic Resources (NBFGR) should lead collaborative projects that bring together diverse expertise and resources (Kumar and Kumar 2020).

15.10.3 Enhancing Community Participation and Stakeholder Engagement

Engaging local communities and stakeholders in the management of fish genetic resources is essential for the success of conservation initiatives. Participatory approaches ensure that the needs and knowledge of local populations are considered, leading to more sustainable and accepted management practices. Studies have shown that community-based management can significantly enhance conservation outcomes and resilience (Berkes 2018). In India, incorporating traditional knowledge and practices through community-based fishery management programs can play a crucial role. Successful examples from Kerala and West Bengal, where local communities are actively involved in sustainable fishery practices, should be scaled up and replicated (Ghosh 2019a, b).

15.10.4 Investing in Capacity Building and Technological Advancements

Investing in capacity building and technological advancements is critical for the long-term sustainability of fish genetic resource management. Training programs for fisheries managers, researchers, and local communities can enhance their ability to effectively implement and monitor conservation strategies. Moreover, adopting advanced technologies such as genomic tools, remote sensing, and data analytics can improve the precision and efficiency of monitoring and management efforts (Allendorf et al. 2010). In India, initiatives by the National Fisheries Development Board (NFDB) and other institutions should focus on enhancing technical skills and providing access to modern tools and technologies (Jena et al. 2020).

15.11 Conclusion

This chapter has provided an extensive review of the national legislations and policies that govern the sustainable management of fish genetic resources in India. The historical context sets the stage by tracing the evolution of fisheries management and highlighting key milestones. Significant national legislations, such as the Indian Fisheries Act (1897), the Wildlife Protection Act (1972) and its 2022 Amendment, the Environment Protection Act (1986), and the National Biodiversity Act (2002), were analyzed for their roles in conserving aquatic biodiversity.

State-specific fisheries acts, both for inland and marine environments, were discussed, emphasizing the need for harmonized implementation across states. Additionally, national policies like the Inland Fisheries and Aquaculture Policy (2020) and supportive frameworks such as the National Fisheries Code and the Coastal Aquaculture Authority Act (2005) were examined.

The chapter also covered various fisheries institutions, including the National Fisheries Development Board (NFDB) and the Marine Products Export Development Authority (MPEDA), and their contributions to fisheries development and conservation. Key challenges in implementing these legislations were identified, such as enforcement issues, stakeholder conflicts, resource constraints, and the need for greater awareness and education.

15.11.1 Importance of a Holistic Approach to Sustainable Management of Fish Genetic Resources

A holistic approach is crucial for the sustainable management of fish genetic resources. Such an approach integrates multiple dimensions, including ecological, socio-economic, and institutional aspects, to ensure that conservation efforts are effective and inclusive.

Ecologically, it involves understanding the intricate relationships within aquatic ecosystems and implementing measures that protect the entire habitat, not just individual species. Socio-economically, it requires balancing the needs of local fishing communities with conservation goals, ensuring that livelihoods are not adversely affected. Institutionally, it necessitates the coordination of various agencies and stakeholders to create a unified and coherent management strategy.

The holistic approach also emphasizes the importance of community involvement and co-management strategies. Empowering local communities and involving them in decision-making processes can lead to more sustainable and culturally appropriate conservation practices (Agardy et al. 2011; Charles et al. 2016).

15.11.2 Future Outlook for Legislative Reforms and Sustainable Practices in India

Looking forward, several areas for legislative reforms and sustainable practices can be identified to enhance the management of fish genetic resources in India.

15.11.2.1 Strengthening Legislative Frameworks

- Revisiting existing laws to address current and emerging challenges in fisheries management.
- Ensuring that national and state laws are harmonized to avoid conflicts and ensure cohesive implementation (FAO 2018).

15.11.2.2 Promoting Interdisciplinary Research and Collaboration

- Encouraging research that integrates ecological, social, and economic perspectives to provide comprehensive solutions.
- Fostering collaborations between governmental bodies, research institutions, and international organizations (Pikitch et al. 2004).

15.11.2.3 Enhancing Community Participation and Stakeholder Engagement

- Implementing policies that actively involve local communities in conservation and management efforts.
- Creating dialogue and conflict resolution platforms among stakeholders to ensure balanced decision-making (Berkes 2009).

15.11.2.4 Investing in Capacity Building and Technological Advancements

- Providing training and resources to fisheries managers and local communities to improve their capacity for sustainable practices.
- Leveraging technology for better monitoring, data collection, and enforcement of regulations (Grafton et al. 2010).

By addressing these areas, India can advance towards more effective and sustainable management of its fish genetic resources, ensuring their conservation for future generations while supporting the livelihoods of those who depend on them.

Appendix

List of fisheries-related institutions/organizations in India and their major activities (Department of Fisheries 2024; ICAR 2024).

Fisheries-related Institutions/ Organizations	Major Activities
Ministry of Fisheries, Animal Husbandry and Dairying, Government of India	– Formulates national fisheries policies and programs – Oversees the implementation of fisheries development schemes
Department of Fisheries	– Develops policies and schemes for inland, marine, and coastal fisheries – Manages Departments at State and Union Territories and fisheries institutes such as FSI, CIFNET, CICEF, NIFPHATT, NFDB, and CAA
Fishery Survey of India (FSI)	– Conducts surveys and assessments of fishery resources in Indian waters – Provides data for sustainable fisheries management
Central Institute of Fisheries Nautical and Engineering Training (CIFNET)	– Provides training in fisheries nautical and engineering disciplines – Supports the development of skilled manpower in the fisheries sector
Central Institute of Coastal Engineering for Fishery (CICEF)	– Undertakes engineering and technical studies for coastal aquaculture – Develops infrastructure for fisheries and coastal development
National Institute of Fisheries Post Harvest Technology and Training (NIFPHATT)	– Develops post-harvest technologies and conducts training programs – Promotes value addition and quality assurance in fisheries products
National Fisheries Development Board (NFDB)	– Promotes sustainable fish production and productivity – Implements programs for the development of infrastructure and modern technologies in fisheries
Coastal Aquaculture Authority (CAA)	– Regulates coastal aquaculture activities to ensure sustainable practices – Issues licenses and monitors compliance with environmental standards
Indian Council of Agricultural Research (ICAR)	– Coordinates fisheries and aquaculture research and academic programs – Supports five resource-specific fisheries research institutes, a deemed university, a bureau, and a directorate
ICAR-Central Marine Fisheries Research Institute (CMFRI)	– Conducts research on marine fisheries resources and ecosystems – Develops strategies for sustainable management of marine fisheries
ICAR-Central Inland Fisheries Research Institute (CIFRI)	– Focuses on research and development of inland fisheries and aquaculture – Provides technical support for the management of inland water bodies
ICAR-Central Institute of Fisheries Technology (CIFT)	– Develops technologies for fish processing and preservation – Ensures quality control and standardization in fisheries products

Fisheries-related Institutions/ Organizations	Major Activities
ICAR-Central Institute of Freshwater Aquaculture (CIFA)	– Conducts research on freshwater aquaculture systems – Develop innovative technologies for sustainable aquaculture practices
ICAR-Central Institute of Brackishwater Aquaculture (CIBA)	– Focuses on research and development of brackishwater aquaculture – Promotes sustainable and eco-friendly aquaculture practices
ICAR-Central Institute of Fisheries Education (CIFE)	– Provides advanced education and training in fisheries and aquaculture – Conducts research to support the development of fisheries science
ICAR-National Bureau of Fish Genetic Resources (NBFGR)	– Conserves and manages fish genetic resources – Conducts research on genetic characterization and improvement of fish species
ICAR-Directorate of Coldwater Fisheries Research (DCFR)	– Conducts research on Coldwater fisheries and aquaculture – Develops strategies for the sustainable management of Coldwater fish resources

References

Agardy T, di Sciara GN, Christie P (2011) Mind the gap: addressing the shortcomings of marine protected areas through large-scale marine spatial planning. Mar Policy 35(2):226–232

Alcala AC, Russ GR (2018) A review of the biological and socio-economic impacts of marine protected areas in The Philippines. Environ Manag 41(1):31–45

Allendorf FW, Luikart G, Aitken SN (2010) Conservation and the genetics of populations. Wiley-Blackwell

Barman B, Nasar S, Roy A (2018) Community-based fishery management in the Sundarbans, West Bengal: a success story. Aquacult Rep 9:50–58

Bavinck M, Chuenpagdee R, Jentoft S et al (2013) Governability of fisheries and aquaculture: theory and applications. Springer

Berkes F (2009) Evolution of co-management: role of knowledge generation, bridging organizations and social learning. J Environ Manag 90(5):1692–1702

Berkes F (2018) Sacred ecology. Routledge

Beveridge MCM, Phillips MJ, Clarke RM (2020) Aquaculture and the environment: lessons from the past, challenges for the future, and emerging issues. Aquac Environ Interact 12:425–439

Blyth RE, Kaiser MJ, Edwards-Jones G et al (2002) Voluntary management in an inshore fishery has conservation benefits. Environ Conserv 29(4):493–508

CAA (2020) Coastal Aquaculture Authority Guidelines. Retrieved from CAA website

CBD (2020) Convention on Biological Diversity: India—Country Profile. Retrieved from https://www.cbd.int/countries/profile/?country=in

Central Inland Fisheries Research Institute (2020) Annual Report 2019-20. CIFRI, Barrackpore

Central Marine Fisheries Research Institute (2019) Marine Fisheries Census 2016 Kochi: CMFRI

Central Marine Fisheries Research Institute (2021) Handbook of Fisheries and Aquaculture Kochi: CMFRI

Charles A, García-Ayllon S, Pita C (2016) Integrating fisheries and coastal management through ecosystem-based spatial planning. Mar Policy 69:73–80

Chuenpagdee R, Jentoft S (2018) Transforming the governance of small-scale fisheries. Marit Stud 17(1):101–115

Chuenpagdee R, Pascual-Fernández JJ, Szeliánszky E et al (2013) Marine protected areas: re-thinking their inception. Mar Policy 39:234–240
Cinner JE, McClanahan TR, MacNeil MA et al (2009) Co-management of coral reef social-ecological systems. Proc Natl Acad Sci 109(14):5219–5222
Cochrane K, Garcia SM (2009) A fishery manager's guidebook. FAO and Wiley-Blackwell
CPCB (2019) Water and air quality monitoring report. Central Pollution Control Board. Retrieved from CPCB website
Department of Fisheries (2017) National policy on marine fisheries. Retrieved from https://dof.gov.in/national-policy-marine-fisheries-2017
Department of Fisheries (2020) Inland fisheries and aquaculture policy. Government of India
Department of Fisheries (2024) Fisheries institutes retrieved from https://dof.gov.in/
Edwin L (2022) Fishing regulations in India. Training manual on bycatch reduction in fisheries: recent advances. Jan. 17–26 2022, ICAR-CIFT: Kochi
FAO (1995) Code of conduct for responsible fisheries. Food and Agriculture Organization of the United Nations. Retrieved from http://www.fao.org/3/v9878e/v9878e00.htm
FAO (2011) Guidelines for Sustainable Mariculture. Food and Agriculture Organization. Retrieved from FAO website
FAO (2016) Agreement on Port State Measures to Prevent, Deter and Eliminate Illegal, Unreported and Unregulated Fishing. Retrieved from http://www.fao.org/port-state-measures/en/
FAO (2018) The state of world fisheries and aquaculture 2018—meeting the sustainable development goals. Food and Agriculture Organization of the United Nations
FAO (2020a) Norway's Fisheries Management Plan. Retrieved from http://www.fao.org/norway-fisheries-management/en/
FAO (2020b) The state of world fisheries and aquaculture 2020. Food and Agriculture Organization of the United Nations. https://doi.org/10.4060/ca9229en
Gadgil M, Rao PRS, Utkarsh G et al (2000) New meanings for old knowledge: the People's biodiversity registers program. Ecol Appl 10(5):1307–1317
Ghosh A (2019a) Community-based fisheries management in India: lessons from traditional systems. Mar Policy 99:361–369
Ghosh S (2019b) Environmental protection and regulation in India: role of the judiciary. J Environ Law 31(1):23–44
Govan H, Tawake A, Tabunakawai K (2009) Community-based marine resource management in the South Pacific. Mar Policy 33(2):182–191
Government of India (2021) Indian Marine Fisheries Bill, 2021. Retrieved from https://www.indiacode.nic.in/handle/123456789/13859
Government of India (2022) The wildlife protection amendment act, 2022. Ministry of Law and Justice, New Delhi
Grafton RQ, Hilborn R, Squires D et al (2010) Handbook of marine fisheries conservation and management. Oxford University Press
Gupta S, Asher S (2013) Environment and pollution laws in India. Oxford University Press
ICAR (2024) Fisheries Institutes retrieved from https://icar.org.in/fisheries-science-institutes
ICCAT (2021) International Convention for the Conservation of Atlantic Tunas. Retrieved from https://www.iccat.int/en/
Jalal KF (2017) Chilika Lake: community-based management for conservation and livelihoods. Aquat Ecosyst Health Manag 20(2–3):183–190
Jena JK, Gopalakrishnan A, Kumar A (2020) Fisheries and aquaculture development in India: current status and future perspectives. Indian J Fish 67(SI):1–10
Jentoft S, McCay BJ (1995) User participation in fisheries management: lessons drawn from international experiences. Mar Policy 19(3):227–246
Jeyabaskaran R (2013) An overview of the wildlife protection act, 1972 and its role in the conservation of marine biodiversity in India. Mar Biodivers 5(3):112–119
Jones PJS, De Santo EM, Qiu W (2021) Governing marine protected areas: getting the balance right. Mar Policy 43:110–118

Kumar R, Chopra AK (2012) Monitoring of environmental pollution and its impact on aquatic ecosystems. Environ Monit Assess 184:161–176
Kumar R, Kumar A (2020) Genetic diversity of fish species: implications for conservation in India. J Aquat Biol Fish 8(1):12–22
Kumar A, Tripathi R (2020) Biodiversity legislation in India: an overview. Environ Policy Law 50(3):229–238
Kumar A, Katiha PK, Joshi P K (2021) Sustainable fisheries and aquaculture for nutritional security in India. National Academy of Agricultural Sciences Policy Paper—8, p 7. https://doi.org/10.1007/s40003-021-00525-3
Makino M (2018) Integrated coastal management in Japan: the satoumi concept. J Environ Manag 93(1):185–193
Meek MH, Wells CP, Tomalty KMH et al (2021) Conservation genomics in the Anthropocene. Annu Rev Anim Biosci 9(1):267–288
Ministry of Environment, Forest and Climate Change (2011) Coastal Regulation Zone Notification. Retrieved from https://moef.gov.in/coastal-regulation-zone-notification-2011/
Mohan V (2018) Coastal aquaculture in India: issues and perspectives. Aquac Inter 26(3):885–899
Nathanson JA (2013) Basic environmental technology: water supply, waste management, and pollution control. Pearson
National Biodiversity Authority (2012) National Biodiversity Act, 2002. Government of India
National Fisheries Development Board (2021) Annual Report 2020–21 Hyderabad: NFDB
Naylor RL, Goldburg RJ, Primavera JH et al (2000) Effect of aquaculture on world fish supplies. Nature 405(6790):1017–1024
Patel A, Mehta P (2020) Collaborative management in the Gulf of Kutch: a case study in marine biodiversity conservation. Mar Policy 115:103840
Patil P, Mude A (2017) Cooperative fisheries management in Maharashtra: a model for sustainable development. Mar Policy 77:48–56
Pikitch EK, Santora C, Babcock EA et al (2004) Ecosystem-based fishery management. Science 305(5682):346–347
Pillai NGK (2012) Fisheries legislation in India: a historical perspective. J Mar Biol Ass India 54(1):1–8
Pomeroy RS, Berkes F (1997) Two to tango: the role of government in fisheries co-management. Mar Policy 21(5):465–480
Pomeroy RS, Rivera-Guieb R (2005) Fishery co-management: a practical handbook. CABI Publishing
Pomeroy RS, Katon BM, Harkes I (2007) Conditions affecting the success of fisheries co-management: lessons from Asia. Mar Policy 25(3):197–208
Pomeroy RS, Parks JE, Balboa CM (2020) Community-based coastal resource management in The Philippines: a review and evaluation of programs and projects, 1984-1994. Mar Policy 20(5):445–464
Pramod G, Nakamura K, Pitcher TJ et al (2014) Estimates of illegal and unreported fish in seafood imports to the USA. Mar Policy 48:102–113
Ramachandran R, Gopikrishna G, Balasubramanian T (2018) Application of remote sensing and GIS in coastal ecosystem management. J Coast Res 34(6):1383–1392
Ramachandran A, Khan H, Biswas N (2022) Coastal aquaculture in India: sustainability challenges and policy implications. Aquac Rep 22:100922
Ramasamy K, Kumar P, Thomas A (2021) Stakeholder participation in the Vembanad Lake fisheries management plan. Fisheries Manag Ecol 28(4):368–378
Rao KS, Ravichandran P (2018) Sustainable aquaculture practices in Andhra Pradesh. Aquac Res 49(4):1123–1133
Reddy AK, Rao V, Mohan M (2019) Genetic diversity of Indian major carps: implications for conservation and management. Fish Res 211:182–192
Salagrama V (2006) Trends in poverty and livelihoods in coastal fishing communities of Orissa State, India. FAO Fisheries Technical Paper, No, p 490

Salim Shyam S, Anuja AR (2022) Marine fisheries policies in India: opportunities and challenges. In: Suvenir 1st Indian Fisheries Outlook: Priming Indian Fisheries in Attaining Sustainable Development Goals. March, 22–24, 2022. ICAR-CIFRI, Barrackpore

Sarkar UK, Pathak AK, Lakra WS (2008) Conservation of freshwater fish resources of India: new approaches, assessment and challenges. Biodivers Conserv 17(10):2495–2511

Sathiadhas R, Panikkar KKP (2015) Fisheries management in India: challenges and opportunities. Indian J Fish 62(2):1–9

Shyam SS (2018) Marine fisheries management in Gujarat: a case study. Indian J Fish 65(3):10–16

Sinha R, Ahmed N (2020) Inland fisheries and aquaculture policy of India: a critique. Fisheries Manag Ecol 27(5):425–433

Sinha RK, Katiha PK (2002) Management of Inland Fisheries Resources of India: issues and options. Inland Fisheries Society of India

Sinha RK, Singh SK, Sharma N (2023) Conservation and management of fish genetic resources in India. J Aquat Ecol 35(2):250–267

Smith ADM, Brown CJ, Bulman CM (2022) The ecosystem approach to fisheries management. ICES J Mar Sci 74(1):16–26

Soto D, Aguilar-Manjarrez J, Hishamunda N. (eds) (2008) Building an ecosystem approach to aquaculture. FAO/Universitat de les Illes Balears Expert Workshop. 7–11 May 2007, Palma de Mallorca, Spain. FAO fisheries and aquaculture proceedings. No. 14. Rome, FAO. 221 p

Srinivasan U (2020) Harmonizing state and central policies in marine fisheries regulation: insights from the MFRM Bill, 2020. Ocean Coast Manag 198:105329

Srinivasan U, Ravikumar M (2017) Sustainable fishing practices in Tamil Nadu: outcomes of the marine fishing regulation act. J Coast Conserv 21(2):175–185

Tyagi LK, Bisht AS, Pal A et al (2013) Functioning of fishing cooperative societies in selected states of India. J Commun Mobilization Sustain Dev 8(1):90–93

Tyagi LK, Bisht AS, Pal A (2014) Performance of fishing cooperative societies in implementing fish conservation measures: case studies from two reservoirs of Orissa. J Commun Mobilization Sustain Dev 9(1):48–51

Tyagi LK, Bisht AS, Pal A (2015) Co-management of reservoir fisheries for sustainable livelihoods: insights for fishery managers and extension professionals from field studies in India. Indian J Ext Educ 51(1&2):45–55

UN (1982) United Nations Convention on the Law of the Sea. Retrieved from https://www.un.org/depts/los/convention_agreements/texts/unclos/unclos_e.pdf

Vivekanandan E, Mohamed KS, Jayasankar J (2017) Stock assessment models for marine fisheries Management in India. Mar Fish Rev 79(2):1–15

Waples RS, Naish KA, Primmer CR (2019) Conservation and management of fish genetic resources: advancing the role of biotechnologies. Trends Ecol Evol 34(10):916–929

World Bank (2019) The Sunken Billions Revisited: progress and challenges in global marine fisheries. World Bank, Washington, DC

The manufacturer's authorised representative in the EU is Springer Nature Customer Service Centre GmbH, Europaplatz 3, 69115 Heidelberg, Germany. If you have any concerns regarding our products, please contact ProductSafety@springernature.com

Printed and bound by CPI Group (UK) Ltd, Croydon, CR0 4YY
01/10/2026
02225857-0007